ANTENNA ENGINEERING
Theory and Problems

ANTENNA ENGINEERING
Theory and Problems

Boris Levin
Holon Institute of Technology
Israel

CRC Press
Taylor & Francis Group
Boca Raton London New York

CRC Press is an imprint of the
Taylor & Francis Group, an **informa** business

A SCIENCE PUBLISHERS BOOK

CRC Press
Taylor & Francis Group
6000 Broken Sound Parkway NW, Suite 300
Boca Raton, FL 33487-2742

First issued in paperback 2020

© 2017 by Taylor & Francis Group, LLC
CRC Press is an imprint of Taylor & Francis Group, an Informa business

No claim to original U.S. Government works

ISBN-13: 978-1-4987-5920-5 (hbk)
ISBN-13: 978-0-367-78248-1 (pbk)

Library of Congress Cataloging-in-Publication Data

Names: Levin, Boris (Electrical engineer), author.
Title: Antenna engineering / author, Boris Levin.
Description: Boca Raton : Taylor & Francis, CRC Press, 2017. | Includes
bibliographical references and index.
Identifiers: LCCN 2016016901 | ISBN 9781498759205 (hardcover : alk. paper)
Subjects: LCSH: Antennas (Electronics)
Classification: LCC TK7871.6 .L44 2017 | DDC 621.382/4--dc23
LC record available at https://lccn.loc.gov/2016016901

Visit the Taylor & Francis Web site at
http://www.taylorandfrancis.com

and the CRC Press Web site at
http://www.crcpress.com

I want to express my deep gratitude

To my teachers—for the transferred to me knowledge of electromagnetic theory,

To employees—for the help in calculations and experiments,
and
To my wife—for the patience and understanding.

Preface

The book is devoted to methods of theoretical and experimental research of antennas and to problems of antenna engineering. The presentation material is based on the electromagnetic (EM) theory.

The book begins with the theory of thin antennas. Thin antennas represent one of the main types of radiators, which are widely used in practice as independent antennas and as elements of complex antennas. Thereby the theory of thin antennas is the basis of the antennas analysis.

Techniques for calculation of the electrical characteristics of thin linear antennas are described and analyzed in the book consistently, beginning with the method of Poynting vector, but the basic attention is given to integral equations for an antenna current, and also to the method of complex potential and to the synthesis of antennas with required characteristics.

Particular attention is paid to the integral equation of Leontovich-Levin. This is the equation for the current along an axis of a thin-walled metal cylinder, it is equivalent to the equation of Hallen with a precise kernel, but unlike him, the equation of Leontovich-Levin allows to accomplish the calculation of the second and subsequent approximations assuming that the current is concentrated on the axis of the radiator, which greatly simplifies the calculations. Two solutions of the equation are considered. The first option allows to obtain for the input impedance of the dipole in the second approximation the well-known result in the form of the set of table functions. The second variant makes it possible to write an expression for the current distribution and hence for the input impedance of the radiator in integral form. This result is compared with the calculation results of the input impedance by using other equations and by means the method of induced electromotive force (emf).

The resulting integral formula for the input impedance of the dipole is identical to the integral formula obtained by the method of induced emf (the second formulation), if the length of the radiator is not close to its length at the frequency of the parallel resonance. This explains the well-known coincidence of the input impedances, calculated by means of both methods in the form of the set of table functions. Application of the theorem about oscillating power for analysis of power transmission between objects allows us to prove the validity of the second formulation of the method of induced emf in comparison with the first formulation, which is based on the equality of reactive powers not having physical meaning.

Serious attention along with an analysis is paid in the book to a synthesis problem. The purpose of synthesis consists in providing high electrical characteristics of antennas —with the aid of development of new variants of radiators, and also by optimizing the electrical characteristics of known radiators.

For a long time engineers tried to solve an optimization problem by finding a law of the current distribution, which provides required electrical characteristics. The task of choosing magnitudes and dimensions of antenna elements in order to optimize antenna characteristics was first staged in relation to the Yagi-Uda antenna. Its solution confirmed rightness of chosen approach.

Later on still two tasks have been considered. The aim of the first task was selecting the shape of the curvilinear monopole, which provides maximal directivity at the given frequency. The second task was more general and in essence dealt with a problem of creating the wide-band radiator. It consisted in determining a type and magnitudes of concentrated loads, which are placed into a linear radiator in series and provide in a given frequency band the required electrical performance, including the good matching of antenna with the signal source, the high efficiency and the necessary directional pattern. The solution of problem was based on understanding advantages of in-phase current distribution and on the hypothesis of Hallen about usefulness of capacitive loads, whose magnitudes are changed along the radiator axis in accordance with linear or exponential law.

Selected approach confirmed the hypothesis of Hallen, demonstrated the rightness of choice of capacitive loads and gave numerical results. Its use helped to solve still three tasks. The first one was selecting loads, which provide in a given range the required current distribution. Thus, in particular, the efforts adopted for finding the current distribution, which creates desired characteristics, have been justified. The second result enabled to determine loads for diminishing distortion of directional pattern of antenna by closely spaced superstructures. Finally, the method was used for selecting concentrated loads, which are placed in series along the wires of V-antenna and significantly expand the frequency range, over which the antenna has high directivity along its axis. V-antenna with curvilinear arms allows to obtain without resistors the equal phases of the fields of all antenna segments in the far region.

These tasks are consistently considered in the book. During the work a reasonable sequence of solving each problem was defined. In the first instance one must propose an approximate method of analysis, in order to use later its result as initial values for the numerical solution of task by methods of mathematical programming. When analyzing the radiator with loads, as approximate methods the method of impedance long line and the method of two-wire line with the concentrated loads were used. At the same time it becomes obvious that by choosing magnitudes of the elements (capacitors, coils of inductance, resistors) and the coordinates of their location as parameters, one can obtain at all frequencies not the given characteristics, but only characteristics maximally close to them. Effectiveness of the methodology also is obvious compared with helpless method of trial and error.

The method of complex potential is widely used for solving of cylindrical (two-dimensional) tasks, for example, for the calculating of electrostatic fields and mutual capacitances of several infinitely long wires located in the homogeneous dielectric medium in parallel to each other with charges uniformly distributed along their lengths. Significant interest is the use of these results for solving three-dimensional tasks with similar mutual placement of metal bodies, since the problem of calculating electric fields of charged bodies is substantially simplified, if the all geometrical dimensions depend only on two coordinates. It is known that in the fullness of time the comparison of conical and cylindrical problems was of great benefit to antenna theory.

In this book the method of complex potential firstly is generalized on the case of piecewise homogeneous media and, secondly, is applied to the three-dimensional structures: conical and parabolic. Comparison of parabolic and cylindrical problems with each other allows to find the equalities relating the replaceable variables. If these equalities are accomplished, the Laplace's equation remains valid in transition from one task to another. In this way, the parabolic problems are reduced to the corresponding cylindrical ones, i.e. parabolic filaments, cylinders, and shells are replaced by parallel filaments, cylinders, and coaxial cylindrical shells. Parabolic structures differ from conic structures in that their equipotential surfaces intersect the axis of symmetry at different points. This allows to use parabolic structures to calculate fields in a phantom in order to determine the influence of its shape and dimensions on a magnitude of created there field.

Separately the principle of complementarities was examined. It is shown that the self-complementary antenna can be located not only on a plane but also on the surface of rotation, in particular conic and parabolic. In the book it is proposed the method of calculating complex flat and three-dimensional self-complementary antennas, including antennas with rotational symmetry, i.e. self-complementary antennas consisting of several metallic and slot radiators.

The obtained results show that a class of self-complementary antennas is considerably wider than that it was considered previously. This class must be complemented, firstly, at the expense of structures, consisting of several metal and slot radiators and, secondly, at the expense of three-dimensional structures, located on surfaces of rotation, in particular on the surfaces of the circular cone or the paraboloid. Closeness of the values of wave impedances of antennas and cables is necessary condition of antennas effectiveness. Known variants of self-complementary antennas do not satisfy this condition, since their wave impedances are substantially higher than wave impedances of standard cables. Antennas, which are regarded in the book, have very different, including sufficiently small values of the wave impedances. That should greatly facilitate the task of matching and expand the scope of using self-complementary antennas.

Problems of antenna engineering are considered in the second half of the book.

Among them there are described results of application of a compensation method for protection of human organism against irradiation and are considered different antenna arrays, including log-periodic, reflector and adaptive. Characteristics of known and offered V-antennas are given.

In the book there are results of studying properties of new types of antennas and methods of their analysis. Among them results of researching transparent antennas are given. In the end of the book the properties and structural features of ship antennas are discussed, including decreasing influence of metal bodies on antenna characteristics and reducing influence of cables on the directional patterns of coaxially placed radiators. Also there are presented the principal circuit, design and characteristics of the antenna-mast with inductive-capacitive load, which was developed with the participation of the author.

The theory of electrically coupled lines allows to calculate characteristics of different antennas, in particular of folded and multi-folded radiators, a multi-level radiator with adjustable directional pattern in a vertical plane, an antenna with meandering load, an impedance folded radiator, etc. Also this theory explains the reasons of appearance of cross talks and in-phase currents in multi-conductor cables.

In the chapter devoted to log-periodic antennas the different ways of reducing their transverse and longitudinal dimensions are considered separately. The main attention is given to new structure of this antenna. It is based on two innovations. Firstly the antenna elements are made in the form of straight and spiral dipoles connected in parallel with each other. Secondly, a two-wire distribution line is replaced by a coaxial cable, and dipoles are replaced by monopoles connected to the central conductor of this cable. As a result asymmetrical coaxial antenna is not in need of rotator and balancing transition from coaxial cable to a symmetrical two-wire line. This version of log-periodic antenna is better than other versions of these antennas with the same dimensions from standpoint of the range width and the radiated power. Its dimensions are smaller than the dimensions of well-known antennas intended for the same frequency range.

In the last chapter a problem of creating antennas for underground radio communication and a question of measuring an antenna gain in a Fresnel zone are discussed.

The proposed book is a natural addition to the known monographs. It is intended for professionals, which are engaged in development, placing and exploitation of antennas. The benefit from this book will be also for lecturers (university-level professors), teachers, students, advisors etc. in the study of fields radiated by antennas. The contents of the book can be used for university courses.

Boris Levin
Israel

Contents

Theory of Thin Antennas

1.1 FIRST STEPS

Ronald King wrote very briefly about the first antennas and first steps in antenna engineering [1]. Through twenty years after Maxwell formulated his famous equations, which have established the foundations of classical electrodynamics [2], Hertz by means of an experiment proved the existence of the wave phenomena predicted by these equations. He used a spark gap for exciting damped oscillations in a wire of length 60 cm with metal plates at the ends [3,4]. Hertz's experiment gave a start to the future rapid development of radio engineering.

The first two Maxwell's equations in differential form are written as

$$curl\,\vec{H} = \vec{j} + \frac{\partial \vec{D}}{\partial t},\, curl\,\vec{E} = -\frac{\partial \vec{B}}{\partial t}, \tag{1.1}$$

where $\vec{H}$ is the vector of magnetic field strength, $\vec{j}$ is the vector of volume density of conduction current, $\vec{D}$ is the electric displacement vector, t is time, $\vec{E}$ is the vector of electric field strength, $\vec{B}$ is the vector of magnetic induction. Hereinafter the International System of Units is used.

Equations (1.1) are to be complemented with the equation of continuity

$$div\,\vec{j} = \frac{\partial \rho}{\partial t}, \tag{1.2}$$

where ρ is the volume density of the electrical charge.

Typically, two more equations are included into the system of Maxwell's equations:

$$div\,\vec{D} = \rho \text{ and } div\,\vec{B} = 0, \tag{1.3}$$

but they follow from equations (1.1) and (1.2) [5].

Equation (1.1) associate the electromagnetic fields and currents in free space. It would be wrong to consider that the left or the right side of an equation is the source of the field and, accordingly, that the other side is the consequence. The electric and magnetic components of fields exist only jointly. And none of these quantities is the cause of appearance of the other.

The field of the antenna is the result of supplying power from a transmitter. In order to take it into consideration, the extraneous currents and fields as the original sources of excitation should be included in the set of equations, in accordance with the Equivalence theorem. They are introduced as summands of quantities $\vec{j}$, $\vec{E}$ and $\vec{H}$. Their nature and placement depend on the model of the segment near a generator, which is commonly called the 'excitation zone'. The total electromagnetic field of an antenna is equal to a sum of the field produced by the excitation zone and the field produced by the currents in the wires, which arise on switching on of the sources. As a rule, on a great distance from the antenna the first field is substantially less than the second one and can be neglected.

Maxwell's equations for electromagnetic field, which are complemented with boundary conditions on the some of another antenna, allow writing the equation for the current in the conductor of the antenna. Solving it and finding the current distribution along the wire, one can determine the electrical characteristics of a radiator. But researchers in the first few decades after the works of Hertz were published, were interested in other matters. Among engineers trying to solve the problem of signal reception, the names of Marconi and Popov are the most known.

In 1894, 23-year-old Rutherford manufactured a device for receiving radio signals, which was based on demagnetization a bunch of needles. He even demonstrated it to Marconi, and the latter undertook to improve it. The invention of radio tubes was very important in order to solve the problem of radio transmission. Then onwards, the power of radio tubes began to grow from year to year.

The article [6], published in 1884, was devoted to calculating the power of radiated signals. This article introduced Poynting's vector

$$\vec{S} = [\vec{E}, \vec{H}]. \tag{1.4}$$

where magnitude $\vec{S}$ is the density of the power flux. Its projection onto the normal to the corresponding part of a closed surface is equal to the density of power flux outgoing from the volume, bounded by the surface.

Using Poynting's vector, one can find the active component of an antenna input impedance. The power, passing through an antenna surface, does not change in free space and is equal to the power flux in the far region. The vectors of electric and magnetic field strengths are mutually perpendicular. Here,

$$|H| = |E|/Z_0,$$

where $Z_0 = 120\pi$ is the wave impedance of free space. Since the field strength of a vertical linear antenna in the spherical coordinate system is

$$E = E_m F(\theta, \varphi),$$

where E_m is the field in the direction of maximal radiation, and $F(\theta, \varphi)$ is the directional pattern, the radiation power of such antenna is equal to an integral of Poynting's vector over a closed surface

$$P_{\Sigma} = \frac{1}{Z_0} \int\limits_{(S)} E_m^2 F^2(\theta, \varphi) dS. \tag{1.5}$$

Integration is performed over the surface of sphere S of a great radius. The surface element is $dS = R_0^2 \sin\theta d\theta d\varphi$. Calculating the ratio of the radiation power to the square of the generator current, we obtain the antenna input resistance

$$R_{\Sigma} = \frac{1}{J^2(0)Z_0} \int\limits_0^{2\pi} d\varphi \int\limits_0^{\pi} E_m^2 F^2(\theta, \varphi) R_0^2 \sin\theta d\theta. \tag{1.6}$$

Let us calculate the electric field in the far region and the directional pattern of a radiator, considering the radiator as a sum of simple electrical dipoles (Hertz' dipoles). The field of such dipole with length b and the current I, located along z-axis, is

$$E_{\theta 0} = j(30kIb/R)exp(-jkR)\sin\theta. \tag{1.7}$$

Here $k = \omega\sqrt{\mu\varepsilon}$ is the propagation constant in the surrounding medium, ω is the circular frequency of the signal, μ is the permeability, $\varepsilon = \varepsilon_r \varepsilon_0$ is the absolute permittivity (ε_r is the relative permittivity and ε_0 is the absolute permittivity of air). The current distribution along a symmetrical dipole with arm length L is determined by the expression

$$J(z) = J(0) \frac{\sin k(L - |z|)}{\sin kL}. \tag{1.8}$$

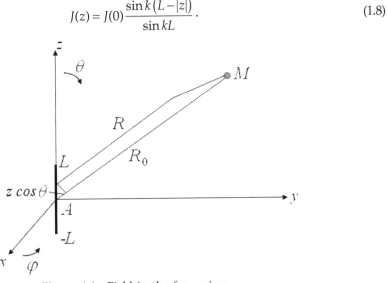

Figure 1.1 Field in the far region.

Putting $R = R_0 - z \cos\theta$ (see Figure 1.1), $I = J(z)$, $b = dz$ and neglecting small magnitudes, we obtain for the field of the symmetrical dipole

$$E_\theta = j\frac{30k \exp(-jkR_0)\sin\theta}{\varepsilon_r R_0} \int\limits_{-L}^{L} J(z)\exp(jkz\cos\theta)dz, \tag{1.9}$$

where R_0 is the distance from the dipole center to the observation point. Substituting (1.8) into (1.9), we get

$$E_\theta = j \frac{60J(0)}{\varepsilon_r \sin kL} \cdot \frac{\exp(-jkR_0)}{R_0} \cdot \frac{\cos(kL\cos\theta) - \cos kL}{\sin\theta}. \tag{1.10}$$

The last factor of this expression defines directional pattern of the dipole. Substituting of (1.10) into (1.6) gives

$$R_\Sigma = \frac{60}{\varepsilon_r^2 \sin^2 kL} \int_0^\pi \frac{[\cos(kL\cos\theta) - \cos kL]^2}{\sin\theta} d\theta. \tag{1.11}$$

Such method of calculating resistance is well-known as the Poynting's vector method. If the dipole length is small ($kL \ll 1$), we find limiting ourselves to the first terms of the function expansion in series ($\cos x$ at small x is equal to $\cos x = 1 - x^2/2$):

$$R_\Sigma = 20k^2L^2/\varepsilon_r^2. \tag{1.12}$$

For comparatively short antennas ($L < 0.3\lambda$, here λ is the wavelength), one can proceed from the expression (1.11)

$$R_\Sigma = 20k^2h_e^2/\varepsilon_r^2, \tag{1.13}$$

where $h_e = \frac{2}{k}\tan\frac{kL}{2}$ is the effective length of the symmetrical dipole.

The next step in the theory of linear radiators was made only in the 20th century. It is known as the method of induced emf. But before proceeding to it let us first consider the field of the conduction current along a filament and a circular cylinder.

1.2   FIELD OF A FILAMENT AND A CIRCULAR CYLINDER

Maxwell's equations require solutions. The solution of the equations is simplified essentially, if we introduce additional functions called potentials. A vector potential (an auxiliary vector field) is introduced by comparing the second equation from (1.3) with the mathematical expression (identity)

$$divcurl\vec{A} = 0,$$

where $\vec{A}$ is an arbitrary vector. This comparison shows that vector $\vec{B}$ can be presented as a curl of some vector $\vec{A}$:

$$\vec{B} = curl\vec{A}, \tag{1.14}$$

Yet, Eqn (1.14) defines vector $\vec{A}$ ambiguously. To define it unambiguously, one should also specify the value of $div\vec{A}$.

Substituting (1.14) with the second equation of the set (1.1) and using the mathematical identity

$$curl\, grad U = 0,$$

where U is an arbitrary scalar function (scalar potential of field), we obtain

$$\vec{E} = -\frac{d\vec{A}}{dt} - grad U. \tag{1.15}$$

Substituting (1.14) and (1.15) into the first equation of the set (1.1) and taking into account the mathematical identity

$$curlcurl\vec{A} = graddiv\vec{A} - \Delta\vec{A},$$

we obtain

$$\Delta\vec{A} = -\mu\varepsilon\frac{d^2\vec{A}}{dt^2} - grad(div\vec{A} + \mu\varepsilon\frac{dU}{dt}) = -\mu\vec{j}. \tag{1.16}$$

Let us define $div\vec{A}$ to simplify the last expression as far as possible. For this purpose, let

$$div\vec{A} = -\mu\varepsilon\frac{dU}{dt}. \tag{1.17}$$

This equality is known as the calibration condition, or Lorentz condition. In accordance with (1.16) and (1.17)

$$\Delta\vec{A} - \mu\varepsilon\frac{d^2\vec{A}}{dt^2} = -\mu\vec{j}. \tag{1.18}$$

For harmonic fields, which depend on the time in accordance with the exponential function $\exp(j\omega t)$, Eq. (1.18) takes the form

$$\Delta\vec{A} + k^2\vec{A} = -\mu\vec{j}. \tag{1.19}$$

Equation (1.19) is called the vector wave equation. Its solution permits to find the vector potential $\vec{A}$, and then the electric and magnetic fields of antenna. Expressions (1.20) are obtained from (1.14), (1.15) and (1.19),

$$\vec{E} = -\frac{\omega}{k^2}(graddiv\vec{A} + k^2\vec{A}), \vec{H} = \frac{1}{\mu}curl\vec{A}. \tag{1.20}$$

If the electromagnetic field sources are distributed continuously in some region V, and the medium surrounding the region V is a homogeneous isotropic dielectric, the solution of the Eq. (1.19) for harmonic field has the form

$$\vec{A} = \mu\int_{(V)} \vec{j}GdV, \tag{1.21}$$

where $G = \exp(-jkR)/(4\pi R)$ is the Green's function.

A similar expression for the scalar potential follows from Eqs. (1.17), (1.21) and (1.2):

$$U = j\frac{1}{\omega\varepsilon}\int_{(V)} Gdiv\vec{j}dV = \frac{1}{\varepsilon}\int_{(V)} \rho GdV. \tag{1.22}$$

It should be noted that region V, where the electromagnetic field sources are located, may be multiply connected (if, e.g., one must regard radiation of several antennas, or metal bodies are located close to the antenna).

Further, consider the special case when the field sources are the electrical currents located in parallel to the z-axis in some region V and having the axial symmetry

$$\vec{j} = j_z\vec{e}_z, j_z = j_z(z) = const(\varphi). \tag{1.23}$$

Here, the cylindrical system of coordinates (ρ, φ, z) is used, with unit vectors $\vec{e}_\rho, \vec{e}_j, \vec{e}_z$ along the axes. As seen from (1.21), the vector potential in this case has only component A_z:

$$\vec{A} = A_z(\rho, z)\vec{e}_z, \tag{1.24}$$

i.e.

$$div\vec{A} = \frac{\partial A_z}{\partial z}, grad\,div\vec{A} = \frac{\partial^2 A_z}{\partial \rho \partial z}\vec{e}_\rho + \frac{\partial^2 A_z}{\partial z^2}\vec{e}_z, curl\vec{A} = -\frac{\partial A_z}{\partial \rho}\vec{e}_\varphi,$$

and in accordance with (1.20)

$$E_z(\rho, z) = -\frac{j\omega}{k^2}\left(k^2 A_z + \frac{\partial^2 A_z}{\partial z^2}\right), E_\rho(\rho, z) = -\frac{j\omega}{k^2}\frac{\partial^2 A_z}{\partial \rho \partial z}, H_\varphi(\rho, z) = -\frac{1}{\mu}\frac{\partial A_z}{\partial \rho}, E_\varphi = H_z = H_\rho = 0. \tag{1.25}$$

Obviously, if the distribution of current $J(z)$ along the radiator is known, one can calculate the electromagnetic field of the current with the help of presented formulas. If the antenna is excited at some point (e.g., $z = 0$) by a generator with concentrated emf e, the antenna input impedance at the driving point is

$$Z_A = e/J(0). \tag{1.26}$$

and in order to determine this impedance, it is enough to know the current magnitude at the corresponding point. When calculating the power absorbed in the load of a receiving antenna, the current magnitude is needed also. So the current distribution along the antenna constitutes a very important characteristic.

As a model of a vertical linear radiator, one can use a straight perfectly conducting filament, coinciding with the z-axis (Figure 1.2a), along which the conduction current flows. Current density $\vec{j}$ is related to this current by

$$\vec{J}(z) = \int_{(S)} \vec{j}dS,$$

where S is the filament cross-section. From (1.21) and (1.25)

$$A_z(\rho, z) = \mu \int_{-L}^{L} J(\varsigma)G_1 d\varsigma, E_z(\rho, z) = \frac{1}{j\omega\varepsilon}\int_{-L}^{L} J(\varsigma)\left(k^2 G_1 + \frac{\partial^2 G_1}{\partial z^2}\right)d\varsigma. \tag{1.27}$$

Here $G_1 = \exp(-jkR_1)/(4\pi R_1)$, distance R_1 from observation point M to integration point P is equal to $\sqrt{(z-\varsigma)^2 + \rho^2}$.

In the considered model the radiator radius is zero. The model of a radiator shaped as a straight circular thin-wall cylinder with radius a (Figure 1.2b) has finite dimensions. Both ends of the cylinder left open, without covers, in order to the current as before had only longitudinal component. The surface density of current along the cylinder is $J_S(z) = J(z)/2\pi a$. Since a volume element in the cylindrical system of coordinates is equal to $dV = \rho d\rho d\varphi dz$, and $\rho = a$ on the cylinder surface, so, in accordance with (1.21) and (1.25),

$$A_z(\rho, z) = \frac{\mu}{2\pi}\int_{-L}^{L} J(\varsigma)\int_{0}^{2\pi} G_2 d\varphi d\varsigma, E_z(\rho, z) = \frac{1}{j\omega\varepsilon}\int_{-L}^{L} J(\varsigma)\left(k^2 G_2 + \frac{\partial^2 G_2}{\partial z^2}\right)d\varsigma, \tag{1.28}$$

where $G_2 = \exp(-jkR_2)/(4\pi R_2)$, and the distance R_2 from observation point M to integration point P is $\sqrt{(z-\varsigma)^2 + \rho^2 + a^2 - 2a\rho\cos\varphi}$. In particular, if the observation point is located on the radiator surface, $R_2 = \sqrt{(z-\varsigma)^2 + 4a^2 \sin^2(\varphi/2)}$.

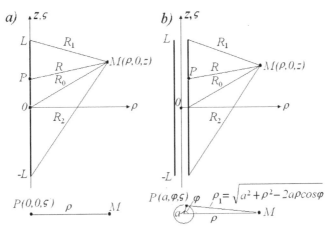

Figure 1.2   The dipole models in the shape of a straight filament (*a*) and a thin-wall circular cylinder (*b*).

Sometimes the dipole model shaped as a filament with finite radius a, i.e. expressions (1.27) are used for A_z and E_z, but distance from the observation point to the integration point is equal to $R_3 = \sqrt{(z-\varsigma)^2 + \rho^2 + a^2}$.

The obtained expressions for the vector potential and the vertical component of the electrical field strength produced by different models of a radiator, confirm the opinion about significance of the current distribution along the radiator. An assumption that this distribution has sinusoidal form played a great role in the antenna theory. It was based partly on results of measurements, but mainly on a simple understanding that the current distribution along wires of two-wire long line does not change if the wires move and diverge from each other. Later on, at derivation and solution of integral equations for currents in radiators, it was rigorously shown that the sinusoidal distribution is the first approximation to the true current distribution. Thus, its use received a reliable justification.

Here it is implied that the sinusoidal distribution may have any phase. In particular, the conduction current on the ends of a dipole and monopole is equal to zero. In this case there is a sinusoidal distribution – see Eq. (1.8). The current distribution along the folded radiator may have an antinode at the upper end, i.e. the current may follow the cosine distribution law.

For a perfectly conducting filament used as a model of a symmetrical radiator, in accordance with (1.28)

$$E_z = \frac{1}{4\pi j\omega\varepsilon}\int_0^L J(\varsigma)\left(k^2 + \frac{\partial^2}{\partial z^2}\right)\left[\frac{\exp(-jkR)}{R} + \frac{\exp(-jkR_+)}{R_+}\right]d\varsigma. \qquad (1.29)$$

This expression takes into account the currents' symmetry in the radiator arms and accordingly the substitution of variable ($-\varsigma$ for ς) is performed at the lower arm, and the designation used is: $R_+ = \sqrt{(z-\varsigma)^2 + \rho^2}$.

Since

$$\frac{\partial R}{\partial \varsigma} = -\frac{\partial R}{\partial z}, \frac{\partial R_+}{\partial \varsigma} = \frac{\partial R_+}{\partial z}, \frac{\partial^2 R}{\partial \varsigma^2} = \frac{\partial^2 R}{\partial z^2}, \frac{\partial^2 R_+}{\partial \varsigma^2} = \frac{\partial^2 R_+}{\partial z^2},$$

then

$$E_z = \frac{1}{4\pi j\omega\varepsilon}\int_0^L J(\varsigma)\frac{\partial^2}{\partial\varsigma^2}\left[\frac{\exp(-jkR)}{R} + \frac{\exp(-jkR_+)}{R_+}\right]d\varsigma + \frac{k^2}{4\pi j\omega\varepsilon}\int_0^L J(\varsigma)\left[\frac{\exp(-jkR)}{R} + \frac{\exp(-jkR_+)}{R_+}\right]d\varsigma.$$

Twice integrating the first term of the expression by parts, we get

$$E_z = \frac{1}{4\pi j\omega\varepsilon}\left\{\int_0^L\left[\frac{d^2 J(\varsigma)}{d\varsigma^2} + k^2 J(\varsigma)\right]\left[\frac{\exp(-jkR)}{R} + \frac{\exp(-jkR_+)}{R_+}\right]d\varsigma + \right.$$
$$\left. J(\varsigma)\frac{\partial}{\partial\varsigma}\left[\frac{\exp(-jkR)}{R} + \frac{\exp(-jkR_+)}{R_+}\right]\Big|_0^L - \frac{dJ(\varsigma)}{d\zeta}\left[\frac{\exp(-jkR)}{R} + \frac{\exp(-jkR_+)}{R_+}\right]\Big|_0^L\right\}. \quad (1.30)$$

If the current along the radiator is distributed in accordance with (1.8), the first factor in the integrand and hence the first term of the expression are zero. As is easy to verify, the second summand is zero too, since the first factor becomes zero at $\varsigma = L$, and the second factor vanishes at $\varsigma = 0$. Derivative of the current is calculated, and auxiliary relation is taken into account $k/(4\pi\omega\varepsilon_0) = 30$:

$$E_z = -j\frac{30 J(0)}{\varepsilon_r \sin kL}\left[\frac{\exp(-jkR_1)}{R_1} + \frac{\exp(-jkR_2)}{R_2} - 2\cos kL\frac{\exp(-jkR_0)}{R_0}\right], \quad (1.31)$$

where $R_1 = \sqrt{(z-L)^2 + \rho^2}$, $R_2 = \sqrt{(z-L)^2 + \rho^2}$, $R_0 = \sqrt{z^2 + \rho^2}$ are the distances from observation point M to the upper end, to the lower end and to the middle of the radiator, respectively (see Figure 1.2a).

Let us present without proof two components of the electromagnetic field for a straight filament:

$$E_\rho = j\frac{30 J(0)}{\varepsilon_r \rho \sin kL}\left[\frac{(z-L)\exp(-jkR_1)}{R_1} + \frac{(z-L)\exp(-jkR_2)}{R_2} - 2z\cos kL\frac{\exp(-jkR_0)}{R_0}\right],$$

$$H_\varphi = j\frac{J(0)}{4\pi\varepsilon_r \rho \sin kL}\left[\exp(-jkR_1) + \exp(-jkR_2) - 2\cos kL\exp(-jkR_0)\right]. \quad (1.32)$$

The rest of the components are zero, see (1.25).

If the model of a symmetrical radiator in the shape of a straight circular cylinder is used, it is necessary to proceed, when calculating the field, from expression (1.28). We obtain instead of (1.31):

$$E_z = -j\frac{30 J(0)}{2\pi\varepsilon_r \sin kL}\int_0^{2\pi}\left[\frac{\exp(-jkR_1)}{R_1} + \frac{\exp(-jkR_2)}{R_2} - 2\cos kL\frac{\exp(-jkR_0)}{R_0}\right]d\varphi, \quad (1.33)$$

where

$$R_1 = \sqrt{(z-L)^2 + \rho^2 + a^2 - 2a\rho\cos\varphi'}, R_2 = \sqrt{(z-L)^2 + \rho^2 + a^2 - 2a\rho\cos\varphi'}, R_0 = \sqrt{z^2 + \rho^2 + a^2 - 2a\rho\cos\varphi'}.$$

Such great attention is paid to the sinusoidal distribution of the current along the radiator because the method of induced electromotive force (method of emf) is based in particular on this distribution.

1.3 THEOREM ABOUT OSCILLATING POWER

Before going to the method of induced emf, it is necessary to consider the theorem about oscillating power.

The theorem and its proof were published for the first time in the book [7]. The book arose on the basis of lectures delivered by the author to undergraduate and graduate students and was devoted to electromagnetic waves of ultra-high frequencies. The reaction of many specialists to the theorem about oscillating power was sharply negative. In their view, the appearance of this theorem was caused by misunderstanding of the sense of the reactive power, although this statement clearly conflicts with the well known postulate, which these experts constantly repeat in articles and lectures. The postulate contends that the reactive power has no physical meaning.

During the years from the date of its first publication, the famous theorem allowed to explain a great many problems.

Let us start with the so-called symbolic method, i.e. with writing equations of the electromagnetic field in a complex form. Widely used electromagnetic fields, time-varying in accordance with the sinusoidal law, are called harmonic or monochromatic fields. Both in the theory of alternating currents and in the field theory it is expedient in mathematical researches of harmonic processes, which are described by linear equations, to introduce complex magnitudes. The transition to these designations is performed in the following way: complex magnitudes denoted as $E(\omega)$ and $H(\omega)$ correspond to magnitudes of electric $\bar{E}(t)$ and magnetic $\bar{H}(t)$ fields at a given point.

Relation between the physical magnitudes and their complex magnitudes is given by the relationships:

$$\bar{E}(t) = \text{Re}[E(\omega)\exp(j\omega t)] \text{ and } \bar{H}(t) = \text{Re}[H(\omega)\exp(j\omega t)], \tag{1.34}$$

where Re A is the real part of a complex vector, located in square brackets and ω is the circular frequency of the investigated process. Complex magnitudes $E(\omega)$ and $H(\omega)$, related with the instantaneous values by the relations of the type (1.34), correspond to two scalar physical magnitudes $E(t) = E \cos \omega t$ and $H(t) = H \cos E \ \omega t$. If $E(\omega)$ and $H(\omega)$, are complex magnitudes:

$$E(\omega) = Ee^{j\alpha} \text{ and } H(\omega) = He^{j\beta}, \tag{1.35}$$

where E and H are the amplitudes, and α и β are the arguments of the complex magnitudes, then

$$E(t) = E \cos(\omega t + \alpha) \text{ and } H(t) = H \cos(\omega t + \beta) \tag{1.36}$$

Thus, the amplitudes of the complex magnitudes are the amplitudes of the corresponding instantaneous values of the physical quantities, and the arguments of the complex magnitudes determine the phases of the instantaneous values of these quantities. Similarly, complex magnitudes are introduced for all physical magnitudes, incoming in the Maxwell equations. Formal coupling of complex equations with the initial equations is simple: in order to obtain complex equations one must replace the differentiation operator $\partial/\partial t$ by the operator of multiplication $j\omega$.

As is well known, energy magnitudes are determined by products (or squares) of instantaneous values of fields and currents. If to create a product

$$a(t)b(t) = \frac{1}{2}AB\big[\cos(\alpha-\beta)+\cos(2\omega t+\alpha+\beta)\big] \qquad (1.37)$$

and to calculate its average value for the period T, one may obtain

$$\overline{a(t)b(t)} = \frac{1}{T}\int_0^T a(t)b(t)dt = \frac{1}{2}AB\cos(\alpha-\beta) = \frac{1}{2}\mathrm{Re}\big[a(\omega)b^{\cdot}(\omega)\big]. \qquad (1.38)$$

Similar expressions are true for vector magnitudes also. These expressions permit to calculate the average value (constant part) of the energy value in accordance with the known complex amplitudes. A similar method can be used to calculate the average value of the variable fraction of the energy (oscillating energy). Indeed, according to (1.37)

$$a(t)b(t) = \overline{a(t)b(t)} + \widetilde{\Theta},$$

where it is natural to assume that the time-dependent second term

$$\widetilde{\Theta} = 0.5AB\cos(2\omega t+\alpha+\beta) \qquad (1.39)$$

is the oscillating fraction of the product $a(t)b(t)$. This part oscillates in time with a frequency 2ω, and its average value is zero. One can rewrite the expression (1.39) as

$$\widetilde{\Theta} = 0.5\,\mathrm{Re}\big[a(\omega)b(\omega)\exp(2j\omega t)\big]. \qquad (1.40)$$

It is seen that half the product of complex amplitudes is the complex amplitude of the oscillating fraction of the product $a(t)b(t)$.

As is well known, the energy conservation law for the electromagnetic field is given by

$$dW/dt + P + \Sigma = 0. \qquad (1.41)$$

Here, W is the electromagnetic energy contained in a volume V, P is an outgoing power (which flows out the volume through its bounding surface), and Σ is the radiation power. Passing from the differential formulation to the integral formulation and using the appropriate complex magnitudes, one can write the theorem about the oscillating power in the form

$$-\widetilde{\Sigma} = \widetilde{P} + 2j\omega\widetilde{W}. \qquad (1.42)$$

In deriving this expression, each term is considered as the sum of the active magnitude (average for the period of oscillation) and the oscillating (variable) fraction. In particular, for an instantaneous value of the power flux one can write according to (1.38)

$$p(t) = \overline{P} + \widetilde{P}, \qquad (1.43)$$

where

$$\overline{P} = 0.5\,\mathrm{Re}(EH^*), \qquad \widetilde{P} = 0.5\,\mathrm{Re}\big[EH\exp(2j\omega t)\big]$$

From here the physical meaning of magnitudes EH^* and EH is clear. The first magnitude is the complex amplitude of the active part of the power flow, equal to its average value. The second magnitude is the complex amplitude of the oscillating part of the power flow. In accordance with the law of energy conservation, if the source of radiation is located inside of a closed surface, then the active (average for the period of oscillation) power, supplied by the source, is equal to the active power passing through a closed surface. It is natural to assume that this equality of powers is true for any time, i.e. the oscillating fraction of the power supplied by the source, is equal to the oscillating fraction of the power passing through a closed surface.

1.4   METHOD OF INDUCED EMF

The induced emf method was proposed in 1922 by Rojansky and Brillouin simultaneously. Klazkin was the first to use it for calculating radiator characteristics. Later on, Pistolkors, Tatarinov, Carter, Brown et al. have contributed to its development. Reference list in the book [8], which is dedicated to regulation and generalization of the results available in the literature, consists of 96 items.

The method of induced emf allows determining both the active and reactive components of the antenna input impedance. Since the active component can be calculated with a similar accuracy by a simpler method of Poynting's vector (see Section 1.1), the method of induced emf actually for practical purposes, as emphasized in [9], is only one of the methods for determining input reactance of antenna.

The theorem about the oscillating power has significantly changed the understanding of the induced emf method, which has been the only way to calculate an antenna input reactance for a long time. The method of induced emf is formulated as follows: A cylindrical radiator of height $2L$ and radius a is placed inside a closed surface. A power, created by the emf source (by a generator), is equal to a complex power passing through this surface. Assume that the closed surface is a circular cylinder of height $2H$ and a radius b, along the axis of which the symmetrical radiator is located (Figure 1.3a). A density of power flux, which leaves a volume, bounded by a closed surface, is determined by the Poynting vector, or rather by its projections onto the normal to the sections surface: to the side surface and to the tops of the cylinders. These projections have the following form:

$$P_\rho = -0.5 E_z H_\varphi^*, P_z = 0.5 E_\rho H_\varphi. \tag{1.44}$$

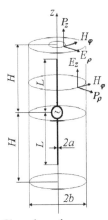

Figure 1.3   Closed surface around a radiator.

Let the cylinder surface coincide with the surface of the radiator, i.e. $H = L$, $b = a$. Then, if the radiator radius is small, power fluxes passing through the upper and lower covers of the cylinder, will also be small. Therefore, power passing through a closed surface is determined by integrating only over the side surface of the cylinder

$$P_1 = \int\limits_{-L}^{L} \int\limits_{0}^{2\pi} P_\rho a d\varphi dz$$

Here P_ρ is determined from (1.44) and does not depend on the coordinate φ, because the field components do not depend on it. Taking into account that $H_\varphi^* = J^*(z)/(2\pi a)$, we obtain

$$P_1 = -0.5 \int\limits_{-L}^{L} E_z J^*(z) dz. \tag{1.45}$$

If a current $J(z)$ is excited by a single generator, located in the middle of the radiator, then the power, created by it, is

$$P_2 = 0.5 |J(0)|^2 Z_A, \tag{1.46}$$

where Z_A is an input impedance of the antenna. Equating the power created by the source of emf, to the power passing through the closed surface, we obtain

$$Z_{AI} = -\frac{1}{|J(0)|^2} \int\limits_{-L}^{L} E_z J^*(z) dz \tag{1.47}$$

The expression (1.47) reveals the essence of the induced emf method. Two other variants of deducing this expression are described in [10] and [11].

If to equate to each other, instead of complex powers, two analogous oscillating powers: the power passing through the closed surface,

$$P_{K1} = -\frac{1}{2} \int\limits_{-L}^{L} E_z J(z) dz \tag{1.48}$$

and the power, created by the generator,

$$P_{K2} = \frac{1}{2} e J(0) = \frac{1}{2} J^2(0) Z_A, \tag{1.49}$$

where e is the emf of the generator, we obtain

$$Z_{AII} = -\frac{1}{J^2(0)} \int\limits_{-L}^{L} E_z J(z) dz. \tag{1.50}$$

After the appearance of an expression (1.50), equation (1.44) has been called the first formulation of the induced emf method. Expression (1.50) was called the second formulation of the induced emf method. This expression was first obtained on the basis of the theorem of reciprocity [12–14]. This theorem holds not only for two separate antennas but also for two points on the same antenna. Using that circumstance and applying the theorem to one radiator, one can obtain the expression (1.50). As is shown here, if to use the concept of oscillating power, then this expression is easily deduced from the energy relations. But despite the fact that the expression (1.50) by means of the theorem about the oscillating power was obtained many years ago [15], most experts kept to argue that it is derived in accordance with the reciprocity theorem by contrast to the expression (1.47) obtained from an equality of powers.

As can be seen from the above, the difference between the first and second formulations is caused by the fact that the first one is based on the equality of complex powers, and the second one - on the equality of the total powers, consisting of the active and oscillating components. Even here an advantage of the second formulation is obvious, since a reactive power unlike the oscillating power has no physical meaning.

Analysis shows that the second formulation is stationary. To verify this, one must show that if the antenna current is changed by the value of the first order infinitesimal, the input impedance will change by the value of the second order. The input impedance Z_{AII} obtained from (1.50) is not be changed in the first approximation for any trial current distribution, which differs from the true current $J^0(z)$ by a small value $\delta J(z)$. This means that if at $J(z) = J^0(z)$ a self-impedance of the radiator is equal to Z_{AII}, then on $J(z) = J^0(z) + \delta J(z)$ the self-impedance is also equal to Z_{AII}. The corresponding proof was given by J.E. Storer and is described in [16]. The stationary property of the second formulation is due to the fact that the integral in this expression is a rough functional of the current function, although an integrand is no rough functional of it [9].

Let a straight, perfectly conducting filament of a finite small radius a, whose axis coincides with z-axis, be located in a lossless medium and be used as a model of a vertical symmetrical dipole with arm length L (see Figure 1.4a). The current distribution along it is determined by the expression

$$J(z) = J(0)\frac{\sin k(L-|z|)}{\sin kL}, \tag{1.51}$$

i.e. a tangential component of the electric field of the filament along a radiator surface is equal to

$$E_z = -j\frac{30J(0)}{\varepsilon_r \sin kL}\left[\frac{\exp(-jkR_1)}{R_1} + \frac{\exp(-jkR_2)}{R_2} - 2\cos kL\frac{\exp(-jkR_0)}{R_0}\right], \tag{1.52}$$

where $R_1 = \sqrt{(z-L)^2 + a^2}$, $R_2 = \sqrt{(z-L)^2 + a^2}$, $R_0 = \sqrt{z^2 + a^2}$ are distances from observation point M to an upper end, to a lower end and to the middle of the radiator, respectively, and ε_r is the air relative permittivity. In this case both formulation of the induced emf method give the same result:

$$R_A = \frac{30}{\sin^2 \alpha}[2(C + \ln 2\alpha - Ci2\alpha) + \sin 2\alpha(Si4\alpha - 2Si2\alpha) + \cos 2\alpha(C + \ln\alpha + Ci4\alpha - 2Ci2\alpha)]$$

$$X_A = \frac{30}{\sin^2 \alpha}[\sin 2\alpha(C + \ln\alpha + Ci4\alpha - 2Ci2\alpha - 2\ln(2L|a)) - \cos 2\alpha(Si4\alpha - 2Si2\alpha) + 2Si2\alpha]. \tag{1.53}$$

Here $Six = \int_0^x (\sin u/u)du$ is sine integral, $Cix = \int_\infty^x (\cos u/u)du$ is the cosine integral, $\alpha = kL$, and $C = 0.5772...$ is the Euler's constant.

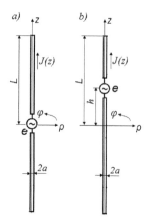

Figure 1.4 Symmetrical (*a*) and asymmetrical (*b*) dipoles.

As can be seen from the expression for antenna reactance, X_A consists of terms of a various order infinitesimal. The great summand is equal to

$$X_{A0} = 30\frac{\sin 2\alpha}{\sin^2 \alpha}\cdot[2\ln(2L/a)-C/2]\approx -120\ln(2L/a)\cot\alpha. \tag{1.54}$$

The value $\chi = 1/\Omega$ is called a small parameter of the thin antennas theory (Ω is a parameter, used by Hallen). The parameter χ is equal to $\chi = 0.5/\ln (2L/a)$. Introducing the notation $W = 60/\chi$, we obtain an expression for the input reactance of an equivalent long line, open at the end: $X_{A0} = -W \cot\alpha$.

In order to calculate losses in antenna conductors (e.g., loss due to skin effect), one must add to a purely real propagation constant a small imaginary value. Calculations show that in this case the second formulation gives positive value of a loss resistance and the first formulation, a negative one. A similar situation occurs during calculating losses in a ferrite shell of the antenna. Thus, if losses exist in the medium or in the antenna, applying of conception of reactive power gives an obvious mistake. The rightness of the second formulation, based on the conception of oscillating power, becomes a fact.

The second formulation of the induced emf method was analyzed when integral equations of Hallen [17] and Leontovich-Levin [18] for the current along a radiator axis were already written and solved. Solutions have been given in the form of expansions into a power series. If we use the formulas presented in [10] and [18], one can show that the solutions of both equations are same [19]. In this case, the coincidence of the results is not only numerical. The results were obtained in an explicit form (in the form of identical tabulated functions).

As already mentioned, solutions obtained by induced emf method for the perfectly conducting filament, using different formulations, gave identical results. They coincide with the solutions of integral equations for different length of radiator, if this length is not close to the parallel resonance when $J(0) \approx 0$. In the latter case, the input impedance, calculated by the induced emf method, becomes infinitely large, and the integral equations give the finite results.

Summarizing, one can say that both formulations of the induced emf method are based on the same two theses. The first thesis assumes the sinusoidal character of the current distribution along the radiator. The second thesis signifies the equality of the source power and the power passing through the closed surface.

Both formulations are useful only in the case the current distribution $J(z)$ along a radiator is known. The selection of the law of the current distribution may be based only on a solution of integral equations for the current, i.e. on a rigorous solution of the problem. Physical base for the selection of another distribution law does not exist. Hence there is no sense in speaking about the accuracy of the induced emf method, excluding artificially the error caused by the inexact current definition. The accuracy of this method is the mutual accuracy of (1.51) and (1.47) or (1.50). The experience in calculations shows that (1.51) gives a quite acceptable approximation, if $\alpha \leq \pi/2$. Therefore, the first thesis is questionable, because this thesis has an approximate nature.

As to the second thesis used for derivation of the first formulation, its inapplicability is obvious, since the reactive power has no physical sense, and the input reactance of antenna is determined as a result of equating two quantities no having physical sense. Equating of two such quantities cannot be justified.

1.5 APPLICATION OF THE INDUCED EMF METHOD TO COMPLICATED ANTENNAS AND TO ANTENNA SYSTEMS

One can use the induced emf method for analyzing more complicated radiators. The expressions (1.47) and (1.50) were obtained without indicating a concrete coordinate of a feeding point. For this reason, they are applicable to the radiator with $h \neq 0$. Really for a radiator with a feed point displaced from the radiator center to point $z = h$ (Figure 1.4b), the flux of an oscillating power through the side cylinder surface by analogy to (1.48) is

$$P_{K1} = -\int_{-L}^{L} E_z(J)J(z)dz .$$ (1.55)

The oscillating power created by one generator by analogy to (1.49) is equal to

$$P_{K2} = eJ(h) = J^2(h)Z_A,$$ (1.56)

By equating the right parts of the expressions, we come to

$$Z_A = -\frac{1}{J^2(h)}\int_{-L}^{L} E_z(J)J(z)dz.$$ (1.57)

Note that in this expression unlike (1.50) not only the denominator other, but also another current distribution $J(z)$ along the radiator and another field $E_z(J)$ of its current.

In the case of a radiator with nonzero surface impedance in (1.50) instead of $E_z(J)$ one should substitute the difference $[E_z(J) - H_\varphi Z(z)]$. Here $Z(z)$ is the surface impedance, i.e. the impedance of the square surface section. Actually, in accordance with the boundary condition on the radiator surface, it is necessary to take into account that a voltage drop along the self radiator makes no contribution to its radiation. Then for an antenna with constant surface impedance Z (Figure 1.5a) in a shape of a straight circular cylinder with radius a we find

$$Z_A = -\frac{1}{J^2(h)}\int_{-L}^{L} [E_z(J) - ZJ(z)/(2\pi a)]J(z)dz.$$ (1.58)

If $h = 0$, the current distribution $J(z)$ coincides in the first approximation with the current distribution along an impedance long line, open at the end:

$$J(z) = J(0)\sin k_1 (L - |z|)/\sin k_1 L. \tag{1.59}$$

Here $k_1 = \sqrt{k^2 - j2k\chi Z/(aZ_0)}$ is the propagation constant of a wave along the impedance line.

For a symmetrical radiator with piecewise constant surface impedance (Figure 1.5b) one can write:

$$Z_A = -1/(J_N^2)\sum_{m=1}^{2N}\int_{b_{m+1}}^{b_m}\left[E_z(J_m) - Z^{(m)}J_m(z)/(2\pi a)\right]J_m(z)dz, \tag{1.60}$$

where m is the segment's number, $2N$ is the total number of segments, $Z^{(m)}$ is the surface impedance on the segment m. Current distribution $J_m(z)$ on the each segment m is sinusoidal. Current distribution $J(z)$ along the radiator coincides in the first approximation, if $h = 0$, with the current distribution along a stepped impedance long line open at the end:

$$J_m(z) = I_m \sin(k_m z_m + \varphi_m), \qquad b_{m+1} \le z \le b_m, \tag{1.61}$$

where

$$I_m = A_m J(0), \qquad \frac{A_m \prod_{p=m+1}^{N} \sin\varphi_p}{\prod_{p=m}^{N}\sin(k_p l_p + \varphi_p)},$$

$$\varphi_m = \tan^{-1}\left\{\frac{k_m}{k_{m-1}}\tan\left[k_{m-1}l_{m-1} + \tan^{-1}\left\langle\frac{k_{m-1}}{k_{m-2}}\tan\left[k_{m-2}l_{m-2} + ... + \tan^{-1}\left(\frac{k_2}{k_1}\tan k_1 l_1\right)...\right]\right\rangle\right]\right\}.$$

In these expressions $z_m = b_m - z$ is the coordinate along the segment m, k_m is the wave propagation constant along this segment, and l_m is its length. The expressions are true for the segment N too, if to adopt that the product $\prod_{p=N+1}^{N}$ is equal to 1.

In the case of a radiator with one concentrated load Z_1 located at point $z = z_1$ (Figure 1.6a), the power is firstly radiated by the antenna:

$$P_{K1} = -\int_{-L}^{L} E_z(J)J(z)dz, \tag{1.62}$$

and secondly it is wasted in the complex load:

$$P_{K2} = J^2(z_1)Z_1. \tag{1.63}$$

The oscillating power produced by the generator is equal to the sum of these powers:

$$P_K = J^2(h)Z_A = P_{K1} + P_{K2}, \tag{1.64}$$

i.e.

$$Z_A = -\frac{1}{J^2(h)}\left\{\int_{-L}^{L} E_z(J)J(z)dz - Z_1 J^2(z_1)\right\}. \tag{1.65}$$

For several loads Z_n, located at points $z = z_n$ of the asymmetrical radiator (Figure 1.6b)

$$Z_A = -\frac{1}{J^2(0)}\left\{\int_0^L E_z(J)J(z)dz - \sum_{n=1}^N Z_n J^2(z_n)\right\}. \qquad (1.66)$$

Free terms in (1.65) and (1.66) are proportional to the square of the current and the magnitude of the concentrated load. It is worth emphasizing that the connection of loads changes the current distribution along the radiator and the field of the current.

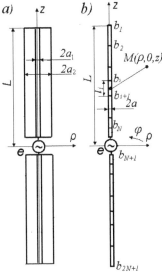

Figure 1.5 Antennas with constant (a) and piecewise constant (b) surface impedances.

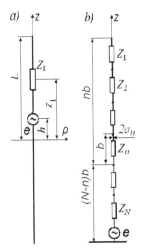

Figure 1.6 Antennas with one (a) and several (b) concentrated loads.

For a folded radiator (Figure 1.7a), which is an example of an antenna consisting of several parallel wires, we obtain

$$Z_A = -\frac{1}{J_g^2}\int_{-L}^L E_z(J)J(z)dz. \qquad (1.67)$$

Here J_g is the generator current, $J(z)$ is the total current of an antenna. The current distribution along the antenna wires coincides in the first approximation with the current distribution along the wires of an equivalent long line and is determined by means of the theory of electrically coupled lines. Generator current J_g of a folded radiator is not always equal to total current $J(z)$ at $z = 0$. If the radiator has a gap at point A then, $J_g = J(0)$. When calculating the field, it is necessary to use the total current, i.e. the sum of the currents of both wires.

Multi-radiator antenna, which is shown in Figure 1.7b, is an example of a radiator consisting of wires with different lengths. The antenna contains the central radiator with complex load Z_1 and side radiators situated around it and connected with it at the base. In this case, one can find the antenna input impedance from (1.65). The current distribution along the antenna wires is found by means of the theory of electrically coupled lines. The equivalent line (Figure 1.7c) consists of three wires. The first wire is equivalent to the central radiator, the second wire is equivalent to the system of identical side radiators, and the third wire is the ground.

Since the wires of the equivalent line have different lengths and the complex load is connected in the central radiator, the line should be divided into three segments. The numbers m of segments are shown in Figure 1.7c. Using the boundary conditions at the segment ends, one can find the current of each wire and the total currents along the segments. Function $J(z)$ is continuous in the entire interval $0 \le z \le l_1$ and behaves as sinusoid along each segment. But its derivative $dJ(z)/dz$ has a jump on the boundaries of the segments. With allowance for jumps of the derivative we obtain instead (1.31):

$$E_z(J) = -j\frac{15}{k}\left\{\frac{2\exp(-jkR_0)}{R_0}\frac{dJ(0)}{dz} - \left[\frac{\exp(-jkR_{11})}{R_{11}} + \frac{\exp(-jkR_{12})}{R_{12}}\right]\frac{dJ(l_1)}{dz} + \right.$$

$$\left. \sum_{m=2}^{3}\left[\frac{\exp(-jkR_{m1})}{R_{m1}} + \frac{\exp(-jkR_{m2})}{R_{m2}}\right]\left[\frac{dJ(l_m+0)}{dz} - \frac{dJ(l_m-0)}{dz}\right]\right\}, \quad (1.68)$$

where $R_0 = \sqrt{a^2+z^2}$, $R_{m1} = \sqrt{a^2+(l_m-z)^2}$, $R_{m2} = \sqrt{a^2(l_m+z)^2}$; a is the radiator radius at point z, and $dJ(l_m+0)/dz$ and $dJ(l_m-0)/dz$ are the values of derivative on the right and on the left of point $z = l_m$.

As it is noted in Section 1.3, the induced emf method does not permit to obtain the finite values of the input impedance at the points of parallel resonance, where $J(h) = 0$, and near these points. The second (integral) variant of solving the Leontovich-Levin integral equation allows in the case of the symmetrical radiator, if $J(h) \approx 0$, to come to the expression

$$Z_A = \frac{e}{\left[2J(0) + \frac{1}{e}\int_{-L}^{L} E_z(J)J(z)dz\right]}. \quad (1.69)$$

One can obtain similar expressions also for more complicated radiators. For example, the input impedance of a radiator with N concentrated loads and with the displaced feed point is equal to

$$Z_A = \frac{e}{\left[2J(h) + \frac{1}{e}\int_{-L}^{L} E_z(J)J(z)dz - \sum_{n=1}^{N} Z_n J^2(z_n)\right]}. \quad (1.70)$$

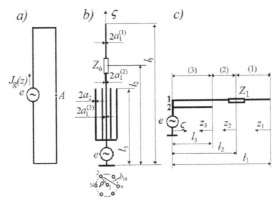

Figure 1.7 Folded antenna (*a*), multi-radiator antenna with the complex load (*b*), and the transmission line equivalent to a multi-radiator antenna (*c*).

These formulas expand essentially the scope of the method of induced emf. Comparison of the results of calculations by these formulas and numerical methods confirms their correctness.

Up to now the subject of discussion was application of the induced emf method for calculating the input impedance of an antenna. But this method is applied widely also for solving another problem – estimating the reciprocal influence of radiators by means of calculating their mutual impedances.

The analysis of two-radiator systems is based on the fact that the current of one radiator creates the field, which has the electrical component tangential to the surface of the second radiator. This component induces the field $E_\varsigma(J_1)d\varsigma$ on the surface of element $d\varsigma$ of the second radiator. In order to execute the boundary condition $E_\varsigma = 0$ on this surface, the own field of the second radiator on its surface must be equal to $-E_\varsigma(J_1)d\varsigma$. The generator of the second radiator must increase the power in the element $d\varsigma$ by $dP = -E_\varsigma(J_1)J_2(\varsigma)d\varsigma$ and, accordingly, the power in the entire radiator by

$$P = -\int_{-L2}^{L_2} E_\varsigma(J_1)J_2(\varsigma)d\varsigma.$$

Power P is equal to the power induced by the first radiator in the second radiator, and the ratio of power P to the square of the current of the second generator determines the magnitude of additional impedance, which the first radiator induced in the second radiator:

$$Z_{21ind} = -\frac{1}{J_2^2(0)}\int_{-L_2}^{L_2} E_\varsigma(J_1)J_2(\varsigma)d\varsigma. \qquad (1.71)$$

The Kirchhoff equation for the second radiator takes the form

$$e_2 = J_2(0)\left[Z_{22} + Z_{21ind}\right] = J_2(0)Z_{22} + J_1(0)Z_{21}, \qquad (1.72)$$

where $Z_{21} = -\int_{-L2}^{L_2} E_\varsigma(f_1)f_2(\varsigma)d\varsigma$ is the mutual impedance of the first and the second radiators,

$f_1(z) = J_1(z)/J_1(0)$, $f_2(\varsigma) = J_2(\varsigma)/J_2(0)$. One can write a similar expression for the first radiator. In the case of Q radiators, it has the follow form for the radiator p:

$$e_p = J_1(0)Z_{p1} + \sum_{q=2}^{Q} J_q(0)Z_{pq}. \tag{1.73}$$

The corresponding circuit for the radiator p is given on Figure 1.8.

The expressions presented in this section are given in the accordance with the second formulation of the induced emf method. The formulas, which allow to calculate the mutual impedances of linear radiators for the different variants of their relative geometrical replacement, are collected in [8].

Figure 1.8   The circuit of radiator p with serial connection of elements.

1.6   LOSS RESISTANCE IN THE GROUND

As already mentioned, the theorem about the oscillating power has significantly changed the understanding of the induced emf method. The losses of asymmetric vertical antenna in an earth and ground are another example of this change. It is presented in this section.

For a long time the calculation of losses in the ground was carried out according to the procedure of Brown [20]. It proceeds from the idea of a high conductivity of the ground, owing to which a magnetic field H_φ at the ground surface (Figure 1.9) is virtually identical to a magnetic field of an antenna, located above a perfectly conducting ground, and its strength is equal to a density of a surface current in the ground:

$$j_\rho(\rho) = H_{\varphi 0}(\rho). \tag{1.74}$$

The surface current has a radial character.

If a resistance per unit area of the earth's surface is equal to R_0, then the resistance of an element in the form of a ring with radius ρ and width $d\rho$ is $dR_g = (R_0/2\pi\rho)d\rho$. The power of losses in this ring is $dP_g = (2\pi\rho|H_{\varphi 0}|)^2 dR_g$. The resistance of losses, referred to the base of the antenna, is found from the expression

$$R_g \frac{1}{|J(0)|^2} \int_a^b dP_g = \frac{2\pi R_0}{|J(0)|^2} \int_a^b |H_{\varphi 0}|^2 \rho d\rho \tag{1.75}$$

Here $J(0)$ is the current in the base. Resistance per unit area is $R_0 = 1/(s\sigma) = 11\pi/\sqrt{\sigma\lambda}$, where s is the depth of current penetration into the ground and σ is the conductivity of the ground.

A lower limit of integration in (1.75) is the antenna radius or the radius of the ground, whose conductivity can be considered infinitely great. The upper limit b must tend to infinity. It is easy to see, however, that in this case, the integral diverges. Indeed, the magnetic field of the monopole in the form of a thin conductive filament, mounted vertically on the perfectly conducting ground, can be written as

$$H_{\varphi 0}(\rho) = j \frac{J(0)}{2\pi\rho \sin kL} \left[\exp\left(-jk\sqrt{\rho^2 + L^2}\right) - \cos kL \exp(-jk\rho) \right]. \tag{1.76}$$

Hence the integrand is

$$\left|H_{\varphi 0}\right|^2 \rho = \frac{J^2(0)}{4\pi^2 \rho \sin^2 kL}\left[1 + \cos^2 kL - 2\cos kL \cos k\left(\sqrt{\rho^2 + L^2} - \rho\right)\right].$$

If $b_2 \gg b_1 \gg L$, the integral

$$\int_{b_1}^{b_2}\left|H_{\varphi 0}\right|^2 \rho d\rho = \frac{J^2(0)(1 - \cos kL)^2}{4\pi^2 \sin^2 kL} \ln\frac{b_2}{b_1}$$

increases unlimitedly with increasing of the upper limit and, consequently, the resistance R_g increases unlimitedly also.

In [21] it was suggested to assume that the upper limit of the integral (1.75) is equal to $\lambda/2$. Outside the boundaries of this area the component of zonal current, which decreases with increasing distance in accordance with the law $1/\rho$, is dominated. This component is taken into account, when the radiation resistance is calculated. Inside the indicated area, the induction component of zonal current, which decreases according to the law $1/\rho^2$, is dominant. It is believed that this component causes losses in the ground. These qualitative considerations were a cause for quantitative evaluation, justifying actually arbitrary choice of the upper limit.

The expression for the additional resistance of the antenna caused by a non-ideal conductivity of the ground (as the resistance of losses in the ground is called commonly), was derived and published in 1954 [22]:

$$Z_g = -\frac{2\pi}{J^2(0)}\int_a^{\infty} E_\rho H_{\varphi 0}\rho d\rho \qquad (1.77)$$

where

$$H_{\varphi 0} = -\frac{1}{2\pi}\frac{\partial}{\partial \rho}\int_0^L J(z)\frac{\exp\left(-jk\sqrt{\rho^2 + z^2}\right)}{\sqrt{\rho^2 + z^2}}dz. \qquad (1.78)$$

Here E_p is a radial component of the electric field on the ground's surface and $J(z)$ is the current along the antenna.

According to the authors' opinion $H_{\varphi 0}$ is a component of the magnetic field on the surface of the perfectly conducting ground.

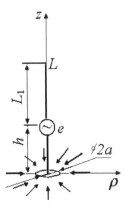

Figure 1.9 Magnetic field near a radiator.

Expression (1.77) is derived by means of an intricate procedure, using direct and inverse Fourier-Bessel transformation. A similar result can be obtained using the theorem about the oscillating power.

This theorem is applied to a volume V, bounded by a hemisphere S_R of a large radius R, a ground surface S_g and an antenna surface S_A (Figure 1.10). Let R tend to infinity. Since in the steady-state mode the energy within the volume is constant, then

$$\int\limits_{(V)} \vec{E}j\,dV = \int\limits_{S=S_R+S_A+S_g} \left[\vec{E},\vec{H}\right]_n dS + j\omega \int\limits_{(V)} \left(\mu\vec{H}^2 + \varepsilon\vec{E}^2\right)dV, \qquad (1.79)$$

where j is a density of an extraneous current, n is an outward normal to the surface S, ω is a circular frequency, and μ and ε are permeability and permittivity of a free space relatively.

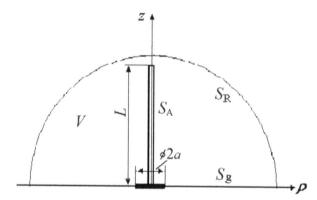

Figure 1.10   Closed surface around a radiator.

The left part of (1.79) is the total (active and oscillating) power associated with the energy of external sources, i.e. with the antenna. It is equal to

$$P_I = \int\limits_0^L E_z J(z)\,dz = J^2(0)(Z_{A0} + Z_g). \qquad (1.80)$$

Here Z_{A0} is the input impedance of the antenna in the case of a perfectly conducting ground. In the absence of losses in the wires, the active component of this impedance is equal to the radiation resistance. Within the limits of solution accuracy, it is supposed that the ground conductivity is high, and the electromagnetic field coincides with the field of an antenna mounted on the ground with infinite conductivity, i.e. losses in the ground do not affect the radiation resistance.

The right part of (1.79) is equal to

$$P_{II} = P_1 + P_2 + P_3 + 2j\omega W. \qquad (1.81)$$

Here, $P_1 = \int\limits_{S_R} [\vec{E},\vec{H}]_n\,dS_R$, $P_2 = \int\limits_{S_A} [\vec{E},\vec{H}]_n\,dS_A$, $P_3 = \int\limits_{S_g} [\vec{E},\vec{H}]_n\,dS_g$ and $W = \frac{1}{2}\int\limits_V \left(\mu\vec{H}^2 + \varepsilon\vec{E}^2\right)dV$ are

total powers of the radiation, of the losses in the wires, of the losses in the ground, as well as the oscillating energy in the volume V. Since electromagnetic fields are considered the same, when the ground conductivity is great or infinite, then

$$J^2(0)Z_{A0} = P_1 + P_2 + 2j\omega W, \quad P_3 = J^2(0)Z_g,$$

i.e.

$$Z_g = \frac{1}{J^2(0)} \int\limits_{(S_g)} \left[\vec{E}, \vec{H}\right]_n dS_g = -\frac{2\pi}{J^2(0)} \int\limits_0^\infty E_\rho H_\varphi \rho \, d\rho. \tag{1.82}$$

Here, E_ρ and H_φ are the field components on the surface of a highly conducting ground. Expression (1.82) differs from (1.77) only by substitution H_φ for $H_{\varphi 0}$. As is seen from (1.82), it is necessary to include in the integrand the field of not perfectly conducting, but of the real ground. Since $J(z)$ is the current of the real antenna, which like the input impedance is distinguished from the current along the antenna mounted on a perfectly conducting ground, then H_φ is the magnetic field on the surface of the ground, which has a finite conductivity. The statement in [22] that it is a component of the magnetic field in the case of the perfectly conducting ground is a mistake. This erroneous argument was repeated by other authors too (see, for example, [23]).

Thus the theorem about the oscillating power allowed not only to obtain simply and clearly the result derived by means of an intricate procedure, but enabled to discover the mistake [24].

Impedance boundary conditions on the ground surface are of the form

$$Z_0 = -E_\rho / H_\varphi, \tag{1.83}$$

where Z_0 is the surface impedance. Here, the minus sign is caused by the fact that the current density $\vec{j} = [\vec{n}, \vec{H}]$ is directed radially toward the origin along the ground. For the ground with high conductivity

$$Z_0 = R_0(1 + j), \tag{1.84}$$

i.e.

$$Z_g = \frac{2\pi R_0(1+j)}{J^2(0)} \int\limits_a^\infty H_\varphi^2 \rho \, d\rho. \tag{1.85}$$

The active component is

$$R_g = \operatorname{Re} Z_g = \frac{2\pi R_0}{J^2(0)} \int\limits_a^b \left(H_{\varphi 1}^2 - H_{\varphi 2}^2 - 2H_{\varphi 1}H_{\varphi 2}\right) \rho \, d\rho, \tag{1.86}$$

where $H_{\varphi 1} = \operatorname{Re} H_\varphi$, $H_{\varphi 2} = \operatorname{Im} H_\varphi$.

The difference between the results of Waite-Pope and Brown is clear from (1.86) and (1.75). If $H_\varphi = H_{\varphi 0}$, when $b_2 \gg b_1 \gg L$,

$$\int\limits_{b_1}^{b_2} \left(H_{\varphi 10}^2 - H_{\varphi 20}^2 - 2H_{\varphi 10}H_{\varphi 20}\right) \rho \, d\rho = J^2(0)(1 - \cos kL)^2 F / (2\pi \sin kL)^2,$$

where

$$F = -\int\limits_{b_1}^{b_2} \left(\frac{\cos 2k\rho}{\rho} + \frac{\sin 2k\rho}{\rho}\right) d\rho = -\sqrt{2} \int\limits_{b_1}^{b_2} \frac{\cos 2k\rho}{\rho} \, d\rho \le \frac{1}{kb_1\sqrt{2}},$$

i.e. integral in (1.86) converges.

Just as expression (1.86) follows from the theorem about the oscillating power, expression (1.75) with an upper limit equal to infinity can be obtained on the basis of the theorem about the complex power. Both formulas follow from the fact that the additional instantaneous power, created by the generator because of losses in the ground, at any given point of time, in accordance with the law of conservation of energy, is equal to the instantaneous power losses in imperfectly conducting ground. Therefore, both expressions should be true.

Integrands in these expressions are different, because Re $(E_\rho H_\varphi) \neq$ Re $(E_\rho H_\varphi^*)$. However, this difference does not exclude equality of integrals. The magnitudes of R_g must be equal to each other upon substituting into integrals a magnetic component H_φ of field on the surface of a real ground. For this, when $\rho \gg L$, the magnitude of $H_{\varphi 2}$ must be equal to zero, i.e. the tangential component of the magnetic field on the ground surface in the far zone must be in phase with the current $J(0)$ in the antenna base.

It is easy to verify that $H_{\varphi 0}$ satisfies this requirement only in the vicinity of the antenna: if $\rho \ll L$, $H_{\varphi 0} = J(0)/2\pi\rho$). When $\rho \gg L$, $H_{\varphi 20}/H_{\varphi 10} = \cot kL$, i.e., $H_{\varphi 0}$ differs substantially from H_φ. The experiment confirms that H_φ coincides with $H_{\varphi 0}$ only at a short distance from the antennas [25]. Let $H_{\varphi 0}$ be different from H_φ on some complex value: $H_{\varphi 0} - H_\varphi = M_1 + jM_2$. Then one can find the difference of integrands in expressions for R_g in the cases of perfectly and imperfectly conducting ground. Using expressions (1.75) and (1.82), we obtain that this difference in the first case is equal to

$$\Delta_1 = |H_{\varphi 0}|^2 - H_\varphi^2 = M_1^2 + M_2^2 + 2H_\varphi M_1,$$

and in the second case it is

$$\Delta_2 = (H_{\varphi 10}^2 - H_{\varphi 20}^2 - 2H_{\varphi 10}H_{\varphi 20}) - H_\varphi^2 = M_1^2 - M_2^2 + 2H_\varphi(M_1 + jM_2).$$

The magnitude Δ_1 depends on the sum of the squares of the real and imaginary components of error, and magnitude Δ_2 depends on their difference. If $M_1 + jM_2$ is a value of an order $\exp(-jk\rho)$ and b_2 grows, an integral $\int_{b_1}^{b_2} \Delta_1 \rho d\rho$ unlimitedly increases, and integral $\int_{b_1}^{b_2} \Delta_2 \rho d\rho$ tends to be zero in proportion to $1/b_1$. The second version is natural, since the tangential component of the magnetic field on the surface of the real ground far from the antenna cannot have the character of no damped spherical electromagnetic wave, i.e. cannot contain summands of order $\exp(-jk\rho)/\rho$ incoming in the expression for $H_{\varphi 0}$.

The authors of [22] attempted to calculate a change of radiated power, caused by the finite conductivity of the ground. They considered that it is equal to a difference between the additional power of the generator determined in accordance with (1.77) and the power of losses in the ground determined by (1.75). This attempt is incorrect, since within the limits of accuracy of the proposed method the radiated power does not depend on the conductivity of the ground, and the power of losses in the ground and the additional power of the generator are identical.

For a radiator, whose feed point is shifted from the middle to point $z = h$ (Figure 1.11).

$$J(z) = \begin{cases} J(0)\cos kz / \cos kh, & 0 \leq z \leq h, \\ J(0)\sin k(L-z)/\sin k(L-h), & h \leq z \leq l. \end{cases} \tag{1.87}$$

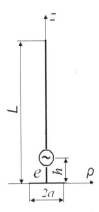

Figure 1.11   A radiator with a shifted feed point.

The magnetic field on the surface of a real ground

$$H_\varphi = j \frac{J(0)}{2\pi\rho \sin k(L-h)} \times$$

$$\left\{ \exp\left(-jk\sqrt{\rho^2+L^2}\right) - \exp\left(-jk\sqrt{\rho^2+h^2}\right)\left[\cos k(L-h) - \tan kh \sin k(L-h)\right]\right\}. \quad (1.88)$$

Substituting (1.88) into (1.86), we find

$$Z_g = \frac{R_0}{4\pi\sin^2 \alpha_2}[F_1 + F_2 + j(F_1 - F_2)], \quad (1.89)$$

where $\alpha_2 = k(L - h)$, F_1 and F_2 for the short radiator with a small radius a of grounding are equal to

$$F_1 = 2\alpha_2^2 \left[\ln(L + h/2a) - 0.5\right] + 4\alpha\,\alpha_0\,\ln(\alpha_1^2/4\alpha\,\alpha_0),\ F_2 = 0. \quad (1.90)$$

Here, $\alpha = kL$, $\alpha_0 = kh$, $\alpha_1 = k(L + h)$. As is seen from expression (1.89) for the short radiator, the reactive component X_g of the loss impedance has an inductive character and is equal in magnitude to the active component R_g. With growth of h the impedance Z_g increases, since $0 \le h \le L$, and a derivative of Z_g with respect to h, is positive always. In the general case F_1 and F_2 have a more complicated character [24].

Results of calculating loss resistance in the ground (water) in *HF* range are presented in Figure 1.12. It is assumed that the magnitude σ is equal to 3 Sm/m. Dimensions are given in meters. Calculations are made in accordance with (1.89): solid lines—in accordance with the general expression, dotted lines—in accordance with (1.90).

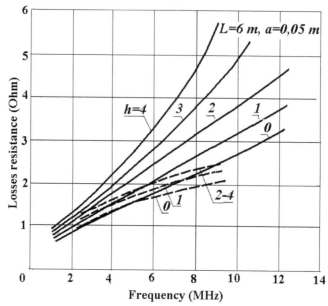

Figure 1.12 Loss resistance in water for a radiator with shifted feed point.

The considered example shows that the theorem about the oscillating power changes significantly the understanding of the processes related with the transfer of power between the objects.

Integral Equation Method

2.1 INTEGRAL EQUATION FOR LINEAR METAL RADIATOR

As shown in Chapter 1, knowledge of the current distribution along a linear radiator allows to determine the electromagnetic field and all electrical characteristics of the radiator. For this reason, calculation of the current distribution is an important problem of the antenna theory.

The current $J(z)$ of a dipole creates an electromagnetic field $E_z(J)$ satisfying the boundary condition

$$E_z(a,z)\big|_{-L \le z \le L} + K(z) = 0 \tag{2.1}$$

Here the cylindrical coordinate system is used. a and L are the radius and the arm length of a dipole, respectively, $K(z)$ is an extraneous emf. Current at the radiator ends is absent:

$$J(\pm L) = 0. \tag{2.2}$$

Expression (2.1) is the mathematical record of the fact that the full field, which is a sum of the extraneous field and the current field, is zero on the surface of a perfectly conducting radiator. The extraneous field is specified usually as the product of potential difference e between the edges of the gap and δ-function. Magnitude $K_1(z) = e\delta(z)$ corresponds to connecting the generator in the radiator middle, at point $z = 0$, and $K_2(z) = e\delta(z - h)$ corresponds to its displacement, i.e. to connecting the generator at point $z = h$.

Equation (2.1) contains as in embryo all the integral equations of the theory of the thin antennas. The external appearance of the equations depends mostly on the selection of function $E_z(J)$. For example, using (1.27), we obtain the integral equation of Hallen for the current along a filament.

$$\int_{-L}^{L} J(\varsigma)G_1 d\varsigma = -\frac{j}{Z_0}\left(C\cos kz + \frac{e}{2}\sin k|z|\right), \tag{2.3}$$

where $G_1 = \exp(-jkR_1)/(4\pi R_1)$, $R_1 = |z - \varsigma|$. Using (1.28), we obtain Hallen's integral equation for the current along a straight thin-wall metal cylinder (the equation with exact kernel)

$$\frac{1}{2\pi}\int_{-L}^{L} J(\varsigma)\int_0^{2\pi} G_2 d\varphi d\varsigma = -\frac{j}{Z_0}\left(C\cos kz + \frac{e}{2}\sin k|z|\right). \tag{2.4}$$

Here $G_2 = \exp(-jkR_2)/(4\pi R_2)$, $R_2 = \sqrt{(z-\varsigma)^2 + 4a^2\sin^2\varphi/2}$. The integral equation for the current along a filament of a finite radius (the equation with approximate kernel) is widely used:

$$\int_{-L}^{L} J(\varsigma)G_3 d\varsigma = -\frac{j}{Z_0}\left(C\cos kz + \frac{e}{2}\sin k|z|\right), \tag{2.5}$$

where $G_3 = \exp(-jkR_3)/(4\pi R_3)$, $R_3 = \sqrt{(z-\varsigma)^2 + a^2}$. Constant C in each equation is found from condition (2.2).

Substituting filament field $E_z(J)$ into (2.1) in accord with (1.27) and replacing R_1 by R_3, one can obtain Pocklington's equation [26]

$$\int_{-L}^{L} J(\varsigma)\left(k^2 G_3 + \frac{\partial^2 G_3}{\partial z^2}\right)d\varsigma = -j\omega\varepsilon K(z), \tag{2.6}$$

which also is the integral equation for the current along a filament of a finite radius.

The first solution of Hallen's equation with an approximate kernel was found by Hallen himself and is described in detail in [10]. The solution uses magnitude $\Omega = 2\ln(2L/a) = 1/\chi$ as the parameter, in inverse powers of which function $J(z)$ is expanded into a series. By means of a successive approximation method (iterative procedure) one can obtain the expression

$$J(z) = j\frac{e}{60\Omega}\cdot\frac{\sin k(L-|z|) + N_1(z)/\Omega + N_2(z)/\Omega^2 + ...}{\cos kL + B_1(L)/\Omega + B_2(L)/\Omega^2 + ...} = J_{0H}(z)/\Omega + J_{1H}(z)/\Omega^2 + ..., \tag{2.7}$$

where $J_{0H}(z) = j\dfrac{e}{60\cos kL}\sin k(L-|z|)$, $J_{1H}(z) = j\dfrac{e}{60}\left[\dfrac{N_1(z)}{\cos kL} - \dfrac{B_1(L)}{\sin k(L-|z|)}\right]$. Functions $N_i(z)$ and $B_i(z)$ are integrals, which can be expressed in terms of integral sine and cosine.

The iterative procedure proposed by King and Middleton [27] yields more accurate results. The common expression for the current in it is similar to (2.7), but expansion parameter Ω is replaced with Ψ. For example, zero approximation instead of $J_{0H}(z)/\Omega$ is given by

$$J_{0KM}(z)\Psi = j\frac{e}{60\Psi\cos kL}\sin k(L-|z|).$$

To find expansion parameter Ψ, magnitude $\psi(z)$ is used. It is calculated as

$$\psi(z) = \int_{-L}^{L} \frac{J_{0KM}(\varsigma)}{J_{0KM}(z)}\cdot\frac{\exp(-jkR)}{R}d\varsigma.$$

By way Ψ, the value of $\psi(z)$ at point $z = z_m$, where the current is maximum or close to maximum, was taken, i.e.

$$\Psi = \begin{cases} \psi(0), & kL \leq \pi/2, \\ \psi(L-\lambda/4), & kL > \pi/2. \end{cases}$$

Such selection of the expansion parameter is caused by the fact that function $\psi(z)$ is proportional to the ratio of vector potential $A_z(z)$ at point z on the antenna surface to current $J(z)$ in the same cross-section. For that reason, function $\psi(z)$ varies slowly along the antenna, or more precisely it is almost constant except for the segments near the wires ends.

2.2 INTEGRAL EQUATION OF LEONTOVICH-LEVIN

The Leontovich-Levin equation [18] played an important part in the progress of the theory of the thin antennas. If the electrical currents parallel to the z-axis and having a circular symmetry are taken by a source of electromagnetic field:

$$j = j_z e_z, j_z = j_z(z) = const(\varphi), \tag{2.8}$$

a vector potential A of a field has only component A_z, which on the surface of the radiator model in a shape of a thin-walled straight metal cylinder with circular cross-section of a radius a, is equal to

$$A_z(\rho,z) = \frac{\mu}{8\pi^2} \int_0^{2\pi} T(z,\varphi)d\varphi, \tag{2.9}$$

where

$$T(z,\varphi) = \int_{-L}^{L} J(\varsigma)\frac{\exp(-jkR)}{R}d\varsigma, R = \sqrt{(z-\varsigma)^2 + \rho_1^2}, \rho = \sqrt{\rho^2 + a^2 - 2a\rho\cos\varphi}, k = \omega\sqrt{\mu\varepsilon}$$

is the propagation constant in the medium surrounding the antenna, ω is the circular frequency, μ and ε are the absolute permeability and permittivity of the medium respectively.

If to integrate $T(z, \varphi)$ by parts and to use successively the circumstance that the radiator radius is small in comparison with its length and the wavelength, i.e. if to neglect by the summands of order of a/L and ka and to keep the summands proportional to the logarithm of these quantities, we obtain:

$$T(z,\varphi) = -2J(z)\ln p\rho_1 - \int_{-L}^{L} \exp\left(-jk|\varsigma-z|\right)\ln 2p|\varsigma-z| \times \left[\frac{dJ(\varsigma)}{d\varsigma} - jkJ(\varsigma)sign(\varsigma-z)\right]d\varsigma,$$

where p is a some constant having the dimensions of inverse length. Since at $\rho > a$,

$$\int_0^{2\pi} \ln p\rho_1 d\varphi \equiv \int_0^{2\pi} \ln(p\sqrt{\rho^2 + a^2 - 2a\rho\cos\varphi})d\varphi = 2\pi\ln p\rho,$$

then

$$A_z(\rho, z) = (\mu/4\pi)[-2J(z)\ln p\rho + V(J, z)], \tag{2.10}$$

where

$$V(J,z) = \int_{-L}^{L} \exp\left(-jk|\varsigma - z|\right) \ln 2p|z - \varsigma| \left[jkJ(\varsigma) + sign(z - \varsigma)dJ(\varsigma)/d\varsigma\right]d\varsigma .$$

The tangential component of the electric field of the antenna is

$$E_z(\rho, z) = -j\frac{\omega}{k^2}\left(k^2 A_z + \frac{\partial^2 A_z}{\partial z^2}\right). \tag{2.11}$$

This expression coincides with the first expression of (1.25). Substituting (2.10) into (2.11) and setting ρ equal to a, we find:

$$E_z(a, z) = \frac{1}{4\pi j\omega\varepsilon}\left[\chi^{-1}\left(\frac{d^2 J}{dz^2} + k^2 J\right) + \frac{d^2 V}{dz^2} + k^2 V\right]. \tag{2.12}$$

Here $\chi = -1/(2 \ln pa)$ is a small parameter of the theory of the thin antennas, used in [18]. As is shown in [28], in the capacity of constant $1/p$, one should choose the distance to the nearest inhomogeneity, i.e. the smallest of three magnitudes: wavelength λ, antenna length $2L$ and the radius R_c of its curvature. In case of a straight radiator, the length of which does not exceed the wavelength, one can consider that $1/p = 2L$, i.e.

$$\chi = 1/[2\ln(2L/a)] \quad \text{or} \quad \chi = 0.5/\ln(2L/ae). \tag{2.13}$$

From (2.11) and (2.12), we obtain the desired equation

$$\frac{d^2 J}{dz^2} + k^2 J = -\chi\left[4\pi j\omega\varepsilon K(z) + \frac{d^2 V}{dz^2} + k^2 V\right]. \tag{2.14}$$

This equation together with the components, which contain the extraneous emf, the current and the current derivative, also has the element incorporating the integral $V(J, z)$ and its derivative. It is known that one concentrated emf cannot create the sinusoidal current along the dipole [29]. The mentioned element is the additional emf, which depends on the current of the antenna. This emf is distributed along the antenna and takes the radiation into account.

   The meaning of manipulations performed during derivation of (2.14), firstly, is that a logarithmic singularity was set off. The function A_z in expression (2.10) including integral $V(J, z)$ is a continuous function everywhere in contrast to the original integral (2.9). Another important advantage of the equation (2.14) is the absence of an argument φ, since the integration with respect to φ has been executed. Nevertheless, this equation is derived for the current along a straight thin-wall cylindrical antenna, and equation (2.14) is equivalent to the equation of Hallen with exact kernel.

   In [18] in order to solve the equation (2.14), the perturbation method is used, i.e. the solution is sought in the form of expansion into a series in powers of the small parameter χ:

$$J(z) = J_0(z) + \chi J_1(z) + \chi^2 J_2(z) + \dots . \tag{2.15}$$

Substituting this series into the equation (2.14) and equating coefficients of equal powers of χ, we obtain, in the case of an untuned radiator [when $J_0(z) = 0$] the set of equations:

$$\frac{d^2 J_1(z)}{dz^2} + k^2 J_1(z) = -4\pi j\omega\varepsilon K(z), \qquad J_1(\pm L) = 0,$$

$$\frac{d^2 J_n}{dz^2} + k^2 J_n = -4\pi j\omega\varepsilon W(J_{n-1}), \quad J_n(\pm L) = 0, n > 1, \tag{2.16}$$

where

$$-4\pi j\omega\varepsilon W(J_{n-1}) = -\left[\frac{d^2 V(J_{n-1}, z)}{dz^2} + k^2 V(J_{n-1}, z)\right].$$

In the first approximation the current at an arbitrary point of the radiator is

$$J_1(z) = j\frac{e}{60\cos kL}\sin k(L - |z|). \tag{2.17}$$

It is easy to make sure that the input impedance of the antenna in this approximation is

$$Z_{A1} = -j60\,\chi^{-1}\cot kL. \tag{2.18}$$

It has only the reactive component. It coincides with the input impedance of the equivalent long line, whose wave impedance is equal to $60/\chi$. The expression for the antenna current in the second approximation, derived using this procedure for an arbitrary point of the radiator, is given in [19].

Another solution of the set of simultaneous equations (2.16), which was published for the first time in [30], is described in the book [19]. This solution gives the opportunity to clarify some questions of the theory of thin antennas. When $n > 1$, if the value $W(J_{n-1})$ is known, one can use the method of variation of constants. The result is:

$$\chi^n J_n(z) = j\frac{\chi}{30\sin 2\alpha} \times$$

$$\left\{\sin k(L+z)\int_z^L W\left(\chi^{n-1}J_{n-1}\right)\sin k(L-\varsigma)d\varsigma + \sin k(L-z)\int_{-L}^z W\left(\chi^{n-1}J_{n-1}\right)\sin k(L+\varsigma)d\varsigma\right\}.$$

We find magnitudes $W(\chi^{n-1}J_{n-1})$ by substituting n first members of the series (2.15) for the current into (2.12):

$$E_z\,(\chi J_1) = -K(z) + W\,(\chi J_1);\; E_z\,(\chi^n\,J_n) = -W(\chi^{n-1}\,J_{n-1}) + W(\chi^n\,J_n),\, n > 1,$$

i.e.

$$E_z\left(\sum_{n=1}^n \chi^m J_m\right) = -K(z) + W(\chi^n J_n).$$

Then

$$\chi^n J_n(z) = j\frac{\chi}{30\sin 2\alpha}\left\{\sin k(L+z)\int_z^L\left[K(\varsigma) + E_\varsigma\left(\sum_{m=1}^{n-1}\chi^m J_m\right)\right]\sin k(L-\varsigma)d\varsigma + \right.$$

$$\left.\sin k(L-z)\int_{-L}^z\left[K(\varsigma) + E_\varsigma\left(\sum_{m=1}^{n-1}\chi^m J_m\right)\right]\sin k(L+\varsigma)d\varsigma\right\}. \tag{2.19}$$

In particular, if $n = 2$,

$$\chi^2 J_2(z) = j\frac{\chi}{30\sin 2\alpha}\left\{\sin k(L+z)\int_z^L \left[K(\varsigma)+E_\varsigma(\chi J_1)\right]\sin k(L-\varsigma)d\varsigma + \right.$$

$$\left. \sin k(L-z)\int_{-L}^z \left[K(\varsigma)+E_\varsigma(\chi J_1)\right]\sin k(L+\varsigma)d\varsigma\right\}. \tag{2.20}$$

Here $\alpha = kL$. The equation (2.19) allows finding term n of the series (2.15) for the current and accordingly nth approximation, if $(n-1)$th approximation is known. Equation (2.20) allows finding the second approximation for the current at any point of the radiator. For this purpose, as it follows from (2.20), it is necessary to calculate the field of the current found in the first approximation.

As is seen from (2.16), when calculating the second and subsequent terms of the series (2.15), one can consider that the current of the radiator is concentrated on its axis. The accuracy level, accepted in derivation of equation (2.14) (accuracy of order of a/L) is retained. This circumstance simplifies essentially the calculation based on the recurrent formulas.

Expression (2.20) allows to calculate the second approximation for the input current. It depends on the first approximation (2.15), which in accordance with (2.17) at a point $z = 0$ is equal to

$$\chi J_1(0) = j\frac{e\chi\tan\alpha}{60}. \tag{2.21}$$

The second term of the series in accordance with (2.20) at this point is

$$\chi^2 J_2(0) = j\frac{\chi}{30\sin 2\alpha}\sin\alpha\int_{-L}^L \left[K(\varsigma)+E_\varsigma(\chi J_1)\right]\sin k(L-|\varsigma|)d\varsigma = \chi^2 J_{21}(0) + \chi^2 J_{22}(0). \tag{2.22}$$

This value consists of two summands corresponding to two elements in square brackets. It is seen that the first summand is equal to $\chi J_1(0)$. The second summand with allowance for (2.21) is equal to

$$\chi^2 J_{22}(0) = \frac{1}{e}\int_{-L}^L E_\varsigma(\chi J_1)\chi J_1(\varsigma)d\varsigma = -\frac{1}{e}\chi^2 J_1^2(0)Z_{A0} \tag{2.23}$$

where the value of Z_{A0} coincides with (1.50), i.e. Z_{A0} is the known expression for the input impedance of the dipole, which was obtained by method of induced emf (second formulation). In order to refine this expression, one must substitute in the integral the field of the current χJ_1 on the radiator surface

$$E_\varsigma(\chi J_1) = \frac{30\chi J_1(0)e}{j\sin\alpha}\left[\frac{\exp(-jkR_1)}{R_1}+\frac{\exp(-jkR_2)}{R_2}-2\cos\alpha\frac{\exp(-jkR_0)}{R_0}\right],$$

where $R_1 = \sqrt{(L-\varsigma)^2+a^2}, R_2 = \sqrt{(L+\varsigma)^2+a^2}, R_0 = \sqrt{\varsigma^2+a^2}$. The result is

$$\chi^2 J_{22}(0) = \frac{2\chi^2 J_1(0)e}{\sin 2\alpha}(Y_1+Y_2-2Y_0\cos\alpha).$$

Here, $Y_m = \int\limits_0^L \dfrac{\exp(-jkR_m)}{R_m} \sin k(L-\varsigma)d\varsigma$. In particular

$$Y_1 = \frac{1}{2j}\left[\int\limits_0^L \frac{\exp\left(jk\langle -R_1+L-\varsigma\rangle\right)}{R_1} - \frac{\exp\left(jk\langle -R_1-L+\varsigma\rangle\right)}{R_1}\right]d\varsigma.$$

Integral Y_1 consists of two integrals. Applying in the first integral the substitution $t = -R_1 + L - \varsigma$ and in the second integral the substitution $u = -R_1 - L + \varsigma$, we obtain

$$Y_1 = \frac{1}{2j}\left[\int\limits_{L-\sqrt{L^2+a^2}}^{-a} \frac{\exp(jkt)}{t}dt + \int\limits_{-L-\sqrt{L^2+a}}^{-a} \frac{\exp(jku)}{u}du\right] =$$

$$\frac{1}{2j}\left[2Ei(-jka) - Ei(-\frac{jka^2}{2L}) - Ei(-j2kL)\right].$$

Results of the integration contains terms with arguments of the order of kL and with small arguments of the order of ka and smaller. If an argument x is small,

$$Ei\,(jx) = Cix + jSix = \ln \gamma x + jx,$$

where $\ln \gamma = C \approx 0.5772...$ is Euler's constant. Summarizing the integration results gives

$$\chi^2 J_{22}(0) = -\frac{\chi^2 J_1(0)e}{\sin 2\alpha}[\Phi(\alpha) + \Delta], \tag{2.24}$$

where $\Phi(\alpha) = e^{2j\alpha}\,[Ei(-4j\alpha) - 2Ei(-2j\alpha) + \ln\gamma\alpha] + 2\ln 2\gamma\alpha - 2Ei(-2j\alpha) - 2j\,\sin 2\alpha\ln(L/a)$, and the small magnitude Δ is equal to

$$\Delta = j\left[-4ka + ka^2/(2L) - \left[2ka - 3ka^2/(4L)\right]\cos 2\alpha - 3ka^2\sin 2\alpha/4L\right]$$

Calculations show that as a rule $\Delta \ll \Phi(\alpha)$, i.e. the value of Δ may be neglected:

$$Z_{A0} = 30\ \Phi(\alpha)/\sin^2\alpha \tag{2.25}$$

The comparison shows that, if the value of Δ is neglected, then one can use the expression (2.23). The reason for this coincidence will be explained later. But at first it is necessary to show the consequences, which are follow from this result. In accordance with (2.15) and (2.20) – (2.25), the input current of the antenna in the second approximation is

$$J(0) = 2\chi J_1(0) - \chi^2 J_1^2(0)\left[Z_{A0} + \frac{30\Delta}{\sin^2\alpha}\right] \tag{2.26}$$

This means that the input impedance of the antenna in this approximation is equal to

$$Z_{A2} = e:\left[2\chi J_1(0) - 30\chi^2\Phi(\alpha)J_1^2(0)/\sin^2\alpha\right] = \frac{30}{\chi^2\,\mathrm{Re}\,\Phi/\left(4\cos^2\alpha\right) + j\left[\gamma\tan\alpha + \chi^2\,\mathrm{Im}\,\Phi/\left(4\cos^2\alpha\right)\right]}. \tag{2.27}$$

2.3   INTEGRAL EXPRESSION FOR CURRENT AND EMF METHOD

Let us return to the expression (2.25). In accordance with (2.15) the input current of no resonant radiator in the second approximation is equal to

$$J(0) = +\chi J_1(0) + \chi^2 J_2(0),$$

and $\chi^2 J_2(0) << \chi J_1(0)$, if $J_1(0) \neq 0$. Taking this inequality into account, one can write in a first approximation:

$$Z_{A2} = \frac{1}{\chi J_1(0)} : \left[1 + \frac{\chi^2 J_2(0)}{\chi J_1(0)}\right] \approx \frac{1}{\chi J_1(0)}\left[1 - \frac{\chi^2 J_2(0)}{\chi J_1(0)}\right]. \tag{2.28}$$

As is shown in Section 2.2 – see (2.22),

$$\chi^2 J_2(0) = \chi J_1(0) + \chi^2 J_{22}(0).$$

Substituting $\chi^2 J_2(0)$ into (2.28), we find

$$Z_{A2} \approx -\chi^2 J_{22}(0) / \left\lfloor \chi^2 J_1^2(0) \right\rfloor, \tag{2.29}$$

where, as it follows from (2.21) and (2.23),

$$\chi^2 J_{22}(0) = j\frac{\chi}{60\cos\alpha}\int_{-L}^{L} E_\varsigma(\chi J_1)\sin k\left(L - |\varsigma|\right)d\varsigma. \tag{2.30}$$

It is seen that the substitution (2.30) into (2.19) gives an expression identical to (1.50), i.e. identical to the second formulation of the method of induced electromotive force.

   Thus, when $J_1(0) \neq 0$, the integral formula for the dipole input impedance derived by the method of emf coincides with the integral formula obtained as a result of solving integral equation of Leontovich-Levin. The identity of integral formulas explains the known fact of coincidence of input impedances (in the second approximation) calculated by the two methods and expressed in terms of tabulated functions.

   As it is known, the solution of Leontovich-Levin equation gives in the area of the parallel resonance of the antenna the finite magnitudes of the active and reactive components of the input impedance. The method of emf in contrast to the integral equation gives infinite magnitude of the input impedance. Really, in accordance with (2.25), if α tends to $n\pi$, where n is natural number, then $\sin \alpha$ tends to zero, and the input impedance Z_{A0} grows indefinitely, i.e. this method leads to incorrect results.

   The opposite situation occurs in the area of the serial resonances of the antenna, when $\alpha = (2n + 1)\pi/2$ It is easy to see that in this case the magnitude of the input impedance including the active component in accordance with the expression (2.27) is equal to zero. In this regard, it is expedient to proceed from (2.25). A reason of that is an approximate nature of expression (2.27) in the area of serial resonance.

   Essentially, both formulas—(2.25) and (2.27)—give more accurate results in one area and have an approximate nature in another area. This allows to offer on the basis of both formulas a general expression that gives high and approximately identical accuracy in both areas:

$$Z_A = \frac{\pi - \alpha}{\pi/2}Z_S + \frac{\alpha - \pi/2}{\pi/2}Z_P. \tag{2.31}$$

Here Z_S is the impedance, which is calculated in accordance with (2.25), Z_P is the impedance, which is calculated in accordance with (2.27), $\pi/2$ is the value of kL at the point of the serial resonance, π is the value of kL at the point of the parallel resonance, and α is the value of kL at an arbitrary point. As can be seen from this expression, when α changes from $\pi/2$ to π the value of the first term uniformly decreases from Z_S to zero, and the value of the second term increases uniformly from zero to Z_P.

Figure 2.1 shows the active and reactive components of the input impedance of a symmetrical cylindrical radiator (dipole) depending on the arm length. Components of the input impedance calculated in accordance with expression (2.31) are marked by number 1, in accordance with (2.25)—by number 2, in accordance with (2.27)—by number 3.

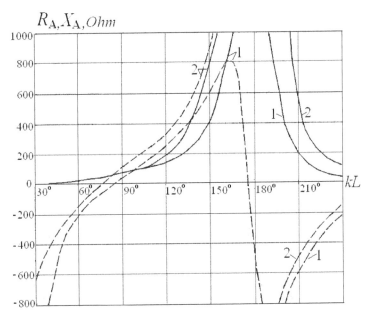

Figure 2.1 Input impedance of a symmetrical cylindrical antenna: curve 1—in accordance with (2.31), curve 2—in accordance with (2.25), curve 3—in accordance with (2.27).

Calculations were made for the radiator with the parameter $\chi = 0.1$, i.e. for the antenna with the rather great transverse dimensions. This choice of a small parameter allows more rigorously analyzing the described procedure. Comparison of the received results with the results of the application of the Moment method shows good agreement of a form of the curves and components of the input impedance. The only difference is that the curves obtained by the Moment method, have a weak shift of resonances to side the lower frequencies.

The calculation results are compared in Table 2.1 with the results presented in [16]. Input impedance is given for four variants of the arm length. It is assumed that the radius of the antenna is equal to 0.01 wavelengths, i.e. it varies with the frequency. Such parameters adopted for ease of comparison of the input impedance with results of other authors collected in [16]. The values presented in this article in accordance with (2.31), are given in Table 2.1 as the results of solving equation of Leontovich-Levin.

The methods used by different authors at different times have been seriously substantiated. The equation of Leontovich-Levin is the most rigorous variant among integral equations of the theory of thin antennas. Unfortunately, for a long time it was in the shadow of other well-known equations. It is necessary to show that this equation, which was created by our predecessors, has not lost its value. Small changes of methods of solving this equation allow to obtain new rigorous results.

The method, which was employed for calculating the input impedance of dipole in the second approximation, allows to calculate the current distribution in this approximation:

$$J(z) = 2\chi J_1(z) + \chi^2 J_{22}(z) = \frac{2\chi J_1(0)}{\sin\alpha}\sin(\alpha - |t|) - \frac{30\chi^2 J_1^2(0)}{\sin^3\alpha}\Phi(t)\sin(\alpha + t),$$

where $\Phi(t) = -Ei(-2j\alpha) + Ei(-2jt) - Ei[-2j(\alpha - t)]$

$+e^{2j\alpha}\{Ei(-4j\alpha) - Ei(-2j\alpha) + Ei(-2jt) - Ei[-2j(\alpha + t)]\} + \ln[2\gamma\alpha(\alpha - t)/t] + e^{-j\alpha}\ln(\alpha + t/2t), (2.32)$

and $t = kz$.

Table 2.1   Input Impedances of Cylindrical Dipole

Method	Arm length			
	$\lambda/8$	$\lambda/4$	$3\lambda/8$	$\lambda/2$
Equation of Hallen, first approximation	19.4–j359	80.4+j35.7	268+j526	1685–j1357
Equation of Hallen, second approximation	16.0–j240	87.3+j35.7	437+j318	559–j599
Solution of King-Middleton, second approximation	14.0–j166	92.5+j38.3	543+j32.2	177–j339
Storer's approximation	11.6–j185	101+j32.8	566+j3.1	290–j363
Method of emf, second formulation	13.4–j391	73.1+j42.5	386+j533	∞
Equation of Leontovich-Levin	13.4–j391	73.1+j42.5	334+j520	1296–j816

2.4   INTEGRAL EQUATION FOR TWO RADIATORS

Generalizing the Leontovich-Levin equation, one can write similar equations for the currents in the system of several radiators, i.e. in antenna array. Consider two parallel symmetrical radiators of different lengths, displaced axially relative to each other (Figure 2.2). In accordance with (1.25) and (2.11), if electrical currents $J_I(\sigma)$ and $J_{II}(\varsigma)$ flow along the radiators, they create the field

$$E_z = -j\frac{\omega}{k^2}\left(k^2 A_z + \frac{\partial^2 A_z}{\partial z^2}\right).$$

In accordance with the superposition principle

$$A_z = A_{z1} + A_{z2}.$$

Model of each radiator is a straight thin-wall circular metal cylinder with radii a_1 and a_2, respectively. The vector potential of the field created by the current of the cylinder is calculated with the help of (2.9), and distances R_1 and R_2 between the observation point with coordinates (ρ_1, φ_0, z) and integration points (a_1, φ, σ) and (a_2, ψ, ς) are calculated in accordance with the explication to this expression. If the observation point is situated near the surface of a first radiator, then at

$$a_1, a_2 \ll d, \tag{2.33}$$

where d is the distance between the axes of the radiators, one can say:

$$A_{z2} = \frac{\mu}{4\pi} \int_{h-L_2}^{h+L_2} J_{II}(\varsigma) \exp(-jkR_2) / R_2 \, d\varsigma, R_2 = \sqrt{(z-\varsigma)^2 + d^2}. \tag{2.34}$$

Vector potential A_{z1}, as in the case of a single radiator, if ρ_1 is small, has a logarithmic singularity. If this singularity was set off, then

$$A_{z1}(a_1, z) = \frac{\mu}{4\pi} \left[\chi_1^{-1} J_1(z) + V(J_1, z) \right], \tag{2.35}$$

Here $\chi_1 = 1/[2 \ln (2L_1/a_1)]$ is a small parameter, and $V(J_1, z)$ is the integral, expression for which is presented in Section 2.2. Vector potential A_{z2} has no such singularity, since, if the assumption (2.33) is true, the distance R_2 is not small at any ς: $R_2 \geq d - \rho_1 - a_2$. Accordingly, the tangential component of the electric field created by current J_I contains a large magnitude of order of χ_1^{-1}:

$$E_z(J_1, a_1, z) = \frac{1}{4\pi j\omega\varepsilon} \left\{ \chi_1^{-1} \left[\frac{d^2 J_1(z)}{dz^2} + k^2 J_1(z) \right] + \frac{d^2 V(J_1, z)}{dz^2} + k^2 V(J_1, z) \right\}, \tag{2.36}$$

and field $E_z(J_{II}, a_1, z)$ created by current J_{II} of the second radiator on the surface of the first radiator does not contain large component for reasons given above.

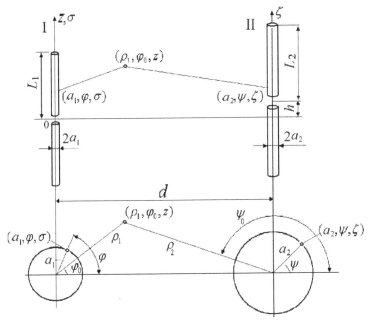

Figure 2.2   System of two parallel radiators.

A boundary condition, similar to (2.1), must be met on the surface of the first radiator:

$$E_z(J_I,a_1,z)+E_z(J_{II},a_1,z)\big|_{-L\leq z\leq L}+K_I(z)=0, \tag{2.37}$$

where $K_I(z)$ is an extraneous emf. Substituting (2.36) into (2.37), we obtain the equation for $J_I(z)$:

$$\frac{d^2 J_I}{dz^2}+k^2 J_I(z)=-4\pi j\omega\varepsilon\chi_1\left[K_I(z)+W(J_I,z)+E_z(J_{II},a_1,z)\right], \tag{2.38}$$

where $4\pi j\omega\varepsilon W(J_I,z)=d^2V(J_I,z)/dz^2+k^2V(J_I,z)$, and $J_I(\pm L_1)=0$.

The right part of this expression contains three components in square brackets: the first component is the exciting emf, the second component is emf which takes the radiation into account, and the third component is emf caused by influence of the second radiator.

While solving the equation (2.21), we present the currents $J_I(z)$ and $J_{II}(\varsigma)$ in the form of series in powers of small parameters χ_1 and χ_2, respectively. Since functionals $W(J_I,z)$ and $E_z(J_{II},a_1,z)$ are linear, they can also be presented in the form of similar series. If χ_1 and χ_2 have the same order of smallness:

$$\chi_1 \sim \chi_2, \tag{2.39}$$

then the equation (2.38) for not resonant radiator reduces to the set of equations, which is a generalization of the set (2.16), written for a single radiator:

$$\frac{d^2 J_{I1}(z)}{dz^2}+k^2 J_{I1}(z)=-4\pi j\omega\varepsilon K_I(z), \qquad J_{I1}(\pm L_1)=0,$$

$$\frac{d^2 J_{In}(z)}{dz^2}+k^2 J_{In}(z)=-4\pi j\omega\varepsilon\left[W(J_{I,n-1})+\left(\frac{\chi_2}{\chi_1}\right)^{n-1}E_z(J_{II,n-1})\right], J_{In}(\pm L_1)=0, n>1. \tag{2.40}$$

As it follows from the first equation of the system (2.40) for the radiator excited by concentrated emf $K_I(z)=e_I\delta(z)$, its current in the first approximation in the presence of the second radiator also has a sinusoidal nature:

$$\chi_1 J_{I1}(z)=j\frac{\chi_1 e_I}{60\cos kL_1}\sin k(L_1-|z|). \tag{2.41}$$

If $n>1$, then, in accordance with (2.40), using the method of variation of constants and considering that magnitudes $W(J_{I,n-1})$ and $E_{II}(J_{II,n-1})$ are known, we obtain

$$\chi_1^n J_{In}(z)=j\frac{\chi_1}{30\sin 2kL_1}\left\{\sin k(L_1+z)\int_z^{L_1}\left[W\left(\chi_1^{n-1}J_{I,n-1}\right)+E_\sigma\left(\chi_2^{n-1}J_{II,n-1}\right)\right]\sin k(L_1-\sigma)d\sigma+\right.$$

$$\left.\sin k(L_1-z)\int_{-L_1}^z W\left[\left(\chi_1^{n-1}J_{I,n-1}\right)+E\sigma\left(\chi_2^{n-1}J_{II,n-1}\right)\right]\sin k(L_1+\sigma)d\sigma\right\}. \tag{2.42}$$

Substituting of the first terms of the series for current $J_I(z)$ into the expression (2.36), allows to find magnitudes $W(\chi_1^{n-1}J_{I,n-1})$:

$$E_z(\chi_1 J_{I1})=-K_I(z)+W(\chi_1 J_{I1})$$

$$E_z\left(\chi_1^n J_{In}\right)=-W\left(\chi_1^{n-1}J_{I,n-1}\right)-E_z\left(\chi_2^{n-1}J_{II,n-1}\right)+W\chi_1^n J_{In}), n>1,$$

i.e.

$$E_z\left(\sum_{m=1}^{n}\chi_1^m J_{I,m}\right) = -K_1(z) - E_z\left(\sum_{m=1}^{n-1}\chi_2^m J_{II,m}\right) + W\left(\chi_1^n J_{In}\right). \qquad (2.43)$$

If to replace n in (2.43) by $(n-1)$, to calculate magnitude $W(\chi_1^{n-1} J_{I,n-1})$ by means of obtained expression, and to substitute it in (2.42), we find the member n of the series for current. In particular, if $n = 2$,

$$\chi_1^2 J_{12}(z) = j\frac{\chi_1}{30\sin 2kL_1}\Bigg\{ \sin k(L_1+z)\int_z^{L_1}\left[K_1(\sigma)+E_\sigma(\chi_1 J_{I1})+E_\sigma(\chi_2 J_{II1})\right]\sin k(L_1-\sigma)d\sigma +$$

$$\sin k(L_1-z)\int_{-L_1}^{z}\left[K_1(\sigma)+E_\sigma(\chi_1 J_{I1})+E_\sigma(\chi_2 J_{II1})\right]\sin k(L_1+\sigma)d\sigma\Bigg\}. \qquad (2.44)$$

Equation (2.44) allows finding the second term of the series for the current at any point of the first radiator. For this purpose, as it follows from (2.44), it is necessary to calculate the fields of the currents in the first approximation. From (2.44), it is see also, as a matter of course, that the magnitude of the second term of the series depends on the geometric dimensions of the second radiator and on the relative position of radiators. In the general case, the expression (2.42), after substituting into it the magnitude $W(\chi_1^{n-1} J_{I,n-1})$ permits to find the member n, if the currents of both radiators are known in approximation $(n-1)$.

From the set of equations (2.40) it follows that, when calculating the second and subsequent terms of the series, one can consider that the current of the first radiator is concentrated on its axis. Also, from (2.34) it follows that the current of the second radiator also is concentrated on the axis. And the accepted in derivation of the equation (2.21) accuracy level (accuracy of order of a_1/L_1) is retained. This circumstance simplifies essentially calculating members of the series for the current based on the recurrence formula, in particular the calculations of the terms n and $(n-1)$, since this formula allows calculating these fields as the fields of the filaments. As a result, calculating the second term of the series for the current of the single radiator, based on using expression (2.44), is simplified, since one can use the expression (2.41) as the first term.

It is interesting to compare the results of solving Leontovich-Levin equation for one and two radiators with solutions obtained by the induced emf method. The input impedance of the first radiator is

$$Z_{AI} = e_I / J_I(0) = \frac{e_1}{\left[\sum_{n=1}^{\infty}\chi_1^n J_{In}(0)\right]}, \qquad (2.45)$$

where $\chi_1^n J_{In}(0) = \dfrac{1}{e_1}\displaystyle\int_{-L_1}^{L_1}\left[K_I(\sigma)+E_\sigma\left(J_I^{(n-1)}\right)+E_\sigma\left(J_{II}^{(n-1)}\right)\right]J_I^{(1)}(\sigma)d\sigma$. Here for simplicity the

following designation is used for the current of first radiator in the approximation n:

$$J_I^{(n)}(z) = \sum_{m=1}^{n}\chi_1^m J_{Im}.$$

The input impedance in the approximation n is equal to

$$J_{AI}^{(n)} = \frac{e_I}{J_I^{(n-1)}(0) + \chi_1^n J_{In}(0)}. \tag{2.46}$$

Let us write the first component of the denominator in the form

$$J_I^{(n-1)}(0) = \frac{1}{e_I} \int_{-L_1}^{L_1} K_I(\sigma) J_I^{(n-1)}(\sigma) d\sigma.$$

Factor $J_I^{(1)}(\sigma)$ in the integrand of the second component of the denominator can be replaced with $J_I^{(n-1)}(\sigma)$, i.e. one can add the terms of higher order to the component of the first order of smallness. Since the polynomial in square brackets of the integrand is a magnitude of $(n-1)$th order of smallness, as is easily seen from (2.43), the addition of terms of higher order does not change the accepted accuracy level. Hence,

$$\chi_1^n J_{In}(0) = J_I^{(n-1)}(0) + \frac{1}{e_I} \int_{-L_1}^{L_1} \left[E_\sigma \left(J_I^{(n-1)} \right) + E_\sigma \left(J_{II}^{(n-1)} \right) \right] J_I^{(n-1)}(\sigma) d\sigma.$$

As a result, we obtain

$$Z_{AI}^{(n)} = \frac{e_I}{\left\{ 2 J_I^{(n-1)}(0) + \dfrac{1}{e_I} \displaystyle\int_{-L}^{L} \left[E_\sigma \left(J_I^{(n-1)} \right) + E_\sigma \left(J_{II}^{(n-1)} \right) \right] J_I^{(n-1)}(\sigma) d\sigma \right\}}. \tag{2.47}$$

One can rewrite the expression (2.28) as

$$Z_{AI}^{(n)} = \frac{e_I}{J_I^{(n-1)}(0)} \Big/ \left[1 + \frac{\chi_1^n J_{1n}(0)}{J_I^{(n-1)}(0)} \right] \approx \frac{e_I}{J_I^{(n-1)}(0)} \left[1 - \frac{\chi_1^n J_{1n}(0)}{J_I^{(n-1)}(0)} \right].$$

If

$$J_I^{(n-1)}(0) \neq 0, \tag{2.48}$$

then

$$Z_{AI}^{(n)} \approx e_I \left[J_I^{(n-1)}(0) - \chi_1^n J_{In}(0) \right] \Big/ \left[J_I^{(n-1)}(0) \right]^2,$$

i.e.

$$Z_{AI}^{(n)} \approx -\frac{1}{\left[J_I^{(n-1)}(0) \right]^2} \int_{-L_1}^{L_1} \left[E_\sigma \left(J_I^{(n-1)} \right) + E_\sigma \left(J_{II}^{(n-1)} \right) \right] J_I^{(n-1)}(\sigma) d\sigma. \tag{2.49}$$

This expression generalizes the expression (1.50), which was presented in Section 1.4 and is called the second formulation of the induced emf method. In (1.50) the sinusoidal distribution of the current along the radiator is used to calculate the input impedance in the second approximation with respect to χ. The $(n-1)$th approximation for the current in the form of (2.49) permits calculating the input impedance in the nth approximation with respect to χ. The equation (1.50) is applicable only to a single radiator, whereas equation (2.49) is true in the presence of the second radiator too. The expressions (2.47) and (2.27) allow to write expressions (1.69) and (1.70).

Comparison of these results with results obtained by the induced emf method allows to draw the following conclusions:

The integral formula of the induced emf method for the radiator input impedance, if the condition (2.48) holds, coincides completely with the integral formula obtained from the solution of the integral equation.

Really, if to take the expression (2.41) as the first term of the series (2.15) and further to perform the transition from the input current to the input impedance of the radiator, which is similar to the transition from (2.45) to (2.49), the result will be identical with the result of calculation performed by the induced emf method. Since the condition (2.48) at the point of a parallel resonance for the sinusoidal distribution of the current is not met, the method of emf gives incorrect results near that point. Resistance and reactance will be increased indefinitely, while the measured values of the input impedances will remain finite.

The derivation of (2.49) uses conditions (2.33) and (2.39). The fulfillment of the conditions is necessary to avoid possible mistakes.

The first formulation of the induced emf method can be reduced to a form similar to expression (2.49):

$$Z_{AI}^{(n)} \approx \frac{1}{\left| J_I^{(n-1)}(0) \right|^2} \int_{-L_1}^{L_1} \left[E_\sigma \left(J_I^{(n-1)} \right) + E_\sigma \left(J_{II}^{(n-1)} \right) \right] J_I^{(n-1)*}(\sigma) d\sigma . \tag{2.50}$$

As it is shown in [19], this expression is obtainable by the direct transition from (2.49). But for that, the equality $J_1^{(1)}(\sigma) = -J_1^{(1)*}(\sigma)$ must be accomplished. In accordance with this equality, the current should be purely reactive, i.e. there should be no losses in the radiator and in the environment.

From the foregoing it follows that in the analysis of the methods of calculating characteristics of the antenna it is necessary to take into account that in the second and subsequent approximation the current along the antenna wire contains not only reactive but also active component. The method of induced emf is equivalent to the analysis of the antenna in the second approximation. The discussion, devoted to the first and second formulations of this method, considered the question of the solution stability in each of these formulations. Stability of the solution using the second formulation was immediately proven. One well-known specialist presented a proof of stability of the solution using the first formulation. The error of the published proof consisted in that the author proceeded from purely reactive magnitude of the current.

2.5   INTEGRAL EQUATIONS FOR COMPLICATED ANTENNAS

The previous sections were devoted to integral equations for the currents along straight metal radiators. Antennas with distributed and concentrated loads are more complicated variants of radiators. An antenna in the form of a metal rod coated by a layer of magneto dielectrics (Figure 1.5a) is an example of a radiator with distributed load. In contrast to (2.1), the boundary condition on the surface of a dipole with distributed load is given as

$$\frac{E_z(a,z) + K(z)}{H_\varphi(a,z)} \bigg|_{-L \le z \le L} = Z(z), \tag{2.51}$$

where $E_z(a, z)$ and $H_\varphi(a, z)$ are the tangential component of the electric field and the azimuthal component of the magnetic field, respectively, and $Z(z)$ is a surface impedance, which is in the general case dependent on coordinate z. The boundary condition of such kind is valid, if the structure of the field inside one medium (e.g., inside a magneto dielectric sheath) is independent of a field structure in another medium (ambient space).

If boundary conditions (2.51) on the antenna surface are valid and the surface impedance substantially changes the distribution of current along the antenna already in the first approximation, the antenna is called an impedance antenna.

In accordance with the equivalence theorem, one can, when calculating the field, replace the radiator by the field on its boundary, and afterwards use only the field as source of the signal. On the other hand, for clearness and simplicity, it is expedient to metallize the antenna surface. Surface density $\vec{j}_S$ of the electric current is related to magnetic field strength $\vec{H}$ as $\vec{j}_S = [\vec{e}_\rho, \vec{H}]$, where $\vec{e}_\rho$ is the unit vector in the ρ direction, i.e.

$$H_\varphi(a,z) = j_z(z) = J(z)/(2\pi a), \tag{2.52}$$

where $J(z)$ is the linear current along a metallized antenna (it is equal to the total radiator current).

The tangential component of the field is determined by the expression (2.36). Substituting (2.36) and (2.52) into (2.51), we obtain the equation for the current along an impedance radiator:

$$\frac{d^2 J(z)}{dz^2} + k^2 J(z) = -4\pi j\omega\varepsilon\chi\left[K(z) + W(J,z) - \frac{J(z)Z(z)}{2\pi a}\right], \tag{2.53}$$

which should satisfy the condition (2.2). Three components in the right part of the equation correspond to the exciter emf, to the radiation, and to the presence of the distributed load, respectively.

As before, we shall seek the solution as a series in powers of small parameter χ, presenting the surface impedance as $2jkZ(z)/(aZ_0) = \chi^{-1}U$. That allows obtaining the set of equations for the not resonant radiator:

$$\frac{d^2 J_1(z)}{dz^2} + k_1^2 J_1(z) = -4\pi j\omega\varepsilon K(z), \qquad J_1(\pm L) = 0,$$

$$\frac{d^2 J_n(z)}{dz^2} + k_1^2 J_n(z) = -4\pi j\omega\varepsilon W(J_{n-1}, z), J_n(\pm L) = 0, n > 1. \tag{2.54}$$

Here $k_1^2 = k^2 - U$. If both components are of the same order of smallness, the surface impedance substantially affects the distribution of current, and one may attribute to the magnitude $k_1 = \sqrt{k^2 - j2k\chi Z(z)/(aZ_0)}$ the meaning of a new wave propagation constant along an antenna. From the first equation of set (2.54) it follows that the current, distributed along the antenna, has in the first approximation sinusoidal character

$$\chi J_1(z) = j\frac{k\chi e}{60k_1 \cos k_1 L}\sin k_1(L-|z|). \tag{2.55}$$

Ratio k_1/k is usually referred the slowing.

Solution of the equation for $J_2(z)$ of the set (2.54) allows to find the current in the second approximation, to determine the active component of input impedance and to define more precisely the magnitude of reactive component. If to use the integral method of solving the equation described in Section 2.2, the additional component

$$Z/(2\pi a)\sum_{m=1}^{N}\chi^m J_1^m(z)$$

will appear in the right part of expression (2.43). If the condition (2.48) holds, then, by analogy to (2.49), we find for the single radiator

$$Z_A^{(n)} \approx -\frac{1}{\left[J^{(n-1)}(0)\right]^2}\int_{-L_1}^{L_1}\left[E_z\left(J^{(n-1)}\right)-\frac{Z}{2\pi a}J^{(n-1)}\right]J_I^{(n-1)}(z)dz. \qquad (2.56)$$

This expression generalizes the expression (1.50) written in accordance with the induced emf method.

A radiator with constant surface impedance is a particular case of a radiator with impedance, changing along the antenna. Let, for example, emf be located in the radiator center, the radiator be symmetrical and consist of $2N$ segments of length l_m. Surface impedance $Z^{(m)}$ is constant in each of them (Figure 1.5b). The equation for current $J_m(z)$ along the segment m of a radiator takes the form

$$\frac{d^2 J_m(z)}{dz^2}+k^2 J_m(z)=-4\pi j\omega\varepsilon\chi\left[K(z)+\sum_{i=1}^{2N}W(J_i,z)-\frac{J(z)Z^{(m)}}{2\pi a}\right],b_{m+1}\le z\le b_m, \qquad (2.57)$$

Considering that the impedance affects essentially the current distribution in the first approximation, one can introduce propagation constant $k_m=\sqrt{k^2-j2k\chi Z^{(m)}/(aZ_0)}$ on each segment and write the current as a series in powers of small parameter χ to obtain the set of equations:

$$\frac{d^2 J_{m1}(z)}{dz^2}+k_m^2 J_{m1}(z)=-4\pi j\omega\varepsilon K(z),$$

$$\frac{d^2 J_{mn}(z)}{dz^2}+k_m^2 J_{mn}(z)=-4\pi j\omega\varepsilon\sum_{i=1}^{2N}W\left(J_{i,n-1},z\right),b_{m+1}\le z\le b_m,n>1. \qquad (2.58)$$

The current and the components of the series for current are continuous along the radiator and absent at its ends. From the first equation, it follows that the current distribution along each antenna segment has in the first approximation a sinusoidal character. In order to find the law of distribution of the current along the entire radiator, it is necessary to complement condition of the current continuity on the segment boundaries by condition of the charge continuity, i.e. by equality of derivatives of the current on the left and the right side of each boundary. This condition means continuity of voltage along the entire radiator, except the point of the generator placement.

The above-mentioned conditions allow expressing the amplitude and phase of the current at any segment through the amplitude and phase of the current of preceding segment, and, therefore, through segments' parameters and one of the currents. The current distribution along the entire radiator coincides in the first approximation with current distribution along a stepped long line open at the end. For the symmetrical radiator excited at the center, the current distribution is determined by the expression

(1.61). If the condition (2.48) holds, the expression for the input impedance in the approximation n with respect to χ takes the form

$$Z_A^{(n)} \approx -\left[J_N^{(n-1)}(0)\right]^{-2} \sum_{m=1}^{2N} \int_{b_{m+1}}^{b_m} \left\{E_z\left[J_m^{(n-1)}\right] - Z_m J_m^{(n-1)}/(2\pi a)\right\} J_m^{(n-1)}(z)dz, \qquad (2.59)$$

i.e. the expression (1.60), obtained by the induced emf method, is generalized.

In the course of researching the radiator with the impedance, which changes along its length, the issue of rational changing the surface impedance along the antenna with the aim of improving matching of the antenna with a cable arises obligatory. The analysis of the problem shows that at a fixed frequency of the first resonance, the surface impedance must be concentrated at a small antenna segment near the generator. A typical wire antenna with an extending coil in the base meets this requirement.

An example of a radiator with concentrated load is given in Figure 1.6a. The integral equation for the current in such antenna is easily derived from the equation for the current in a metal dipole. The connection of concentrated complex impedance Z_n in a wire (at point $z = z_n$) is equivalent to connection of additional concentrated emf $e_n = -J(z_n)Z_n$, which produces the extraneous field

$$E_n = -J(z_n)Z_n \delta(z - z_n). \qquad (2.60)$$

The boundary condition for the electric field on the radiator surface with N loads will has the form

$$E_z(a,z)\Big|_{-L\leq z\leq L} + K(z) - \sum_{n=1}^{N} J(z_n)Z_n \delta(z - z_n) = 0, \qquad (2.61)$$

i.e.

$$\frac{d^2 J(z)}{dz^2} + k^2 J(z) = -4\pi\omega\varepsilon\chi\left[K(z) + W(J,z) - \sum_{n=1}^{N} J(z_n)Z_n \delta(z - z_n)\right]. \qquad (2.62)$$

If the radiator is symmetric and loads Z_n placed in both arms are identical and located at identical distances z_n from the coordinates origin, it follows from (2.61) that

$$E_z(a,z)\Big|_{-L\leq Z\leq L} + K(z) - \sum_{N=1}^{N/2} J(z_n)Z_n\left[\delta(z - z_n) + \delta(z + z_n)\right] = 0. \qquad (2.63)$$

For example, Hallen's equation (2.3) for the current along a filament takes the form

$$\int_{-L}^{L} J(\varsigma)G_1 d\varsigma = -\frac{j}{Z_0}\left\{C\cos kz + \frac{e}{2}\sin k|z| - \frac{1}{2}\sum_{n=1}^{N/2} J(z_n)Z_n\left[\sin k|z - z_n| + \sin k|z + z_n|\right]\right\}.$$

This equation was used in paper [31]. If the radiator has only one load Z_1 connected in the wire (at point $z = z_n$), it follows from (2.62) that

$$\frac{d^2 J(z)}{dz^2} + k^2 J(z) = -4\pi j\omega\varepsilon\chi\left[K(z) + W(J) - J(z_1)Z_1\delta(z - z_1)\right]. \qquad (2.64)$$

Three components in the right part of the expression correspond to the exciting emf, the radiation and to the presence of the load, respectively. We seek the solution as a series in powers of small parameter χ, which allows obtaining the set of equations

$$\frac{d^2 J_1(z)}{dz^2} + k^2 J_1(z) = -4\pi j\omega\varepsilon\left[K(z) - \chi J_1(Z_1)Z_1\delta(z-z_1)\right], J_1(\pm L) = 0,$$

$$\frac{d^2 J_n(z)}{dz^2} + k^2 J_n(z) = -4\pi j\omega\varepsilon\left[W(J_{n-1}) - \chi J_n(z_1)Z_1\delta(z-z_1)\right], J_n(\pm L) = 0, n > 1. \quad (2.65)$$

The equations were written provided that Z_1 has the magnitude of order of $1/\chi$, i.e. it is comparable with the antenna wave impedance. The solution of the first equation for the particular case when the antenna feed point is displaced from the center, i.e., $K(z) = e\delta(z - h)$, takes the form

$$\chi J_1(z) = j\frac{\chi e}{30\sin 2kL}\sin k(L+\gamma_2 h)\sin k(L-\gamma_2 z)$$

$$+ \frac{\chi^2 e}{900\sin^2 2kL}\frac{Z_1 Z_2}{Z_1 + Z_2}\sin k(L+\gamma_1 z_1)\sin k(L+\gamma_3 z_1)\sin k(L-\gamma_3 h)\sin k(L-\gamma_1 z), \quad (2.66)$$

where

$$Z_2 = -j\frac{30\sin 2kL}{\chi\sin k(L+z_1)\sin k(L-z_1)}, \gamma_1 = \begin{cases} +1, z_1 \le z, \\ -1, z_1 > z, \end{cases} \gamma_2 = \begin{cases} +1, h \le z, \\ -1, h > z, \end{cases} \gamma_3 = \begin{cases} +1, h \ge z_1, \\ -1, h < z_1. \end{cases}$$

The solution of the equations for $J_n(z)$ at $n > 1$ may be found by replacing magnitude $K(\varsigma)$ in equation for $J_1(z)$ by $W(J_{n-1})$. If we take into account (2.66), we get at the excitation point:

$$\chi^n J_n(h) = \frac{1}{e}\int_{-L}^{L} W\left(\chi^{n-1} J_{n-1}\right)\chi J_1(\varsigma)d\varsigma$$

If to use an equality of the type (2.43), in which the additional component in the form

$$\sum_{m=1}^{n} \chi^m J_m(z_1)Z_1\delta(z-z_1)$$

appears in accordance with (2.64) due to the concentrated load Z_1, then from this equality one can find magnitude $W(\chi^{n-1} J_{n-1})$ and substitute it in $\chi^n J_n(h)$. If the condition (2.48) holds, then, by analogy to (2.49),

$$Z_A^{(n)} \approx -\frac{1}{\left[J^{(n-1)}(h)\right]^2}\left\{\int_{-L}^{L} E_\sigma\left(J^{(n-1)}\right)J^{(n-1)}(\sigma)d\sigma - Z_1\left[J^{(n-1)}(z_1)\right]^2\right\}. \quad (2.67)$$

This expression at $Z_1 = 0$, $h = 0$ coincides with (2.49) in the absence of the second radiator. As is seen from (2.67), if a concentrated load is connected in the antenna wire, then a free member, proportional to the impedance magnitude and the square of the current at the point of connection, appears in the formula for Z_A together with the integral. The addition of such member does not contradict the logic of the induced emf method. The expression (2.67) generalizing the expression (1.65), corresponds to this method. In the case of several (N) loads with magnitudes Z_n, located at points $z = z_n$ (see Figure 1.6b), we come to the expression generalizing the formula (1.58).

Therefore, the solution of the integral equations for currents in antennas of different types confirms and defines more precisely the results determined by the method of induced emf when its second formulation is used. The conclusion is true also for radiators made of several parallel wires. They are considered in Chapters 3 and 4. The results of using the theory of the impedance antennas and of the antennas with concentrated loads are considered in Chapter 5.

2.6　INTEGRAL EQUATIONS FOR A SYSTEM OF RADIATORS

In Section 2.4 the system consisting of two radiators was analyzed with the help of the integral equation. The expression (2.49) of this Section shows clearly that an input impedance of a radiator in the system is equal to the sum of the self-impedance $Z_{II}^{(n)}$ in the approximation n with respect to χ and the additional impedance, equal to a product of the mutual impedance $Z_{III}^{(n)}$ of radiators and the ratio of currents at centers of radiators. By analogy, in the case of several (Q) radiators, the strength of the electric field on the surface of radiator p is

$$E_p = \sum_{q=1}^{Q} E_p(J_q),\qquad(2.68)$$

where $E_p(J_q)$ is the field along the radiator p, created by current $J_q = J_q(0)f_q(\sigma)$ of the radiator q, $f_q(\sigma)$ is the current distribution in the radiator q. The oscillating power, created by the radiator p with current $J_p = J_p(0)f_p(\sigma)$ in all radiators, is

$$P_p = -\sum_{q=1}^{Q} \int_{-L_p}^{L_p} E_p(J_q)J_p(0)f_p(\sigma)d\sigma,$$

i.e. the input impedance of the radiator p is

$$Z_p = \frac{P_p}{J_p^2(0)} = \sum_{q=1}^{Q} J_q(0)\frac{Z_{qp}}{J_p(0)},\qquad(2.69)$$

where $Z_{qp} = -\int_{-L_p}^{L_p} E_p(f_q)f_p(\sigma)d\sigma$ is the mutual impedance of the radiators q and p.

In the notation system adopted in Section 2.4, where the order n of smallness is taken into account, this expression is given in the form:

$$Z_p^{(n)} = -\frac{1}{J_p^{(n-1)}(0)}\sum_{q=1}^{Q} J_q^{(n-1)}(0)\int_{-L_p}^{L_p} E_\sigma\left[f_q^{(n-1)}\right]f_p^{(n-1)}(\sigma)d\sigma.$$

Multiplying the current of the source into the input impedance of radiator, we obtain the magnitude of emf, located at the center of radiator p:

$$e_p = J_p(0)Z_p = \sum_{q=1}^{Q} J_q(0)Z_{qp}, p = 1,2,\cdots Q.\qquad(2.70)$$

This is Kirchhoff's equation for a close circuit. According to the equation, the emf in the circuit is the sum of the voltage drops on the elements. Since the equality is true for each radiator, then really the complete system of equations is written by one formula (2.70). The expression (2.70) is identical to the equation (1.73), written in accordance with the logic of the induced emf method.

Note that (2.70) corresponds to the connection of circuit elements with each other in series (see Figure 1.8). The circuit of connection in series is employed widely in the analysis of radiators system. The input impedance of each radiator is calculated usually in accordance with expression of the type (1.50). For this reason, the connection in series is true for the system of radiators with the arm length smaller than 0.4λ. At higher frequencies near the parallel resonance it is expedient to use the connection of circuit elements in parallel. Here, the input impedance is calculated in accordance with an expression of the type (1.69).

In spite of seeming diversity of the described methods, they have a common essential disadvantage. They are developed for specific radiators and possess no flexibility and freedom for the analysis of arbitrarily constructed radiators. The method, which allows analyzing a wire structure consisting of straight segments located arbitrarily and connected partially with each other, offers in this context much greater possibilities (Figure 2.3a). It is considered that current flows along thin, perfectly conducting filaments. Two segments of a filament are shown in Figure 2.3b. The distance from point O_p of the segment p to element ds of the segment s is

$$R = \left[\vec{r}_p + p\vec{e}_p - \vec{r}_s - s\vec{e}_s \right]_{p=0}, \tag{2.71}$$

where $\vec{r}_p$ and $\vec{r}_s$ are radii vectors from the coordinates origin to points O_p and O_s of the corresponding segments, p and s are coordinates measured along the segments and $\vec{e}_p$ and $\vec{e}_s$ are the unit vectors, directions of which coincide with wire axes.

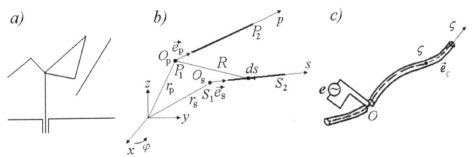

Figure 2.3 An antenna, consisting of several straight segments (a), two straight segments (b) and curvilinear variant (c).

Let us write for the current along the segment S: $\vec{j}_s = \vec{j}_s(s)\vec{e}_s$. According to (1.21),

$$\vec{A}_s(\vec{j}_s) = A_s(j_s)\vec{e}_s = \mu\vec{e}_s \int_{S_1}^{S_2} J(s')G_3 ds,$$

where S_1 and S_2 are the coordinates of the beginning and end of the segment s on the s-axis. In order to find the vector potential of the total field, one has to sum up the vector potentials of fields of all segments:

$$\vec{A} = \sum_{n=1}^{N} \vec{A}_{sn}\left(\vec{j}_s\right) = \sum_{n=1}^{N} A_{sn}\left(j_{sn}\right)\vec{e}_{sn},\tag{2.72}$$

where n is the segment's number, N is the number of segments, and

$$A_{sn}\left(j_{sn}\right) = \mu \int_{Sn1}^{Sn2} J(s_n)G_3 ds_n.$$

In accordance with (1.20) and (1.21), the field of the segment S at point O_p is

$$\vec{E}_s(O_p) = \frac{1}{j\omega\varepsilon}\int_{S_1}^{S_2} J(s)\left[k^2 G_3 \vec{e}_s + graddiv(G_3\vec{e}_s)\right]ds.$$

The differentiation in the last term is performed with respect to the coordinates of the observation point. Since in the rectangular coordinate system the distance between the observation point and the integration point is equal to

$$R = \sqrt{(x_p - x_s)^2 + (y_p - y_s)^2 + (z_p - z_s)^2},$$

where x_p, y_p, z_p are the coordinates of the observation point, and x_s, y_s, z_s are the coordinates of the integration point, then as a result of symmetry $grad_p G = -grad_s G$ (here the differentiation is performed with respect to coordinates p and s). Taking into account that in accordance with the gradient definition $\vec{e}_s grad_s G = \partial G/\partial s$ and using the mathematical identity $div(G\vec{e}_s)\vec{e}_s grad G$, we find $div_p(G\vec{e}_s) = -\partial G/\partial s$, i.e.

$$\vec{E}_s(O_p) = \frac{1}{j\omega\varepsilon}\int_{S_1}^{S_2} J(s)\left[k^2 G_3 \vec{e}_s + grad_p\left(\frac{\partial G_3}{\partial s}\right)\right]ds.$$

The projection of the field of wire S onto direction p is calculated as a product of $\vec{E}_s(O_p)$ and $\vec{e}_p$:

$$E_{ps} = \vec{E}_s\left(O_p\right)\vec{e}_p = \frac{1}{j\omega\varepsilon}\int_{S_1}^{S_2} J(s)\left[k^2 G_3 \vec{e}_3\vec{e}_p - \frac{\partial^2 G_3}{\partial p\partial s}\right]ds,\tag{2.73}$$

and the projection of the complete field is the sum of the field's projections of all segments:

$$E_p = \sum_{n=1}^{N} E_{pn} = \frac{1}{j\omega\varepsilon}\sum_{n=1}^{N}\int_{S_{n1}}^{S_{n2}} J_n(S_n)\left[k^2 G_3 \vec{e}_n\vec{e}_p - \frac{\partial^2 G_3}{\partial p\partial s_n}\right]ds_n.$$

Substituting this field in (2.1), we get the equation generalizing the Pocklington's equation (2.6):

$$\sum_{n=1}^{N}\int_{S_{n1}}^{S_{n2}} J_n(S_n)\left[k^2 G_3 \vec{e}_n\vec{e}_p - \frac{\partial^2 G_3}{\partial p\partial s_n}\right]ds_n = -j\omega\varepsilon K_p(p).\tag{2.74}$$

If $N = 1$, Eq. (2.74) converts to (2.6),

$$\int\limits_{-L}^{L} J(\varsigma)\left[k^2 G_3 \vec{e}_\varsigma \vec{e}_z - \frac{\partial^2 G_3}{\partial z \partial \varsigma}\right]d\varsigma = -j\omega\varepsilon K(z). \tag{2.75}$$

To this end, first the replacement of variables is performed: $p \to z, s_n \to \varsigma$. Furthermore, one should take into account that the Green's function is symmetrical relative to the coordinates of the points of observation and integration: $\partial G_3 / \partial z = -\partial G_3 / \partial \varsigma$.

Let the wire antenna have a shape of a polygonal line, along which the coordinate ς is postponed, and the lengths of straight segments tend to be zero. Then from (2.74) we obtain the integral equation of Pocklington for the current in a curvilinear wire (Figure 2.3c):

$$\int\limits_{(L)} J(\varsigma)\left[k^2 G_3 \vec{e}_\varsigma \vec{e}_z - \frac{\partial^2 G_3}{\partial z \partial \varsigma}\right]d\varsigma = -j\omega\varepsilon K(z). \tag{2.76}$$

Here, $\vec{e}_z$ and $\vec{e}_\varsigma$ are unit tangent vectors at the points of observation and integration. If the curvilinear wire is symmetrical relative to some middle point, the form of this equation completely coincides with the form of (2.75).

2.7 GENERALIZED INDUCED EMF METHOD

An analytical solving problem of antenna radiation has been obtained for a small number of the simplest variants of radiators. As a rule, the small-scale radiators situated in free space were considered. This is explained by the difficulty of the problem. In this connection, numerical methods allowing reducing the problem to solution of a set of linear algebraic equations became frequent practice in solving integral equations for the antenna current. These methods permit to find characteristics of complex antennas of great dimensions (in comparison with a wavelength), and also to take into account the influence of nearby antennas and metal bodies.

Integral equation reduces to a set of algebraic equations with the help of Moment method. In the general case the integral equation for the current in a wire antenna has the form

$$\int\limits_{(l)} J(\varsigma) K(z,\varsigma) d\varsigma = F(z), \tag{2.77}$$

where $J(\varsigma)$ is the sought function (the current distribution along a wire), $K(z, \varsigma)$ is the kernel of the equation, which depends on coordinate Z of the observation point and on coordinate ς of the integration point. $F(\varsigma)$ is a known function, it is determined by extraneous sources of the field. The terms proportional to the current may enter into this function, for example in the case of antenna with loads. Here this is of no great importance. The integral is taken over an all wire length. It is easy to verify that the equations considered earlier are particular cases of the equation (2.77).

Unknown current $J(\varsigma)$ is expressed in the form of a sum of linearly independent function $f_n(\varsigma)$, which are called by the basis functions:

$$J(\varsigma)\sum_{n=1}^{N} I_n f_n(\varsigma), \tag{2.78}$$

where I_n are unknown coefficients, which in the general case are complex. Substituting (2.78) into (2.77), we obtain:

$$\sum_{n=1}^{N} I_n \int_{(l)} f_n(\varsigma)K(z,\varsigma)d\varsigma = F(z). \tag{2.79}$$

Often the second system of linearly independent functions $\varphi_p(z)$ is introduced. They are called by the weight functions. If to multiply both parts of equation (2.79) by $\varphi_p(z)$ and to integrate over entire wire length and then to repeat the operation at different p, we shall obtain the set of equations:

$$\sum_{n=1}^{N} I_n \int_{(l)} \varphi_p(z) \int_{(l)} f_n(\varsigma)K(z,\varsigma)d\varsigma dz = \int_{(l)} \varphi_p(z)F(z)dz, \quad p=1,2...N. \tag{2.80}$$

Obviously number N of equation (2.70) must coincide with the number N of unknown magnitudes. The integration result of each expression is its moment. From this the method's name comes.

If the system of weight functions coincides with the system of basis functions, such a variant of the Moment method is known as Galerkin's method. In this case

$$\sum_{n=1}^{N} I_n \int_{(l)} f_p(z) \int_{(l)} f_n(\varsigma)K(z,\varsigma)d\varsigma dz = \int_{(l)} f_p(z)F(z)dz, \quad p=1,2...N. \tag{2.81}$$

One can rewrite this set of equations as

$$\sum_{n=1}^{N} I_n Z_{np} = U_p, \quad p=1,2...N, \tag{2.82}$$

where

$$Z_{np} = \int_{(l)} f_p(z) \int_{(l)} f_n(\varsigma)K(z,\varsigma)d\varsigma dz, \; U_p = \int_{(l)} f_p(z)F(z)dz.$$

Equation (2.82) is true also for the set of equations (2.80), if one replaces $f_p(z)$ with $\varphi_p(z)$ in formulas for Z_{np} and U_p.

Expression (2.82) is the set of linearly independent algebraic equations with N unknown I_n, having the dimensionality of the current. Coefficients Z_{np} and U_p have the dimensionalities of the impedance and voltage; they can be calculated, e.g. by means of numerical integration. Accordingly, one can interpret the expression (2.82) as Kirchhoff's equation for the contour p with current I_p and emf U_p, which enters into the system of N coupled contours. Here Z_{pp} is the own impedance of the contour element, and Z_{np} is the mutual impedance of the contours n and p.

The set of equations (2.82) can be solved on the computer with the help of standard software. If to write down the set in a matrix form:

$$[I][Z] = [U], \tag{2.83}$$

where $[Z]$ is the impedance matrix, $[I]$ and $[U]$ are a current and a voltage vectors, then one can say that the solution is obtained by means of the standard method of matrix inversion:

$$[I] = [Z]^{-1}[U]. \tag{2.84}$$

Substitution of values I_n into (2.78) allows to calculate current distribution $J(\varsigma)$, and afterwards all electrical characteristics of the radiator.

In practice the calculation of matrix elements Z_{np} may prove to be difficult, since it is connected with the double numerical integration. To alleviate the difficulties, one can use δ-functions in the capacity of weight functions: $\varphi_p(z) = \delta(z - z_p)$. Then, the double integral in the calculation of Z_{np} becomes a simple integral, the calculation of U_p requires no integration, and the expression (2.80) takes the form

$$\sum_{n=1}^{N} I_n \int_{(l)} f_n(\varsigma) K\left(z_p, \varsigma\right) d\varsigma = F(z_p), p = 1,2 \ldots N.$$

One can obtain this equation directly from (2.77) and (2.78), if the left and right parts of the equation (2.77) are equated to each other at isolated points. Their number N corresponds to that of the obtained equations. For this reason, the variant of the Moment method is known as the point-matching technique or the collocation method (see, e.g., [31]).

The collocation method ensures an exact equality of the left and right parts of the equation (2.77), at N points at least. In the intervals between the points the difference between the two parts of the equation may increase sharply. When using the Moment method with weight functions of other type, the equality may not take place in all points of the interval of z changing. But equating of both moments of function (integration with some weight) minimizes the difference between the left and right parts at whole interval of z changing. This property in the final analysis is almost always more important than the exact equality at isolated points. Therefore, Galerkin's method allows providing, as a rule, an essentially more accurate solution than the collocation method. Yet, sometimes the collocation method is useful too.

The choice of basis functions is of great importance for using the Moment method, since the successful selection of the system permits to decrease the amount of calculation under given accuracy or increases the accuracy under the same calculation time. For that end, as a rule, the basis functions must correspond to the physical sense of the problem, i.e. must coincide, in the first approximation, with the actual distribution of the current along a radiator or its elements.

Basis functions are subdivided into two types: entire domain functions, which are other than zero along the entire radiator length, and functions of sub-domains, which are other than zero along segments of radiator. In the capacity of basis functions of the first type, one can use, for example, terms of Fourier series and polynomials of Tchebyscheff or Legendre. Their field of application is limited mainly by solitary radiators of a simple shape. Basis functions of sub-domains are typically employed for an antenna of a complex shape. In particularly, such approach is expedient, if the antenna consists of arbitrarily situated segments of straight wires partially connected with each other. A straight radiator may also consist of physically isolated segments, if concentrated loads are located in the conductor of the radiator at given distances from each other. Piecewise-constant (impulse) functions (Figure 2.4a), piecewise-linear functions (Figure 2.4b), and piecewise-parabolic functions (Figure 2.4c) are shown at Figure 2.4 for illustration of basis functions of sub-domains. These basis functions are special cases of a wider class of basis functions – of polynomials. A simplest variant of approximation with the help of a polynomial is proposed in [32]:

$$J(\varsigma) = \sum_{m=0}^{M_n} I_{nm} (\varsigma - \varsigma_n)^m, \quad \varsigma_n < \varsigma < \varsigma_{n+1}.$$

Here, M_n is the selected degree of the polynomial on the segment n, and I_{nm} are unknown coefficients. Comparing this expression with (2.78), we obtain:

$$I_n = I_{n0}, f_n(\varsigma) = \sum_{m=0}^{M_n} \frac{I_{mm}}{I_{n0}} (\varsigma - \varsigma_n)^m \text{ at } \varsigma_n \leq \varsigma \leq \varsigma_{n+1} \text{ and } 0 \text{ elsewhere.}$$

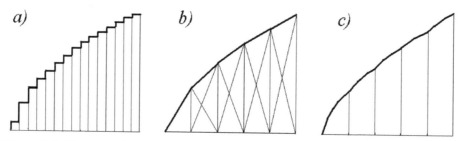

Figure 2.4 The curve line as the sum of pulsed (a), piecewise-linear (b) and piecewise-parabolic (c) basis functions of sub-domains.

One can use terms of Fourier series as basis functions of sub-domains. A particular case of such functions are piecewise-sinusoidal functions:

$$J(\varsigma) = \frac{I_p \sin k\left(\varsigma_{p+1} - \varsigma\right) + I_{p+1} \sin k\left(\varsigma - \varsigma_p\right)}{\sin k\left(\varsigma_{p+1} - \varsigma_p\right)}, \quad \varsigma_{p-1} \leq \varsigma \leq \varsigma_{p+1}. \tag{2.85}$$

Comparing this expression with (2.78) and choosing a simpler variant, one can write:

$$f_p(\varsigma) = \begin{cases} \sin k\left(\varsigma - \varsigma_{p-1}\right) / \sin k\left(\varsigma_p - \varsigma_{p-1}\right), \varsigma_{p-1} \leq \varsigma \leq \varsigma_p, \\ \sin k\left(\varsigma_{p+1} - \varsigma\right) / \sin k\left(\varsigma_{p+1} - \varsigma_p\right), \varsigma_p \leq \varsigma \leq \varsigma_{p+1} . \\ 0 \ elsewhere \end{cases} \tag{2.86}$$

Application of expression (2.78) with the basis functions in the form of (2.86) is equivalent to dividing of wire onto short dipoles with overlapped arms and with centers at points ς_p, wherein I_p is the current at the center of dipole p. In this sense, expressions (2.78) and (2.86) are the generalization of expression (1.8). When lengths of short dipoles are decreased, piecewise-sinusoidal basis functions are converted to piecewise-linear functions. Figure 2.4b permits to visualize how the basis functions of sub-domains form the curve line corresponding to distribution of the current along an antenna.

In [33] it is proposed to use the functions in the form (2.86) as the basis and weight functions. Such variant of the Moment Method has two advantages. First, a rapid convergence of results is ensured, i.e. dimension of the matrix [Z] is small in comparison with dimensions of the matrixes when using other basis and weight functions. This means that application of piecewise-sinusoidal functions as the basis and weight functions corresponds to the physical content of the problem. Second, expressions containing sine integrals and cosine integrals can be used to calculate many matrix elements.

If to substitute the current distribution (2.78) with weight functions (2.86) into the equation (2.74) for the complicated wire radiator and to multiply in accordance with Galerkin's method, both parts of the equation to weight function $f_s(z)$ and after that to integrate along the entire wire length, then we obtain, repeating this operation for different s, a set of p equations of type (2.82) with p unknown magnitudes I_p and with the coefficients

$$Z_{ps} = -\frac{1}{j\omega\varepsilon}\int_{z_{s-1}}^{z_{s+1}} f_s(z) \int_{\varsigma_{p-1}}^{\varsigma_{p+1}} f_p(\varsigma)\left[k^2 G_3 \vec{e}_\varsigma \vec{e}_z - \frac{\partial^2 G_3}{\partial z \partial \varsigma}\right]d\varsigma dz, U_s = \int_{z_{s-1}}^{z_{s+1}} f_s(z)K_s(z)dz. \qquad (2.87)$$

Comparing (2.87) with expression (1.50), where magnitude E_{ps} is taken from (2.73), it is easy to verify that the formula for Z_{ps} corresponds to the mutual impedance between dipoles p and n, calculated by the induced emf method. As seen from (2.87), the dipoles are considered as isolated, i.e. the current of each dipole follows the sinusoidal law. Substituting extraneous field $K_s(z)$ into (2.87), we see that magnitude U_s is the emf of the generator connected at the center of the dipole s. Therefore, the set of equations (2.82) with coefficients Z_{ps} and U_s is the set of Kirchhoff equations for the set of dipoles constituting the wire antenna.

Thus, the variant of Galerkin's method, which was proposed by Richmond for calculating the current distribution in a complicated antenna, is equivalent to dividing of the radiator onto isolated dipoles. Their self- and mutual impedances are calculated by the induced emf method. For this reason, Richmond's method can be named by the generalized induced emf method.

It is expedient to divide the antenna wire with connected in it concentrated loads onto short dipoles so that to place each load in the center of a dipole. Then, in accordance with (2.59), one can generalize the set of equations (2.82) and write it in the form:

$$\sum_{p=1}^{p} I_p Z_{ps} = U_s - I_s Z_s, s = 1, 2 \ldots N, \qquad (2.88)$$

or in matrix form

$$\left\lfloor I_p \right\rfloor \left\lfloor Z_{ps} \right\rfloor = [U_s] - [I_s][Z_s]. \qquad (2.89)$$

The accuracy of the induced emf method for calculating a dipole as is known decreases when the dipole length increases. The accuracy of calculation is acceptable at dipole arm length $L \leq 0.4\lambda$. The advantage of the generalized induced emf method consists in the fact that one can divide the long dipole onto several short dipoles, e.g., with the arm length no greater than 0.2λ. That allows ensuring the required exactness.

Calculation of the coefficients Z_{ps} requires the double numerical integration. But the problem is simplified essentially, if the method described in [8] is used for calculating the mutual impedance of two arbitrarily situated dipoles. Here, the double integrals are reduced to ordinary integrals, and each integral is a sum of alternating series. The components of series are calculated by means of recurrence formulas, almost as quickly as the components of the power series.

From all the above it follows that the induced emf method is a constant companion and satellite of the integral equation method. Also it is inseparable from the concept of an equivalent long line open at the end with the the sinusoidal current distribution coinciding with the current distribution along the symmetrical dipole. In the case of the usual line of metal wires, the propagation constant of a wave along the line is equal to the propagation constant of a wave in the air.

The generalized induced emf method in substance is the basis of all programs of calculation used in modern computers. This allows us to stop talking about the strict theory of thin linear radiators. But before that, we should say a few words on the cross-sectional shape of the radiators.

The radiator's models considered in the first two chapters, are shaped like a straight circular cylinder. But a cross-section of the dipoles may have an arbitrary shape. In practice, the circular (see Figure 1.4) and rectangular cross-sections are encountered most often. The circular cross-section is the usual cross-section of metal radiator. Slot radiators in a metal sheet have a rectangular cross-section. Appearance of printed circuits caused an interest in thin dipoles of rectangular cross-section. They are printed on dielectric substrates and excited by strip lines. In order to provide a distinction between dipoles of circular and rectangular cross-section the latter dipoles often are called strip dipoles.

Calculations as a rule are limited by the variants, in which the dipoles consist of conductors with perfect conductivity. Maximal linear dimension of the cross-section is $u \ll L$. The currents flow along conductors' surface. This implies that a potential difference, created by a generator between edges of a gap, is not a function of a coordinate along the perimeter of the cross-section. But the field of antenna depends on the shape and dimensions of its cross-section.

This question is considered in [16]. The author studies the dependence of the vector-potential of the field on the antenna shape. The solution is based on the first expression of (1.28), in which the integral is taken along the perimeter of the antenna cross-section, particularly along the perimeters of the gap and adjacent segments. Using the mean value theorem in order to simplify the problem, it is possible to calculate the vector potential for the cross-section of an elliptical cylinder, and compare it with the vector potential of a circular cylinder. They are equal to each other, if the radius of the circular cross-section is equal to $a_e = 0.5(a_e + b_e)$. Here $2a_e$ and $2b_e$ are large and small axes of the ellipse.

The similar result is given in [13]. Here it is assumed that the equivalent radius of the flat antenna with a width b is equal to $a_e = b/2$. That permits to find the self capacitance of the flat antenna with this equivalent radius. It is equal to $C_{01} = 2\pi\varepsilon/\ln(2L/a_e) = 2\pi\varepsilon/\ln(4L/b)$. The other self-capacitance of a rectangular plate with the length $2L$ and the width b per unit of its length is given in [34]. When $2L/b \geq 10$, $C_{02} = 2\pi\varepsilon/\ln(2.4L/b)$. The formulas obtained for C_{01} and C_{02} lie within the limits of accepted accuracy.

$$3$$

Folded Antennas of Metal Wires

3.1 ELECTRICALLY RELATED LONG LINES PARALLEL TO METAL SURFACE

Antennas from the parallel wires (symmetric and asymmetric), spaced on a distance, which is small in comparison with a length of a wave and a length of a wire, are widespread side by side with conventional linear radiators. The simplest version of such radiators is the symmetric antenna of two thin wires of the identical diameter, excited in the middle of one of the wires (Figure 3.1). This antenna is called the symmetrical folded radiators. Two asymmetrical variants of this radiator are shown in Figure 3.2. In the first variant, the second (unexcited) wire is shorted to the ground (Figure 3.2a), in the second variant there is a gap between this wire and the ground (Figure 3.2b). In the general case the diameters of wires forming the radiator are not the same (Figure 3.2c). The upper ends of the wires of an asymmetric radiator are connected or not connect with each other and may be positioned at different heights (Figure 3.2d). The radiator may be formed in the shape of coaxial structure (Figure 3.2e).

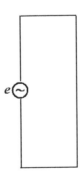

Figure 3.1 Symmetric folded radiator.

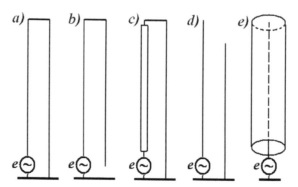

Figure 3.2 Asymmetric folded radiators: (*a*) with the ground connection of the second wire, (*b*) with a gap in the second wire, (*c*) with wires of different diameter, (*d*) with wires of different length, (*e*) in a form of a coaxial structure.

An advantage of antennas consisting of parallel conductors is in the first place in the fact that it is substantially shorter than the linear antennas intended for operating at the same frequency. Selecting variant and dimensions of such antennas provides additional degrees of freedom for obtaining the desired electrical characteristics, for example, for improving matching with signal source (with generator or cable). To sum up, we can conclude that the folded radiators combine the functions of radiation and matching.

The two-wire long line open at the end is a useful analog of a conventional linear radiator. This line allows finding the current distribution along an antenna wire. An analog of folded radiator is an equivalent asymmetric line of two wires located above an infinitely large metal surface—above the conductive ground (Figure 3.3). In the general case, the number of wires can be significantly greater than two. They form the structure from electrically related lines located above the ground. Theory of coupled lines designed by A.A. Pistolkors [35] makes a basis of the analysis of antennas consisting of parallel wires.

This theory permits to analyze antennas and cables as multi-wire lines. In particular, it permits to study structures of N closely spaced parallel wires with allowance of the ground. An asymmetrical line of two wires (see Figure 3.3) is the simplest example of such structure. This line is equivalent to a folded radiator with a gap (see Figure 3.2*b*). Telegraph equations for the current and potential of wires 1 and 2 in this line take the form:

$$-\frac{\partial u_1}{\partial z} = jX_{11}i_1 + jX_{12}i_2, \qquad u_1 = j\frac{1}{k^2}\left(X_{11}\frac{\partial i_1}{\partial z} + X_{12}\frac{\partial i_2}{\partial z}\right),$$

$$-\frac{\partial u_2}{\partial z} = jX_{22}i_2 + jX_{12}i_1, \qquad u_2 = j\frac{1}{k^2}\left(X_{22}\frac{\partial i_2}{\partial z} + X_{12}\frac{\partial i_1}{\partial z}\right). \qquad (3.1)$$

Here u_i is the potential of the wire i relative to the ground, i_i and is the current along the wire i, and $X_{ik} = \omega\Lambda_{ik}$ is the self- or mutual inductive impedance per unit length.

The two left equations of the set (3.1) are based on the fact that the decrease of potentials at segment dz of each wire is the result of the emf influence. The emf's are induced by the self-currents and by the currents of the adjacent wires. The other two equations are written on the basis of the electrostatic equations relating charges and potentials in accordance with the equation of continuity.

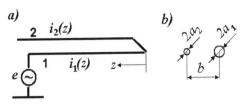

Figure 3.3 The asymmetrical line, which is equivalent to a folded radiator with the gap: (*a*) circuit, (*b*) cross-section.

Dependence of the current on coordinate z is adopted in the form $\exp(\gamma z)$, where γ is the propagation constant. Differentiation of the right equations and substituting them into the left equations brings to a set of uniform equations, which shows that propagation constant γ in the system of two metal wires is equal to k. We search the solution of the set of equations in the form of $U_1 = A_i \cos kz + jB_i \sin kz$. Assuming $z = 0$ in order to determine constant quantities A and B, we obtain

$$i_{1(2)} = I_{1(2)} \cos kz + j \left[\frac{U_{1(2)}}{W_{11(22)}} - \frac{U_{2(1)}}{W_{12(21)}} \right] \sin kz, U_{1(2)} = \cos kz + j \sum_{s=1}^{2} \rho_{1(2)s} I_s \sin kz, \quad (3.2)$$

where $I_{1(2)}$ and $U_{1(2)}$ are the current and the potential at the beginning of wire 1 or 2 (at point $z = 0$), $W_{1(2)s}$ and $\rho_{1(2)s}$ are the electrostatic and the electrodynamics wave impedances between wire 1 or 2 and wire S.

In the general case, when the system consists of N parallel metal wires located above the ground, expressions for the current and the potential of wire n take the form

$$i_n = I_n \cos kz + j \left[\frac{2U_n}{W_{nn}} - \sum_{s=1}^{N} \frac{U_s}{W_{ns}} \right] \sin kz, u_n = U_n \cos kz + j \sum_{s=1}^{N} \rho_{ns} I_s \sin kz, \quad (3.3)$$

where I_n and U_n are the current and potential at the beginning of wire n (at point $z = 0$) respectively and W_{ns} and ρ_{ns} are the electrostatic and electrodynamics wave impedances respectively between wire n and wire s:

$$\rho_{ns} = \frac{\alpha_{ns}}{c}, W_{ns} = \begin{cases} 1/(c\beta_{ns}), n = s, \\ -1/(c\beta_{ns}), n \neq s. \end{cases} \quad (3.4)$$

Here, α_{ns} is the potential coefficient (with due account of a mirror image in the perfectly conducting ground surface), β_{ns} is the coefficient of electrostatic induction, and c is the light velocity. The coefficients β_{ns} and α_{ns} are related as follows:

$$\beta_{ns} = \Delta_{ns}/\Delta_N, \quad (3.5)$$

where $\Delta_N = |\alpha_{ns}|$ is the $N \times N$ determinant, and Δ_{ns} is the cofactor of the determinant Δ_N. For an asymmetrical line of two wires, we can write

$$\frac{1}{W_{11}} = \frac{\rho_{22}}{\rho_{11}\rho_{22} - \rho_{12}^2}, \frac{1}{W_{22}} = \frac{\rho_{11}}{\rho_{11}\rho_{22} - \rho_{12}^2}, \frac{1}{W_{12}} = \frac{\rho_{12}}{\rho_{11}\rho_{22} - \rho_{12}^2}, \quad (3.6)$$

i.e.

$$\left(\frac{1}{W_{12}} - \frac{1}{W_{11}} \right) : \left(\frac{1}{W_{12}} - \frac{1}{W_{22}} \right) = \frac{\rho_{22} - \rho_{12}}{\rho_{11} - \rho_{12}} = g. \quad (3.7)$$

Finally, if the wires of an asymmetrical line have unequal lengths, or if concentrated loads are connected into them, one must divide the line to segments. The expressions for the current and potential of wire n at segment m take the form:

$$i_n^{(m)} = I_n^{(m)} \cos kz_m + j\left(\frac{2U_n^{(m)}}{W_{nn}^{(m)}} - \sum_{s=1}^{N} \frac{U_s^{(m)}}{W_{ns}^{(m)}}\right) \sin kz_m ,$$

$$u_n^{(m)} = U_n^{(m)} \cos kz_m + j\sum_{s=1}^{N} \rho_{ns}^{(m)} I_s^{(m)} \sin kz_m , \tag{3.8}$$

where $b_{ns}^{(m)}$ and $U_n^{(m)}$ are the current and potential of wire n at the beginning of segment m (at point $z_m = 0$), respectively, M is the number of wires in segment m, and $W_{ns}^{(m)}$ and $\rho_{ns}^{(m)}$ are the electrostatic and electrodynamics wave impedances between wires n and s at segment m respectively. Equation (3.8) generalizes the expressions (3.3).

In order to solve each set of equations, the boundary conditions are used. They establish the absence of currents at the free ends of the wires, the continuity of the current and potential along each wire, the abrupt changes in potential at the points of connecting loads and generator e. If to calculate the current magnitude $J(0)$ at the feed point, one can find the input impedance of the asymmetrical line,

$$Z_l = e/J(0). \tag{3.9}$$

It is equal approximately to the reactive impedance of the antenna, whose equivalent is the given asymmetrical line. One can find the antenna impedance more accurately, if the antenna is treated as a linear radiator, the current along which is equal to the total current along the line.

When calculating the antenna input impedance, one needs, as a rule, to find field E_ς at antenna surface. And it should be kept in mind that, while current function $J(\varsigma)$ is continuous along the entire length of the antenna and sinusoidal at each segment, function $dJ/d\varsigma$ may have a jump near the segment boundaries.

Equation (3.8) use wave impedances $W_{ns}^{(m)}$ and $\rho_{ns}^{(m)}$, equation (3.3) use similar magnitudes. The magnitudes of the wave impedances, as is seen from (3.4), are determined by the potential coefficients. The coefficients are found by the method of mean potentials in accordance with the actual position of antenna wires. The simplest variant of this method is the Howe's method. It easily shows that the mutual potential coefficient of two parallel wires of equal lengths, which dimensions and position are presented in Figure 3.4, is given as

$$\alpha_{ns} = \alpha(L,l,b)/(2\pi\varepsilon), \tag{3.10}$$

where

$$\alpha(L,l,b) = \frac{1}{2L}\left[(L+l)sh^{-1}\frac{L+l}{b} + (L-l)sh^{-1}\frac{L-l}{b} - 2Lsh^{-1}\frac{l}{b} - \sqrt{(L+l)^2+b^2} - \sqrt{(L-l)^2+b^2} + 2\sqrt{l^2+b^2}\right],$$

i.e. $\rho_{ns} = \alpha_{ns}/c = \alpha(L,l,b)/(2\pi\varepsilon_0\varepsilon_r c) = 60\alpha(L,l,b)/\varepsilon_r.$

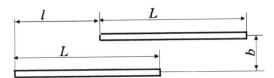

Figure 3.4 The mutual location of two wires.

If the parallel conductors are located upright, i.e. perpendicular to conductive ground, then the self-potential coefficient of conductor n at segment m with the account of the mirror image is

$$\alpha_{nn}^{(m)} = \alpha\left[l_m - l_{m+1}, 0, a_n^{(m)}\right] - \alpha\left[l_m - l_{m+1}, l_m + l_{m+1}, a_n^{(m)}\right], \qquad (3.11)$$

where l_m and l_{m+1} are the boundary coordinates of segment m, $a_n^{(m)}$ is the radius of wire n at segment m. The mutual potential coefficient between wires n and s at segment m is

$$\alpha_{ns}^{(m)} = \alpha\left[l_m - l_{m+1}, 0, b_{ns}^{(m)}\right] - \alpha\left[l_m - l_{m+1}, l_m + l_{m+1}, b_{ns}^{(m)}\right]. \qquad (3.12)$$

Here, $b_{ns}^{(m)}$ is the distance between the axes of wires n and s at segment m. As an example of such structure the circuit of one of possible variants of a multi-radiator antenna is presented in Figure 3.5a. The circuit of an equivalent long line is given in Figure 3.5b.

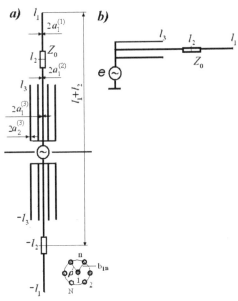

Figure 3.5 Multi-radiator antenna (a) and equivalent long line (b).

It is necessary to emphasize the important detail of the method history. The method was at first proposed for calculating the electrical characteristics of related lines which are parallel to each other and to the ground. But this method allows to find the current distribution along the parallel wires, which are perpendicular to the ground, taking into account the mirror image in its conductive surface, i.e. it permits to determine the electrical characteristics of multi-wire vertical structures.

Summing up, it is important to note the general principle underlying the theory of related lines. The theory allows to find, in the first approximation, the current distribution along each wire in order to use later this distribution for calculating the active component and for defining more precisely the reactive component of the input impedance with the help of the induced emf method. As is known, this principle was used in calculating the input impedances of linear radiators by means of the induced emf method, since the current distribution along the linear radiator coincides in the first approximation with the current distribution along a uniform long line.

A similar approach is used in calculating the electrical characteristics of impedance antenna, i.e. of the radiator with nonzero boundary conditions on its surface. One can obtain the current distribution along the radiating structure by analyzing the integral equation for the current. Here, the laws of a current distribution along the wires of the radiator and along the wires of the equivalent line (or of system of lines) are identical. The advantage of equivalent lines is the maximal simplicity in finding the law and the efficiency of applying the obtained results to designing radiators with required characteristics.

As already mentioned, the asymmetrical line of two wires situated above the ground is an equivalent of a folded antenna. The theory of related lines is a base for the analysis of antennas, consisting of parallel wires.

3.2 FOLDED ANTENNAS, PERPENDICULAR TO A METAL SURFACE

We begin to consider method of calculating folded antennas, using as an example an asymmetric folded radiator with a gap in an unexcited wire (see Figure 3.2b). At first in this gap between the free end of the antenna and the ground we connect in parallel two generators of a current of equal magnitude (mJ) and opposite sign (Figure 3.6). Also we divide a main generator (with a current J) onto two parallel generator of identical sign and different magnitude: mJ and $(1 - m)J$.

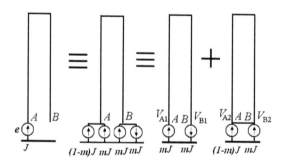

Figure 3.6 Calculation of the folded radiator with a gap.

The voltage at the point A in accordance with the principle of superposition is equal to the sum of the voltages created in it by all generators. Therefore, as it is shown in Figure 3.6, the circuit of the folded radiator may be split into two circuits, with two generators in each circuit. The voltages at point A, created in each of these circuits, are calculated and summed up. Let the currents of the wires of the first circuit be in anti-phase, i.e. equal in magnitude and opposite in direction. This means that the first circuit is the two-wire long line short-circuited at the end. Let the currents of the wires of the second circuit be in phase, i.e. the potentials of the points of both wires located at the same height (including the points near the antenna base) are identical. This means that the second circuit is the linear radiator (monopole), excited in the base.

As shown in Section 3.1, depicted in Figure 3.3 asymmetric line of two wires located above the ground is equivalent to folded radiator. Currents and potentials in the line wires are defined by the set of equations (3.2). Electrostatic and electrodynamics wave impedances included in these equations are defined by equalities (3.4). From the first equation of (3.2) follows that in considered line there are only anti-phase currents $(i_1 = -i_2)$, if the following conditions are met:

$$I_1 = -I_2, \ U_1/W_{11} - U_2/W_{12} = -(U_2/W_{22} - U_1/W_{12}),$$

where from in accordance with (3.6) $U_1 = -U_2/g$. From the second equation of (3.2) and the equality (3.6) it follows that the ratio of potentials of points located in the different wires of the same section is equal to $u_1/u_2 = U_1/U_2 = -1/g$. This means that the voltages in points A and B of short-circuited two-wire line are related by the equality

$$V_{A1}/V_{B1} = -1/g. \tag{3.13}$$

If only in-phase currents exist in asymmetric line, the potentials of wires in the same cross-section of line should be equal along the entire length of the line $(u_1 = u_2)$, i.e.

$$U_1 = U_2, \ \rho_{11}I_1 + \rho_{12}I_2 = \rho_{12}I_1 + \rho_{22}I_2,$$

where from $I_1 = gI_2$. This means that the currents of wires in each cross-section of the monopole (including the base of the monopole) are related by

$$i_1/i_2 = J_{A2}/J_{B2} = g. \tag{3.14}$$

From (3.13) and (3.14), we obtain

$$(V - V_{A1})/V_{A1} = (J - J_{B2})/J_{B2} = g.$$

Here $V = V_{A1} - V_{B1}$ is the voltage at the input of the line, and $J = J_{A2} + J_{B2}$ is the current in the base of the folded radiator, which is equal to the total current (to the current of the generator). From here

$$V_{A1}/V = J_{B2}/J = m = 1/(1 + g). \tag{3.15}$$

From (3.15) and Figure 3.6 it is clear that m is a fraction of the in-phase current in the right wire of the monopole.

In the first circuit (in the short-circuited two-wire line) the voltage at the point A is

$$V_{A1} = mV = m^2 \ JZ_l$$

where $Z_l = jW_l \tan kL$ is the input impedance of the line, W_l is its wave impedance. In the second circuit (monopole)

$$V_{A2} = JZ_e \ (a_e)$$

where $Z_e \ (a_e)$ is the input impedance of the asymmetric linear radiator of height L with an equivalent radius a_e. Dividing the total voltage at the point A on the current of the generator, we find the input impedance of the folded radiator with a gap:

$$Z_A = \frac{V_{A1} + V_{A2}}{J} = Z_e(a_e) + jm^2 W_l \tan kL. \tag{3.16}$$

As it follows from this expression, from point of view of the input impedance, the folded radiator with a gap is a serial connection of the monopole and the short-circuited two-wire long line.

Similarly, the input impedance of the folded radiator with the ground connection of the second wire is a parallel connection of the monopole and the short-circuited long line:

$$Y_A = \frac{1}{jW_l \tan kL} + \frac{p^2}{Z_e(a_e)}. \tag{3.17}$$

Here $p = 1 - m = g/(1 + g)$ is the fraction of the in-phase current in the left wire of the radiator. In order to obtain (3.17), one must connect in the right wire of the closed folded radiator in series two generators of voltage, which are equal in magnitude (pe) and opposite in sign. Also one must divide the main generator (with electromotive force e) on two consistently connected generators with the same direction and with electromotive forces pe and $(1 - p)e$. Further the circuit of the folded radiator is divided into two circuits—the short-circuited two-wire line and the monopole, and the currents in the base of the left wire created in each of these circuits are calculated and summed.

Expression (3.17) was obtained in [36], where procedure of calculating the folded radiator with wires of different diameters and the ground connection of the second wire is given. If the folded radiator is formed by two identical thin wires with radii $a = a_1 = a_2 \ll b$ (here b is the distance between the axes of the wires—see Figure (3.3), then

$$m = p = 0.5, \qquad g = 1, \qquad a_e = \sqrt{ab}, \qquad W_l = 120 \ln(b/a).$$

For wires with different radii we have

$$p = 1 - m = \frac{\ln(b/a_2)}{2\ln\left(b/\sqrt{a_1 a_2}\right)}, \ a_e = \exp\frac{\ln a_1 \ln a_2 - \ln^2 b}{\sqrt{a_1 a_2}/b}, \ a = \frac{\ln(b/a_2)}{\ln(b/a_1)}, \ W_l = 120\ln\frac{b}{\sqrt{a_1 a_2}}. \tag{3.18}$$

In the general case

$$m = C_{22}/(C_{11} + C_{22}), \tag{3.19}$$

where C_{22} is the self-capacitance of the right wire and C_{11} is the self-capacitance of the left wire. From (3.15) and (3.19) it follows that the currents in the wires of the monopole are proportional to the self-capacitances of the wires, and potentials of the wires of the long line are proportional to the capacitive reactances $X_n = -1/(\omega C_{nn})$ between the wires and the ground.

The limiting case ($m = 1$, $C_{11} = 0$) is shown in Figure 3.2e. Here the folded radiator is designed as a segment of the coaxial line, which is open below and closed at the top. According to (3.16),

$$Z_A = Z_e(a_e) + jW_l \tan kL$$

where $W_l = 60 \ln(a_2/a_1)$. If the outer conductor of the coaxial line is shorted to the ground, then according to (3.17),

$$Y_A = -j/(W_l \tan kL)$$

Thus, in extreme cases, expressions (3.16) and (3.17) give a sufficiently obvious result.

As can be seen from expressions (3.16) and (3.17), the replacement of a linear radiator by a folded radiator changes significantly the input impedance of the antenna. Folded radiator with a gap, if the length of this radiator is less than a quarter of a wave length, contains the inductive reactance of a short-circuited line. This inductive reactance is

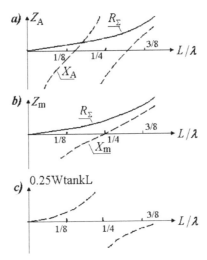

Figure 3.7 Input impedance of folded radiator with a gap (*a*), of the monopole (*b*) and of the long line (*c*).

connected in series with the input impedance of the monopole and compensates its capacitive reactance. Therefore, when the height of a linear and a folded radiator are the same, the frequency of a first parallel resonance of the folded radiator is close to the frequency of the first serial resonance of the linear radiator. Accordingly, the frequency of the first serial resonance of the folded radiator is substantially lower than the frequency of the first serial resonance of a linear radiator, approximately twice. Figure 3.7 shows typical active and reactive components of the input impedance of the folded radiator with a gap and the wires of equal diameter (*a*) and also the input impedances of its elements: of the monopole (*b*) and of the long line (*c*).

Input impedance of folded radiator with the ground connection of the second wire in contrast to folded radiator with a gap is a parallel connection of the input impedances of the monopole and a short-circuit long line, i.e. it has a more complex character. However, this option allows transforming an active component of a monopole input impedance. Indeed, according to expression (3.17) the input admittance of the folded radiator at the resonant frequency of a monopole is

$$Y_A = \frac{R_e(a_e) + jp^2 W_l \tan kl}{jR_e(a_e)W_l \tan kL}$$

i.e. neglecting the first term of the numerator as against the second term and substituting the value p in accordance with the (3.18), we obtain

$$R_A \approx \frac{R_e(a_e)}{p^2} = R_m(a_e)\left[1 + \frac{\ln(b/a_1)}{\ln(b/a_2)}\right]^2. \qquad (3.20)$$

Selecting the radii of the wires permits increasing a level of matching such antenna with a cable or a generator.

It is necessary to say a few words about the physical meaning of the obtained results. It is obvious that in the free space two closely spaced parallel wires, along which equal currents flow in opposite directions, do not radiate the signal, since the fields of wires are mutually cancel each other. The conductive metallic surface (ground), on which antenna is installed, causes appearance of displacement currents between the wires and ground (they are shown by dotted lines in Figure 3.8) and decreasing the conduction currents in parallel wires with increasing distance from the ground. Displacement currents are returning to the generator pole along the ground surface. They create the radiation and cause the emergence of the summand in the input impedance of the antenna, analogous to the resistance of the monopole.

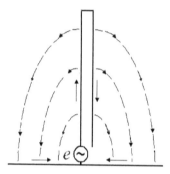

Figure 3.8 Displacement currents and currents in the ground near the folded radiator.

3.3 FOLDED RADIATOR WITH WIRES OF DIFFERENT LENGTH

The previous sections were devoted to folded radiators and equivalent long lines made of parallel wires of equal length. Procedure of calculating a current distribution along wires of line is based on the theory of electrically coupled lines. In order to determine the input impedance of a folded antenna, its circuit is divided into a linear radiator and a two-wire long line. If an asymmetrical folded radiator is formed by wires of different lengths (see Figure 3.2d), the structure may be similarly split into the line and the radiator, but the method of calculating the line and the radiator is not obvious. This is not the sole task that requires determining electrical characteristics of a line and a radiator with wires of different lengths.

An example of a line with wires of different lengths is shown in Figure 3.9a. As can be seen from the figure, it is distinguished from the line shown in Figure 3.3 and consists of two segments. The lengths of the upper and lower segments are equal respectively to $l = l_1 - l_2$ and $L = l_2$. Here l_1 is the length of a longer wire, and l_2 is the length of a shorter wire. The lower segment is made in the form of two parallel wires

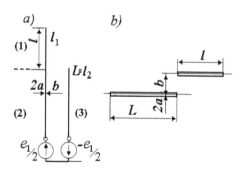

Figure 3.9 Long line with wires of different length (*a*) and its capacitive load (*b*).

of the same length with the same cross-section, for example, circular cross-section with the radius a. Capacitance per unit length of the wire between two thin wires located in a homogeneous medium with permittivity ε is equal to

$$C_0 = \pi\varepsilon/\ln (b/a) \tag{3.21}$$

where b is the distance between the axes of the wires. This capacitance determines the wave impedance of the lower segment of the two-wire long line.

Influence of the upper segment of the line on its input impedance will be taken into account by means of calculating the capacitance between the upper segment of the longer wire (with length l) and the short wire. We shall define an input impedance of the line with wires of different lengths as the input impedance of the line with short wires and capacitive load at the end. The electrostatic structure in this case consists of three conducting elements designated as (1), (2) and (3) in Figure 3.9a. C is the capacitance between elements (2) and (3) in the presence of element (1). It is not equal to the capacitance between isolated elements (2) and (3) in the absence of element (1). For this reason let us find the necessary capacitance as the difference of two capacitances:

$$C = C_1 - C_0 L. \tag{3.22}$$

Here, C_1 is the total capacitance between the long and short wire, and $C_0 L$ is the capacitance between the wires of the lower segment. At that C_1 is the capacitance of an electrically neutral system consisting of two conductors (see, for example, [34]):

$$C_1 = (\alpha_{11} + \alpha_{22} - 2\alpha_{12})^{-1}, \tag{3.23}$$

where α_{ik} are potential coefficients, calculated by the following formulae:

$$\alpha_{11} = \frac{1}{2\pi\varepsilon L}\left\{\ln\left[L/a + \sqrt{1+(L/a)^2}\right] + a/L - \sqrt{1+(a/L)^2}\right\},$$

$$\alpha_{22} = \frac{1}{2\pi\varepsilon(L+l)}\left\{\ln\left[(L+l)/a + \sqrt{(L+l)^2/a^2 + 1}\right] + a/(L+l) - \sqrt{a^2/(L+l)^2 + 1}\right\},$$

$$\alpha_{12} = \frac{1}{4\pi\varepsilon(L+l)}\left\{\ln\left[\left(L+\sqrt{L^2+b^2}\right)/b\right] + (L+l)\ln\left[\left(L+l+\sqrt{(L+l)^2+b^2}\right)/b\right]/L\right.$$
$$\left. -\sqrt{L^2+b^2}/L + \sqrt{l^2+b^2}/L - l\ln\left[\left(l+\sqrt{l^2+b^2}\right)/b\right]/L + b/L - \sqrt{(L+l)^2+b^2}/L\right\}.$$

Since L/a, $l/a \gg 1$, then

$$\alpha_{11} = \frac{1}{2\pi\varepsilon L}\left(\ln\frac{2L}{a} - 1\right), \quad \alpha_{22} = \frac{1}{2\pi\varepsilon(L+l)}\left[\ln\frac{2(L+l)}{a} - 1\right].$$

Sometimes inequalities L/b, $l/b \gg 1$ are true. In this case the expression for α_{12} gets simplified:

$$\alpha_{12} = \frac{1}{4\pi\varepsilon(L+l)}\left[\ln\frac{2L}{b} + \ln\frac{2(L+l)}{b} + \frac{l}{L}\ln\frac{L+l}{l} - 2\right].$$

Calculations show that the capacitance C is small as compared with the capacitance $C_0 L$. In particular, if wires are located in the air, i.e. $\varepsilon = 1/(36\pi \cdot 10^9)$, and $L = 7.5$, $b = 1.0$, $2a = 0.05$, when the excess length l changes from 1 to 4 (all dimensions are in centimeters), we have $C_0 L = 7.07$ pF, and C changes from 0.05 to 0.1 pF. Thus the excess length l creates the capacitive load at the end of the two-wire line. This load is equivalent to lengthening of the line by a value l_0:

$$l_0 = (1/k) \cot^{-1}[1/(\omega C W_l)]. \qquad (3.24)$$

where W_l is the wave impedance of the line. The results of calculating capacitances C, and equivalent lengths l_0 for the above mentioned dimensions at 1 GHz are given in Table 3.1.

It is easily convinced that the capacitance between the elements (2) and (3) in the absence of the element (1) is significantly greater than the capacitance presented in Table 3.1.

Table 3.1 Capacitive Loads and Values of Lengthening

l, cm	$2a = 0.05$ cm			$2a = 0.2$ cm		
	l_0, cm	l_{01}, cm	C, pF	l_0, cm	l_{01}, cm	C, pF
0.0	0	0	0.020	0	0	0.047
0.5	0.22	0.19	0.037	0.21	0.15	0.073
1.0	0.41	0.39	0.050	0.37	0.30	0.093
1.5	0.56	0.52	0.063	0.49	0.45	0.108
2.0	0.69	0.86	0.073	0.58	0.61	0.119
2.5	0.80	1.10	0.081	0.65	0.79	0.128
3.0	0.90	1.38	0.089	0.71	1.00	0.135
3.5	0.98	1.66	0.095	0.75	1.24	0.140
4.0	1.05	1.94	0.101	0.78	1.48	0.144
4.5	1.12	2.17	0.107	0.81	1.64	0.148

These calculations were verified by simulations with the help of CST program. The model of structure, which was applied at this simulation, is shown in Figure 3.10, where e is a generator with output impedance $R = 50$ Ohm. The simulation results for the value l_{01}, of lengthening are also presented in Table 3.1. Since the distance b between the wires is finite, then the dimensions l_0 and l_{01} for $l = 0$, based on the two-wire line approximation, differ from 0. The cause of this circumstance is the self-capacitance of the wires. In order to clearly demonstrate how the excess length l has an effect on lengthening of the line, dimensions l_0 and l_{01} are decreased by their values at $l = 0$.

As it is seen from Table 3.1, the calculation and simulation results agree well for $l \le 0.1\lambda$. It turns out that the input impedance of a line with wires of unequal lengths differs comparatively weakly from the input impedance of a two-wire line with such length as the shorter wire.

Similar results at $2a = 0.2$ cm are given in Table 3.1.

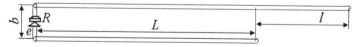

Figure 3.10 Simulation model for a two-wire long line with wires of different lengths.

In accordance with the obtained results one can write the current distributions along the wires of line as:

$$i_1(z) = \begin{cases} I_0 \sin kl_0 \sin k(L+l-z)/\sin kl, L \le z \le L+l, \\ I_0 \sin k(L+l_0 - z), \quad 0 \le z \le L, \end{cases}$$

$$i_2(z) = \begin{cases} 0, & L \leq z \leq L+l, \\ -I_0 \sin k(L+l_0 - z), & 0 \leq z \leq L, \end{cases} \qquad (3.25)$$

where I_0 is the current of a generator. A long line line with equal length of wires located in free space can radiate only in the case when the distance between the wires is not too small compared with the wave length. In case of wires of unequal lengths, the excess lengths l of the longer wire radiates, as it follows from expressions (3.25) for the currents.

The obtained results allow to consider another problem—calculating the input impedance of a linear radiator (monopole) composed of two parallel wires with different lengths (Figure 3.11a). Figure 3.11b shows an equivalent asymmetric line for such a structure.

In this case it is necessary to divide the equivalent line into two segments, as shown in Figure 3.11b. The segment 1 has one wire; the segment 2 consists of two wires. The segment number is indicated in parentheses, the number of wire is indicated in its base. The currents and potentials along the mth segment of the nth wire of the asymmetric line are given by (3.8), where $n = 1, 2, m = 1, 2$. If the distance between the wires is small in comparison with the wires lengths, one can consider that

$$\rho_{nn}^{(m)} = \text{const}(n) = \rho_{nn}^{(m)}, \rho_{ns}^{(m)}\big|_{n \neq s} = \text{const}(n) = \rho_2^{(m)}, W_{nn}^{(m)} = \text{const}(n) = W_1^{(m)}, W_{nS}^{(m)}\big|_{n \neq s} = \text{const}(n) = W_2^{(m)}.$$

The zero currents at the ends of the wires and the continuity of the current and potential along each wire permit to write the boundary conditions:

$$i_1^{(1)}\big|_{z_1=0} = i_2^{(2)}\big|_{z_2=0} = 0; \; i_1^{(1)}\big|_{z_1=l} = i_1^{(2)}\big|_{z_2=0} \; ;$$

$$u_1^{(1)}\big|_{z_1=l} = u_1^{(2)}\big|_{z_2=0} \; ; \; u_1^{(2)}\big|_{z_2=l_2} = u_2^{(2)}\big|_{z_2=l_2} = e \cdot$$

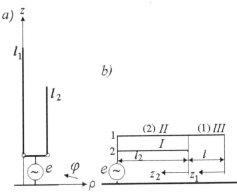

Figure 3.11 Monopole formed by the wires of different length (a), and an equivalent long line (b).

From these boundary conditions we get

$$I_1^{(1)} = I_2^{(2)} = 0; I_1^{(2)} = j\frac{U_1^{(1)}}{W_1^{(1)}}\sin kl; U_1^{(2)} = U_1^{(1)}\cos kl;$$

$$U_2^{(2)} = U_1^{(1)}\left[\cos kl - \frac{\rho_1^{(2)} - \rho_2^{(2)}}{W_1^{(1)}}\sin kl \cdot \tan kl_2\right];$$

$$U_1^{(1)} = \frac{e}{\cos kl \cos kl_2}\left[1 - \frac{\rho_1^{(2)}}{W_1^{(1)}}\sin kl \tan kl_2\right]^{-1}.$$

The current distribution along the first segment of the longer wire is given by

$$i_1^{(1)} = j\frac{U_1^{(1)}}{W_1^{(1)}}\sin k(l_1 - z).\tag{3.26}$$

The current along the second segment is

$$i_1^{(2)} = jU_1^{(1)}\left\{\frac{\sin kl \cos kz_2}{W_1^{(1)}} + \cos kl\left[\frac{1}{W_1^{(2)}} - \frac{1}{W_2^{(2)}}\left(1 - \frac{\rho_1^{(2)} - \rho_2^{(2)}}{W_1^{(1)}}\right)\tan kl \tan kl_2\right]\sin kz_2\right\}\tag{3.27}$$

and the current along the shorter wire is

$$i_2^{(2)} = jU_1^{(1)}\cos kl\left\{\frac{1}{W_1^{(2)}} - \frac{1}{W_2^{(2)}}\left[1 - \frac{\rho_1^{(2)} - \rho_2^{(2)}}{W_1^{(1)}}\right]\tan kl \tan kl_2\right\}\sin kz_2.\tag{3.28}$$

The total current along the second segment is

$$i_1^{(2)} + i_2^{(2)} = jU_1^{(1)}\left\{\frac{\sin kl \cos k(l_2 - z)}{W_1^{(1)}} + \cos kl\left[\frac{1}{W_1^{(2)}} - \frac{1}{W_2^{(2)}}\left(1 - \frac{\rho_1^{(2)} - \rho_2^{(2)}}{W_1^{(1)}}\right)\tan kl \tan kl_2\right]\sin k(l_2 - z)\right\}$$

$$\tag{3.29}$$

These expressions show that the current distribution along each segments of the monopole is sinusoidal, i.e. it is similar to the current distribution along a monopole consisting of two segments with different wave impedances (for example, with different wire diameters).

Let us write the expression for the total current along the monopole in the form

$$J_{Am}(z) = \sum_{n=1}^{M} i_n^{(m)}(z), l_{m+1} \le z \le l_m,\tag{3.30}$$

where $i_n^{(m)}(z) = A_{nm}\cos k(l_m - z) + jB_{nm}\sin k(l_m - z)$. In accordance with (3.26)–(3.28)

$$A_{11} = A_{21} = A_{22} = B_{21} = 0, B_{11} = \frac{U_1^{(1)}}{W_1^{(1)}}, A_{12} = j\frac{U_1^{(1)}}{W_1^{(1)}}\sin kl,$$

$$B_{12} = B_{22} = U_1^{(1)}\cos kl\left[\frac{1}{W_1^{(2)}} - \frac{1}{W_2^{(2)}}\left(1 - \frac{\rho_1^{(2)} - \rho_2^{(2)}}{W_1^{(1)}}\right)\tan kl \tan kl_2\right].\tag{3.31}$$

The input reactance of the monopole is equal to the input impedance of the equivalent line:

$$jX_A = Z_l = \frac{e}{J_A(0)} = -j\frac{\left[W_1^{(1)} - \rho_1^{(2)}\tan kl \tan kl_2\right]\cos^2 kl_2}{\tan kl \cos^2 kl_2 + DW_1^{(1)}\sin 2kl_2},\tag{3.32}$$

where

$$D = \frac{1}{W_1^{(2)}} - \left[1 - \frac{\rho_1^{(2)} - \rho_2^{(2)}}{W_1^{(1)}}\right]\frac{\tan kl \tan kl_2}{W_2^{(2)}}.$$

The radiation resistance of the monopole is equal to

$$E_\Sigma = 40k^2 h_e^2. \tag{3.33}$$

where h_e is the effective height of the monopole given by

$$h_e = \frac{1}{kJ(0)}\sum_{m=1}^{2}\left\{(A_{1m}+A_{2m})\sin k(l_m - l_{m+1}) + j(B_{1m}+B_{2m})\left[1-\cos k(l_m - l_{m+1})\right]\right\} =$$

$$\frac{jU_1^{(1)}}{kJ(0)W_1^{(1)}}\left\{1+\sin kl \sin kl_2 + \cos kl\left[2DW_1^{(1)}(1-\cos kl_2)-1\right]\right\}. \tag{3.34}$$

These expressions define the currents along each wire of the asymmetric radiator and allow calculate more accurately its input impedance. Considering that an antenna is a linear radiator and the current along it is equal to a total current of both wires, it is possible to find the input impedance, for example, by the method of induced emf (second formulation). During calculating the tangential component of the electric field, one must take into account the discontinuity of the current derivative on the segment boundaries.

3.4 LOSS RESISTANCE IN THE GROUND

Application of folded radiators largely depends on their losses, particularly losses in the ground. Each of the elements, of which antenna consists (i.e., monopole and a long line), has losses in the ground. Loss resistance R_{ge} for a monopole is calculated in the usual manner. With regard to a long line, its loss resistance R_{gl} in the ground is also non-zero. Of course, magnetic fields, created by opposite currents of two closely spaced parallel wires of a line, cancel each other, and the radius of an area, in which the full compensation is absent, is relatively small (the center of this area is located at the middle point between wires). However, it is necessary to take into account that, when an observation point is approaching to a conductor with the current, the magnetic field increases.

Let J_1 and J_2 be the currents in left and right wire of the line:

$$J_1 = -J_2 = mJ\frac{\sin k(l+z)}{\sin kl}.$$

Here mJ is the current in the base of the wire and $l = \lambda/4 - L$ is the wire length (Figure 3.12a). If the origin of a rectangular coordinate system is placed in the middle of the interval between the wires, the vector potential of the electromagnetic field produced by the current J_1 at an arbitrary point on the ground surface with coordinates $(x, y, 0)$ located at the distance $\rho = \sqrt{x^2+y^2}$ from the axis of the first wire in view of mirror image is equal to

$$A_{z1} = \frac{mJ\mu}{2\pi\sin kL}\int_0^L \sin k(l+z)\frac{\exp\left(-jk\sqrt{\rho^2+z^2}\right)}{\sqrt{\rho^2+z^2}}dz.$$

The tangential components of the magnetic field are

$$H_{x1} = \frac{\partial A_{z1}}{\mu \partial y} = \frac{mJ}{2\pi j \sin kL} y A_0(L,l,\rho), H_{y1} = -\frac{\partial A_{z1}}{\mu \partial x} = \frac{mJ}{2\pi j \sin kL} x A_0(L,l,\rho), \quad (3.35)$$

where

$$A_0(L,l,\rho) = -\frac{1}{\rho^2}\left[jL\frac{\exp(-kR)}{R}\sin k(l+L) + \exp(-jkR)\cos k(l+L) - \exp(-jk\rho)\cos kl \right]$$

Here $R = \sqrt{\rho^2 + L^2}$. In calculating the integral, the substitution $R = \sqrt{\rho^2 + L^2}$ was used.

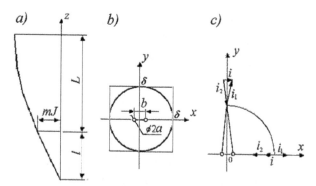

Figure 3.12 Current distribution along the wire (*a*), area of losses (*b*) and currents on a ground surface in the outer area (*c*).

The current density in the ground coincides with the strength of the magnetic field created by wires of the long line. Its components are equal to

$$i_x = -H_y = \frac{mJ}{2\pi j \sin kL}\left[(x+b/2)A_0(L,l,R_1) - (x-b/2)A_0(L,l,R_2)\right],$$

$$i_y = -H_x = \frac{mJy}{2\pi j \sin kL}\left[A_0(L,l,R_1) - A_0(L,l,R_2)\right],$$

where $R_{1(2)} = \sqrt{(x\pm b/2)^2 + y^2}$.

Introducing a value δ satisfying the inequality

$$b/2 << \delta << L, \lambda/2, \quad (3.36)$$

one can divide the area of losses (Figure 3.12*b*) into two areas, the boundary between which is a circumference of radius δ. In the outer area the distance from the long line to the observation point is large in comparison with the distance between the wires. Here the fields produced by the currents J_1 and J_2 cancel each other (Figure 3.12*c*). Thus, for example, when $x = 0$, $i = i_1 b/\rho \approx bH_{x1}/\rho$, and taking (3.36) into account, we find

$$|i| \le \frac{1,5mJ}{\pi \sin kL}\frac{b}{\rho^2}.$$

Power losses in the outer area

$$P_I = R_0 \int_0^{2\pi}\int_0^{\infty}|i|^2 \, \rho d\rho d\varphi \le \pi R_0 \left(\frac{1,5mJb}{\pi\delta\sin kL}\right)^2,$$

where $R_0 = 1/(s\sigma)$ is the resistance per unit area of the ground surface. In accordance with (3.36), this value is small and can be neglected. In the inner area $\rho << L,\ \lambda/2$ expressions for the components of the surface current are greatly simplified:

$$i_x = \frac{mJ}{2\pi}\left[\frac{x-b/2}{(x-b/2)^2+y^2}-\frac{x+b/2}{(x+b/2)^2+y^2}\right],$$

$$i_y = \frac{mJy}{2\pi}\left[\frac{1}{(x-b/2)^2+y^2}-\frac{1}{(x+b/2)^2+y^2}\right]. \tag{3.37}$$

Without wasting adopted accuracy, we calculate the losses not in a circle of radius δ, but in the square of side 2δ, (see Figure 3.12b). Power of losses in the inner area is

$$P_{II}=4R_0\int_0^\delta\int_0^\delta\left(i_x^2+i_y^2\right)dxdy=\frac{R_0(mJb)^2}{\pi^2}\int_0^\delta\int_0^\delta\frac{dxdy}{\left[(x-b/2)^2+y^2\right]\left[(x+b/2)^2+y^2\right]},$$

from which

$$R_{gl}=\frac{P_{II}}{(mJ)^2}=\frac{R_0}{\pi^2}\sum_{i=1}^{2}Q_1,$$

where

$$Q_1=-\int_0^1\left[\cot^{-1}(t-b/2\delta)+\cot^{-1}(t+b/2\delta)\right]\frac{dt}{t},$$

$$Q_2=\int_0^1\left[\frac{\cot^{-1}(t-b/2\delta)}{t-b/2\delta}+\frac{\cot^{-1}(t+b/2\delta)}{t+b/2\delta}\right]dt.$$

One can show that

$$\sum_{i=1}^{2}Q_i=\pi\ln(b/2a)+2G,$$

where $G=\int_0^1\tan^{-1}z\frac{dz}{z}\approx 0{,}916$ is the Catalan's constant. The value δ, as would be expected, in this expression, is not included. Taking into account that $b/a >> 1$, we find

$$R_{gl}=\frac{R_0}{\pi}\ln(b/a)=\frac{11}{\sqrt{\sigma\lambda}}\ln(b/a). \tag{3.38}$$

This result can also be obtained using known analogy between electric field in a conductive medium and an electrostatic field. The rightness of this approach follows from (3.37): a magnetic field and current density in the ground in the area of the losses are determined only by the currents in the base of wires and have a quasi-static nature. Similarly, one can obtain for the coaxial line (see Figure 3.2e)

$$R_{gl}=\frac{R_0}{2\pi}\ln\left(a_2/a_1\right)=\frac{5.5}{\sqrt{\sigma\lambda}}\ln(b/a). \tag{3.39}$$

As can be seen from (3.38) and (3.39), the loss resistance in the ground for vertically located long line (both two-wire and coaxial) does not depend on its length L. This resistance depends only on the ratio of the distance between the wires to the wire radius. Figure 3.13 shows the dependence of loss resistance R_{gl} of a two-wire line on the frequency at different values b/a and δ (of sea water). The calculation shows that ignoring the losses in the water is irregular.

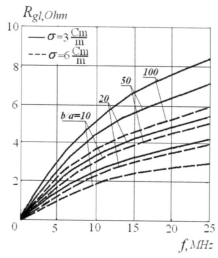

Figure 3.13   Loss resistance of the vertical two-wire long line in the water.

One must emphasize that the loss resistance R_{ge} of the monopole in the ground should be connected in series with the input impedance of the radiator itself, and the loss resistance R_{gl} of the line in the ground should be connected in series with the input impedance of the line. This is easily seen, proceeding from Figure 3.14, where the contour along which the current flows, is shown for both antenna elements. Taking into account losses in the ground, the expression for the input impedances of folded radiator with a gap and the expression for the input admittance of folded radiator with ground connection of the second wire take the form:

$$Z_A = Z_e(a_e) + R_{ge} + jm^2 W_l \tan kL + m^2 R_{gl}, \quad Y_A = \frac{1}{R_{gl} + jW_l \tan kL} + \frac{p^2}{R_{ge}Z_e(a_e)}. \quad (3.40)$$

Loss resistances of folded radiators in the wires are considered in Section 4.2.

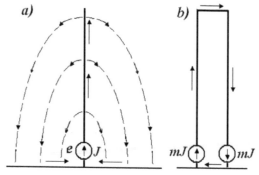

Figure 3.14   Direction of currents in monopole and ground (*a*), and in two-wire line and ground (*b*).

3.5 IMPEDANCE FOLDED RADIATORS

Folded radiators, on whose surface in contrast to metal radiator nonzero boundary conditions are performed, are called impedance folded radiators. Nonzero boundary conditions outside an excitation zone have the form:

$$\frac{E_z(a,z)}{H_\varphi(a,z)}\Big|_{-L\leq z\leq L}=Z(z). \qquad (3.41)$$

Here E_z (a, z) and H_φ (a, z) are the tangential component of the electric field and the azimuthal component of the magnetic field respectively, $Z(z)$ is the surface impedance, which in the general case depends on the coordinate z and substantially changes the distribution of current along an antenna in the first approximation. As mentioned already, the boundary conditions of this type are valid, if the structure of a field in one of a media (for example, inside the magneto dielectric sheath of the antenna) does not depend on a field structure in another medium (ambient space). Using the surface impedance (or concentrated loads) creates an additional degree of freedom and permits to expand opportunities of the antenna [28].

Two asymmetrical variants of the impedance folded radiator are shown in Figure 3.15. In the first variant there is a gap between the second (unexcited) wire and the ground (Figure 3.15a). In the second variant this wire is shorted to the ground (Figure 3.15b).

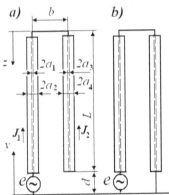

Figure 3.15 Asymmetric impedance folded radiators with a gap in the second (unexcited) wire (a), and with the ground connection of this wire (b).

The method of calculating an impedance folded radiator, similarly to calculation of a metal folded radiator, is based on the theory of asymmetric lines [37]. For two-wire impedance long lines, consisting of two wires with impedance coating (see Figure 3.3) and located above ground, telegraph equations (3.1) are valid:

$$-\frac{\partial u_1}{\partial z}=j(X_{11}+Q_1)i_1+jX_{12}i_2, \quad u_1=j\frac{1}{k^2}\left(X_{11}\frac{\partial i_1}{\partial z}+X_{12}\frac{\partial i_2}{\partial z}\right),$$

$$-\frac{\partial u_2}{\partial z}=j(X_{22}+Q_2)i_2+jX_{12}i_1, \quad u_2=j\frac{1}{k^2}\left(X_{22}\frac{\partial i_2}{\partial z}+X_{12}\frac{\partial i_1}{\partial z}\right). \qquad (3.42)$$

Here, as in Section 3.1, u_i is the potential of the wire i relative to ground, i_i is the current along the wire i, $X_{ik} = \omega\Lambda_{ik}$ is the self- or mutual inductive impedance per unit length.

Besides that, $jQ_i = Z_i/(2\pi a_i)$ is the additional impedance per unit length due to surface impedance Z_i.

The two left equations of the set (3.42) are based on the fact that the decrease of potentials at segment dz of each wire is the result of the emf influence. The emf's are induced by the self-currents and by the currents of the adjacent wires. The other two equations were written on the basis of the electrostatic equations relating charges and potentials in accordance with the equation of continuity. Dependence of the current on coordinate z is adopted in the form exp (γz), where γ is the propagation constant.

Differentiation of the right equations and substituting them into the left equations brings to a set

$$\left(X_{11} + Q_1 + \frac{\gamma^2}{k^2}X_{11}\right)i_1 + X_{12}\left(1 + \frac{\gamma^2}{k^2}\right)i_2 = 0,$$

$$X_{12}\left(1 + \frac{\gamma^2}{k^2}\right)i_1 + \left(X_{22} + Q_2 + \frac{\lambda^2}{k^2}X_{22}\right)i_2 = 0. \tag{3.43}$$

This system of equations has a solution, if the condition is satisfied

$$\gamma_{1(2)}^2 = -k^2 - \frac{\omega}{2}\left[G_1\beta_{11} + Q_2\beta_{22} \mp \sqrt{(Q_1\beta_{11} - Q_2\beta_{22})^2 + 4Q_1Q_2\beta_{12}^2}\right]. \tag{3.44}$$

Here $\beta_{11} = \dfrac{k^2 X_{22}}{\omega(X_{11}X_{22} - X_{12}^2)}$, $\beta_{12} = -\dfrac{k^2 X_{12}}{\omega(X_{11}X_{22} - X_{12}^2)}$, $\beta_{11} = \dfrac{k^2 X_{11}}{\omega(X_{11}X_{22} - X_{12}^2)}$, are coefficients

of electrostatic induction. Thus a system of two non metallic wires located above the ground has two different propagation constants. If the impedance is a purely reactive, they are equal

$$k_{1(2)} = \sqrt{-\gamma_{1(2)}^2}.$$

We seek a solution in the form

$$u_1 = A \cos k_1 z + jB \, \text{Sin} \, k_1 z + C \cos k_2 z + jD \, \text{Sin} \, k_2 z. \tag{3.45}$$

The ratio of the currents obtained from (3.43) is substituted in the first equation of (3.42):

$$i_1 = j\frac{k^2 + \gamma^2}{\gamma^2 Q_1}\frac{du_1}{dz} = a_1\left(B\cos k_1 z + jA\sin k_1 z\right) + a_2\left(D\cos k_2 z + jC\sin k_2 z\right), \tag{3.46}$$

where $a_i = (k^2 - k_i^2)/(k_i Q_i)$. From (3.46) and similar relationship between i_2 and u_2 one can obtain

$$\frac{du_2}{dz} : \frac{du_1}{dz} = \frac{Q_2 i_2}{Q_1 i_1}, \tag{3.47}$$

i.e.

$$u_2 = b_1(A \cos k_1 z + jB \, \text{Sin} \, k_1 z) + b_2(C \cos k_2 z + jD \sin k_2 z). \tag{3.48}$$

and $b_i = \dfrac{Q_2}{Q_1}\dfrac{k^2\left(X_{11} + Q_1\right) - k_i^2 X_{11}}{(k_i^2 - k^2)X_{12}}$. Finally from (3.47) and (3.46) it follows that

$$i_2 = a_1 c_1 (B \cos k_1 z + jA \sin k_1 z) + a_2 c_2 (D \cos k_2 z + jC \sin k_2 z), \tag{3.49}$$

where $c_i = b_i Q_1/Q_2$.

Putting $z = 0$, we find: $A = U_{11}$, $D = I_{11}/a_1$, $C = U_{12}$, $D = I_{12}/a_2$. Here U_{11}, U_{12}, I_{11}, I_{12} are fractions of voltages and currents at the beginning of the first wire (near load), which correspond to phase constant k_1 and k_2. Considering that the positive current is the current flowing from the generator to the load, i.e. in the direction of negative z, we rewrite (3.45), (3.46), (3.48) and (3.49), taking into account the calculated coefficients:

$$u_1 = U_{11}\cos k_1 z + U_{12}\cos k_2 z + j(W_1 I_{11}\sin k_1 z + W_2 I_{12}\sin k_2 z),$$

$$i_1 = I_{11}\cos k_1 z + I_{12}\cos k_2 z + j\left(\frac{U_{11}}{W_1}\sin k_1 z + \frac{U_{12}}{W_2}\sin k_2 z\right),$$

$$u_2 = b_1 U_{11}\cos k_1 z + b_2 U_{12}\cos k_2 z + j(b_1 W_1 I_{11}\sin k_1 z + b_2 W_2 I_{12}\sin k_2 z),$$

$$i_1 = c_1 I_{11}\cos k_1 z + c_2 I_{12}\cos k_2 z + j\left(c_1\frac{U_{11}}{W_1}\sin k_1 z + c_2\frac{U_{12}}{W_2}\sin k_2 z\right). \qquad (3.50)$$

One must emphasize the important conclusion, which follows directly from (3.50). Currents and potentials of both wires are connected by rigid relations depending not on the details of the connecting antenna in a circuit (not in accordance with so-called boundary conditions), but on the wires, diameters and the surface impedance. Therefore, it is impossible, changing only the boundary conditions (for example, changing the point of connecting emf, magnitudes and points of connecting loads), to obtain in the wires of the impedance folded radiator purely in-phase or anti-phase currents (by contrast to purely metallic folded radiators). Accordingly, the input impedance of such impedance radiators cannot be presented as an aggregate of impedance lines and radiators, connected with each other in parallel or in series. Only identical wires, as it will be shown below, are an exception.

We apply these results to the calculation of the input impedance of a folded radiator with a gap in the second (unexcited) wire (see Figure 3.15a). The boundary conditions for this variant have the form:

$$i_1(0) + i_2(0) = 0,\ u_1(0) = u_2(0),\ i_2(L) = 0,\ u_1(L) = e. \qquad (3.51)$$

Substituting (3.50) into (3.51), we find

$$I_{12} = -I_{11}\frac{1+c_1}{1+c_2},\ U_{12} = U_{11}\frac{1-b_1}{b_2-1},\ I_{11} = -jd_1\frac{U_{11}}{W_1},$$

$$e = U_{11}\left(\cos k_1 L - \frac{1-b_1}{1-b_2}\cos k_2 L\right) + I_{11}\left(W_1\sin k_1 L - W_2\frac{1+c_1}{1+c_2}\sin k_2 L\right), \qquad (3.52)$$

and

$$d_1 = \tan k_1 L\left[1 - \frac{W_1(1-b_1)c_2\sin k_2 L}{W_2(1-b_2)c_1\sin k_1 L}\right]:\left[1 - \frac{c_2(1+c_1)\cos k_2 L}{c_1(1+c_2)\cos k_1 L}\right].$$

Then the input impedance of impedance line is

$$X_{il} = \frac{e}{ji_1(L)} = -W_1\cot k_1 L\frac{1 - \dfrac{(1-b_1)\cos k_2 L}{(1-b_2)\cos k_1 L} + d_1\tan k_1 L\left[1 - \dfrac{W_2(1+c_1)\sin k_2 L}{W_1(1+c_2\sin kL_1)}\right]}{1 - \dfrac{W_1(1-b_1)c_2}{W_2(1-b_2)c_1} - d_1\cot k_1 L\left[1 - \dfrac{(1+c_1)\cos k_2 L}{(1+c_2)\cos k_1 L}\right]}. \qquad (3.53)$$

Expression (3.53) makes it possible to determine the approximately reactive component of the input impedance of the impedance folded radiator (similarly to the fact as formula for the input impedance of an equivalent long line allows to determine a reactive component of an input impedance of the line radiator). The antenna input impedance can be found more precisely by the method of induced electromotive force. Equating the oscillating part of the power passing through a closed surface surrounding the antenna and the oscillating part of the power passing through the source of emf, we obtain (for the asymmetrical radiator)

$$Z_A = -\frac{1}{J_1^2(0)}\int_0^L E_y J(y)dy.$$ (3.54)

Here E_y is a field on the antenna surface, $J_1(0) = i_1(L)$ is the current of the generator and $J(y) = J_1(y) + J_2(y)$ is a total current of an antenna as function of coordinate $y = L - z$ (see Figure 3.15a).

Expression (3.54) is a generalization of the second formulation of the method of induced emf as applied to the folded radiator. For the folded radiator with a gap between the second (unexcited) wire and the ground a total input current of the antenna coincides with a generator current. For the folded radiator with the ground connection of this wire this coincidence is absent.

In the vicinity of parallel resonance, where the method of induced emf gives the wrong result, for the folded radiator with a gap, one must use the expression

$$Z_A = e/\left[2J_1(0) + \frac{1}{e}\int_0^L E_y(y)J(y)dy\right].$$ (3.55)

and the total current of this antenna is equal to

$J(y) = j(1 + c_1)U_{11}/W_1 \times$

$$\left\{\sin k_1\left(L-|y|\right) - \frac{W_1(1-b_1)(1+c_2)}{W_2(1-b_2)(1+c_1)}\sin k_2\left(L-|y|\right) - d_1\left[\cos k_1\left(L-|y|\right) - \cos k_2\left(L-|y|\right)\right]\right\}.$$ (3.56)

The field in the far region with allowance for a mirror image is

$$E_\theta = \frac{60k(1+c_1)U_{11}\exp(-jkr)\sin\theta}{W_1 r}\left[\frac{\Theta_1(\cos\theta)}{k^2\cos^2\theta - k_1^2} - \frac{\Theta_2(\cos\theta)}{k^2\cos^2\theta - k_2^2}\right],$$ (3.57)

where

$$\Theta_i(\cos\theta) = k_i e_i[\cos(kL\cos\theta) - \cos k_i L] + d_1[k\cos\theta\sin(kL\cos\theta) - k_i\cos k_i L],$$

$$e_1 = 1, e_2 = \frac{W_1(1-b_1)(1+c_2)}{W_2(1-b_2)(1+c_1)}.$$

An effective length of asymmetric radiator is

$$h_e = \frac{k_2(1-\cos k_1 L) - e_2 k_1(1-\cos k_2 L) - d_1(k_2\sin k_1 L - k_1\sin k_2 L)}{k_1 k_2\left[\sin k_1 L - e_2\sin k_2 L - d_1(\cos k_1 l - \cos k_2 L)\right]}.$$ (3.58)

Thus, the calculation of the folded radiator with nonzero boundary conditions is divided into two stages. First, the current distribution along the antenna wires is determined using the theory of coupled lines, afterwards electrical characteristics of the antenna are calculated. In order to calculate the far field, the total current of antenna is used. Input impedance is calculated by the method of induced electromotive force, or by solving the integral equation. Coefficients W_i, b_i, c_i, k_i depend on the inductive impedances $X_{ik} = \omega p_{ik}/c^2$ per unit length, where α_{ik} are the potential coefficients, which are determined by a method of an average potential (for example, by method of Howe), in accordance with the actual location of the antenna wires.

Practically, important special cases, when the surface impedance on one wire of the folded radiator is equal to zero, are of particular interest. Main characteristics of folded antennas with a gap, if one or the other wire is purely metallic, are given in Table 3.2. One must note that in the calculation of the difference $k_1^2 - k^2$, it is necessary to expand it into the series of Maclaurin.

Table 3.2   Characteristics of Folded Radiators with a Gap

Characteristic	$Q_2 = 0$	$Q_1 = 0$
k_1	k	k
k_2	$k\sqrt{1+\dfrac{Q_1 X_{22}}{X_{11}X_{22}-X_{12}^2}}$	$k\sqrt{1+\dfrac{Q_2 X_{11}}{X_{11}X_{22}-X_{12}^2}}$
$k_1^2 - k^2$	$Q_2\dfrac{k^2}{X_{22}}$	$Q_1\dfrac{k^2}{X_{11}} - Q_1^2\dfrac{k^2 X_{12}^2}{Q_2 X_{11}^3}$
X_{il}	$-\dfrac{k_2 Q_1}{k_2^2-k^2}\times\dfrac{F_1+F_2}{F_3+F_4}\cot kL,$ where $F_1 = 1+\dfrac{(X_{22}-X_{12})\cos k_2 L}{X_{12}\cos k_2 L},$ $F_2 = d_1\tan kL\left\{1-\dfrac{k_2 k Q_1 \sin k_2 L}{[(k_2^2-k^2)(X_{12}-X_{11})+k^2 Q_1]\sin kL}\right\}.$ $F_3 = \dfrac{(X_{22}-X_{12})\sin k_2 L}{X_{12}\sin kL},$ $F_4 = d_1\tan kL\left\{\dfrac{k_2 k Q_1 \cos k_2 L}{[(k_2^2-k^2)(X_{12}-X_{11})+k^2 Q_1]\sin kL}\right\}$	$-X_{11}\dfrac{1+d_1\tan kL}{k(1-d_1\tan kL)}\cot kL$
d_1	$\dfrac{1+\dfrac{X_{22}-X_{12}}{k_2 k Q_1 X_{12}}[(k^2-k_2^2)X_{11}+k^2 Q_1]\dfrac{\sin k_2 L}{\cos kL}}{1+\dfrac{[(k_2^2-k^2)X_{11}-k^2 Q_1]\cos k_2 L}{[(k_2^2-k^2)(X_{12}-X_{11})+k^2 Q_1]\cos kL}}$	$-\dfrac{(k_2^2-k^2)(X_{11}-X_{12})^2}{k_2 k Q_2 X_{11}}\cot kL$
e_2	$\dfrac{X_{12}-X_{22}}{k_2 k Q_1 X_{12}}[(k_2^2-k^2)(X_{12}-X_{11})+k^2 Q_1]$	$-\dfrac{(k_2^2-k^2)(X_{11}-X_{12})^2}{k_2 k Q_2 X_{11}}$

If the radiator is made up of two identical wires ($Q_1 = Q_2$, $a_2 = a_4$), then

$$k_1 = \sqrt{k^2 + \omega Q_1\left(\beta_{11} + \beta_{12}\right)} = k\sqrt{1 + Q_1/\left(X_{11} + X_{12}\right)},$$

$$k_2 = \sqrt{k^2 + \omega Q_1\left(\beta_{11} - \beta_{12}\right)} = k\sqrt{1 + Q_1/\left(X_{11} - X_{12}\right)},$$

where from, $b_1 = c_1 = 1$, $b_2 = c_2 = -1$, $d_1 = 0$, i.e. expressions (3.50) take the form

$$u_1 = U_{11}\cos k_1 z + U_{12}\cos k_2 z + j\left(W_1 I_{11}\sin k_1 z + W_2 I_{12}\sin k_2 z\right),$$

$$i_1 = I_{11}\cos k_1 z + I_{12}\cos k_2 z + j\left(\frac{U_{11}}{W_1}\sin k_1 z + \frac{U_{12}}{W_2}\sin k_2 z\right),$$

$$u_2 = U_{11}\cos k_1 z - U_{12}\cos k_2 z + j\left(W_1 I_{11}\sin k_1 z - W_2 I_{12}\sin k_2 z\right),$$

$$i_2 = I_{11}\cos k_1 z - I_{12}\cos k_2 z + j\left(\frac{U_{11}}{W_1}\sin k_1 z - \frac{U_{12}}{W_2}\sin k_2 z\right) \tag{3.59}$$

This means that in this particular case, irrespective of the boundary conditions for the currents and voltages, their components with the propagation constant k_1 are equal in magnitude and opposite in sign (anti-phase wave). Accordingly, the input impedance of the folded radiator with a gap (see Figure 3.15a) can be presented as an aggregate of input impedances of two-wire line and monopole:

$$X_A = -W_m \cot k_1 L + 0,25 W_l \tan k_2 L \tag{3.60}$$

where

$$W_m = W_1/2 = \frac{2k_1}{2kc\left(\beta_{11} + \beta_{12}\right)} = \frac{k_1}{2kcC_{11}}$$

is the wave impedance of the impedance linear radiator consisting of two parallel wires, and

$$W_l = 2W_2 = \frac{2k_2}{kc\left(\beta_{11} - \beta_{12}\right)} = \frac{k_2}{kc\left(C_{12} + C_{11}/2\right)}$$

is the wave impedance of an impedance long line also consisting of two wires located symmetrically relatively surface of zero potential (ground). Magnitudes C_{11} and C_{12} in these expressions are partial capacitances.

For the folded radiator with the ground connection of the second wire (see Figure 3.15b), instead of the third boundary condition (3.51) we have

$$u_2(L) = 0. \tag{3.61}$$

Therefore, instead of the third equation of the set (3.52), we obtain

$$I_{11} = -jd_2\frac{U_{11}}{W_{11}}, \tag{3.62}$$

where

$$d_2 = -\cot k_1 L\left[1 + \frac{b_2(1-b_1)\cos k_2 L}{b_1(1-b_2)\cos k_1 L}\right]\left[1 - \frac{W_2 b_2(1+c_1)\sin k_2 L}{W_1 b_1(1+c_2)\sin k_1 L}\right]^{-1},$$

and d_2 will take the place of the coefficient d_1 in expressions for electrical characteristics of the radiator. When $Q_2 = 0$, coefficient d_2 is equal to $d_2 = -\cot kL$. When $Q_1 = 0$,

$$d_2 = -\cot kL \left[1 + \frac{(X_{11} - X_{22})\cos k_2 L}{X_{12} \cos kL} \right] \left[1 - \frac{k_2 k Q_2 X_{11} \sin k_2 L}{(k_2^2 - k^2) X_{12} (X_{11} - X_{12}) \sin k_1 L} \right]^{-1}.$$

When $Q_1 = Q_2$, $a_2 = a_4$,

$$Y_A = \frac{1}{W_l \tan k_2 L} - \frac{1}{4W_m \cot k_1 L}. \tag{3.63}$$

When $Q_1 = Q_2 = 0$, i.e. $k_1 = k_2 = k$, equalities (3.60) and (3.63) become by expressions (3.16) and (3.17).

As an example, in Figure 3.16 the model of the impedance folded radiator is presented. One wire of this radiator is made in the form of a rod with a ferrite coating (relative magnetic permeability of the coating is 10), and the other wire is made of a metal tube. Dimensions of a model are given in millimeters.

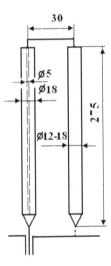

Figure 3.16 Model of the impedance folded radiator.

The calculated curves and experimental values of active R_A and reactive X_A components of the input impedance of the presented radiator for the different variants of its connection to the generator and the ground, as well as for different diameters of the tube are given in Figures 3.17 and 3.18. The coincidence of the calculated and experimental results is quite satisfactory. As can be seen from the figures, the radiator characteristics are substantially changed, if one or the other wire is excited. Using slowing coating allows to decrease resonant frequencies in 2-2.5 times.

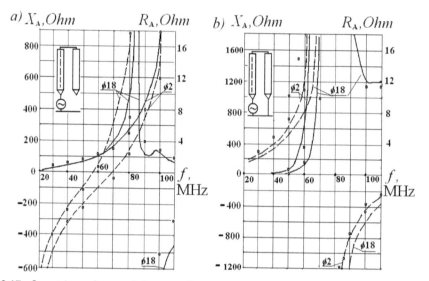

Figure 3.17 Input impedance of the impedance folded radiator with excited impedance wire: with a gap (*a*), with a shorting to the ground (*b*).

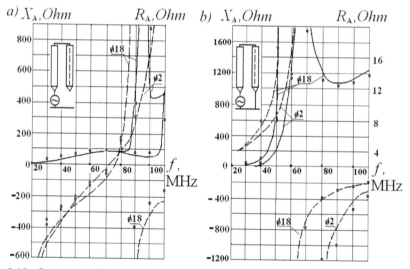

Figure 3.18 Input impedance of the impedance folded radiator with excited metal wire: with a gap (*a*), with the ground connection of the second wire (*b*).

4

Multi-Folded and Multi-Level Antennas

4.1 METHOD OF CALCULATING MULTI-FOLDED RADIATORS

As is shown in Section 3.1, asymmetric folded radiators consisting of two parallel wires, upper ends of which are connected with each other, combine the functions of radiation and matching. In a folded radiator with a gap, a length of which is less than a quarter of a wave length, a capacitive impedance of a linear radiator (of a monopole) is compensated by an inductive impedance of a long line with shorting at the end. In a folded radiator with a ground connection of an unexcited wire a long line, connected in parallel with an input impedance of a linear antenna, transforms its resistance.

A multi-folded radiator (Figure 4.1) gives more opportunities in order to obtain useful results in this direction. This radiator is a group of parallel wires connected in pairs at the top and bottom so that there is formed a system of coupled and connected in series elongated loops (of two-wire long lines). In the particular case when the number of wires is two, this antenna is converted into a folded radiator.

Figure 4.1 Multi-folded radiator.

If the transverse dimensions of a multi-folded radiator are small in comparison with its height L and the wavelength λ, then, as is shown in the article [38], devoted to a research of electromagnetic oscillations in systems of parallel thin wires, the current in each wire of such system can be divided into in-phase and anti-phase components, and the entire system may be reduced to an aggregate of linear radiator and non-radiating long lines. The method of calculating the characteristics of such antenna may be considered using an example of a two-folded radiator with a gap (Figure 4.2).

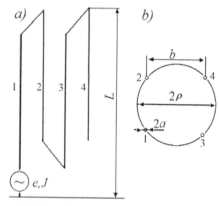

Figure 4.2   Two-folded radiator with a gap: circuit (*a*), cross-section (*b*).

At first we must divide the two-folded antenna onto a radiator and long lines. For that, into the gap between the free end of the antenna and the ground we include two parallel generators of the current, which are equal in magnitude (mJ) and opposite in sign (Figure 4.3). Here J is the current of the main generator. The main generator is also divided onto two generators of currents, identical in direction and different in magnitude: mJ and $(1 - m)J$. The total current of the generator as a result of such operation is not change; the total current in the gap is zero as before.

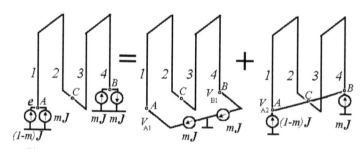

Figure 4.3   Calculating the two-folded radiator with a gap.

According to the superposition principle a voltage at point A is equal to a sum of voltages, produced by all generators. Therefore, as it is shown in Figure 4.3, one can divide the circuit of the two-folded antenna onto two circuits, with two generators in each one, and then add up the voltages at the point A, created in each of these circuits.

In the first auxiliary circuit the generators are identical and connected in series. Therefore, the voltage between the point A and the ground is

$$V_{A1} = V/2 = 0.5mJZ_l, \tag{4.1}$$

where $V = V_{A1} - V_{B1}$ is the input voltage, and Z_l is the impedance of a complicate long line. More precisely, this is two coupled lines with the same distance $b\sqrt{2}$ between wires and the equal wave impedances. $W_1 = 120 \ln(b\sqrt{2}/a)$. One of these lines is a load for the other line. In the first approximation it could be considered that it is a united two-wire line, bended in the middle at an angle $180°$. If to use the theory of electrically coupled lines, W_1 in this expression will be replaced by the value $W_2 = W_3 W_4 / (W_4 - W_3)$, where

$$W_3 = 60 \frac{4\ln^2(b/a) - 2\ln^2\sqrt{2}}{\ln\left[b^3/\left(a^3\sqrt{2}\right)\right]}, W_4 = 240\ln\left\lfloor b/\left(a\sqrt{2}\right)\right\rfloor.$$

It is easy to make sure that for small radius of wires$(b \gg a)$

$$W_2 \approx W_1 \approx 120 \ln(b/a).$$

The points A and B in the second auxiliary circuit are connected with each other as equipotential points. It means that $m = 1/2$, if the wires diameters are identical. The second circuit is the folded radiator, in which each "conductor" consists of two parallel wires (1 and 4, 2 and 3 respectively). This circuit can also be divided into two ones: the four-wire line of length L, with shorting at the end, and the asymmetric radiator (the monopole) of height L. The wave impedance W_l of the line is equal to $W_5 = 60\ln\left\lfloor b/(a\sqrt{2})\right\rfloor$. Equivalent radius of the monopole with four wires (of radius $b/\sqrt{2}$), located along the cylinder generatrices, is equal to $a_e = \sqrt[4]{ab^3\sqrt{2}}$. Input impedance of the second circuit is

$$Z_{A2} = Z_m(a_e) + jm^2 W_l \tan kL.$$

This impedance is calculated by means of the same procedure of dividing the initial circuit onto two circuits (the monopole and the long line), and adding up the input voltages. For the voltage between point A and the ground in the second circuit, one can write

$$V_{A2} = J[Z_m(a_e) + j0.25W_5 \tan kL]. \tag{4.2}$$

Hence, the input impedance of the entire antenna

$$Z_A = Z_m(a_e) + j0.25W_2 \tan 2kL + j0.25W_5 \tan kL. \tag{4.3}$$

This result is illustrated by Figure 4.4, which shows the input impedances of the antenna and its components. From the point of view of the input impedance the considered antenna is a series connection of the monopole and two lines of length L and $2L$, with shorting at the end. The wave impedances of these lines are close to values

$$W(n) \approx (120/n)\ln(b/a), \tag{4.4}$$

where n is the number of wires in each "conductor" of the line.

From (4.4) and Figure 4.4 it follows that the radiation resistance of the two-folded antenna with a gap is equal to the radiation resistance of the monopole of the same height. The reactive component of the input impedance has additional resonances, and the first parallel resonance is caused by the parallel resonance of the long line with length $2L$, i.e. its frequency is half the frequency of the first serial resonance of an ordinary monopole. The frequency of the first serial resonance of the antenna is even smaller (but not necessarily twice).

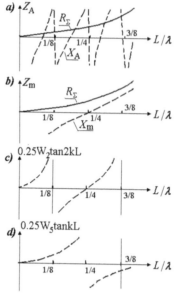

Figure 4.4 Impedance of two-folded antenna (a) and of its components: monopole (b), line of length $2L$ (c), line of length L (d).

Due to increase of the wires number and a corresponding increase of the length of the total antenna wire, the number of resonances increases in a concrete frequency range. For example, if the loop's number is equal to $N = 2^n$, then for the antenna with a gap we obtain similarly to (4.3)

$$Z_A = Z_m(a_e) + j0.25 \sum_{m=0}^{n} W\left(2^m\right) \tan\left(NkL / 2^m\right). \tag{4.5}$$

Here $a_e = \sqrt[2N]{2Na\rho^{2N-1}}$ is the equivalent radius of the monopole, consisting of N loops, which are located along the generatrices of the cylinder with the radius ρ (if N grows, the equivalent radius tends to be ρ). The frequency of the first parallel resonance of the antenna (i.e. of its second resonance) is N times lower than the frequency of the first serial resonance of an ordinary monopole. Such character of the input impedance allows, firstly, to use multi-folded antenna in the range of longer waves, and secondly, when it is necessary, to tune the antenna onto several frequencies.

If N-folded antenna with the ground connection of the unexcited wire consists of the wires of identical diameter (see Figure 4.1, dotted line), its input admittance is

$$Y_A = \frac{1}{j120 \ln(b/a) \tan NkL} + \frac{1}{4Z_{N/2}}, \tag{4.6}$$

where $Z_{N/2}$ is the input impedance of the $N/2$-folded antenna with a gap and with "conductors" of two wires. This result generalizes expression (3.17) and is obtained similarly.

For the odd number of antenna wires, the calculation becomes more complicated. For example, a radiator of three wires (Figure 4.5) may be divided onto a three-wire line and a monopole with a height L (Figure 4.6). Potentials of all wires of the monopole in each cross-section must be the same. So in accordance with (3.19) the magnitude m depends on the relation of capacitances of two branches of folded antenna. The right branch consists of two wires, and its capacitance is twice as much. Therefore one derives that $m = 2/3$. One can show, using the theory of electrically coupled lines that an impedance of a three-wire line of the identical three wires (of a first circuit) is equal to

$$Z_l = j80 \ln (b/a) \frac{1}{\cot 2kL + \cot kL}. \tag{4.7}$$

In this circuit there are only anti-phase currents, and their sum is equal to zero in each cross-section. The sum of potentials in an arbitrary cross-section also is zero, i.e.

$$V_{A1} = -2V_{B1} = 2V/3,$$

where $V = V_{A1} - V_{B1}$ is the voltage at the input of the line. From here the impedance of the three-wire antenna is

$$Z_A = \frac{V_{A1} + V_{A2}}{J} = Z_m(a_e) + j80 \ln (b/a) \frac{1}{\cot 2kL + \cot kL}. \tag{4.8}$$

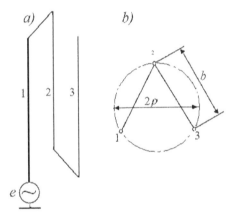

Figure 4.5   Three-wire antenna: circuit (a), cross-section (b).

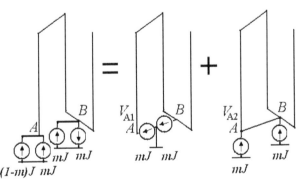

Figure 4.6   To the calculation of three-wire folded antenna with a gap.

Equivalent radii of the three-wire monopole and the three-folded radiator are equal accordingly to $a_{em} = \sqrt[3]{ab^2}$ and $a_{er} = \sqrt[6]{6ab^5}$. If the number of loops is $N = 3 \cdot 2^n$, then

$$Z_A = Z_m\left(a_e\right) + j0.25\sum_{m=0}^{n} W\left(2^m\right)\tan\left(NkL/2^m\right) + j0.33W\left(2^n\right)\frac{1}{\cot 2kL + \cot kL}. \tag{4.9}$$

The value of a_e is given earlier.

4.2 ELECTRICAL CHARACTERISTICS OF MULTI-FOLDED RADIATORS

The first section of this chapter is devoted to the method of analysis of multi-folded antennas and to their input impedances. Here we shall briefly talk about other properties of these radiators.

The directional patterns of a multi-folded antenna do not differ from the directional patterns of an ordinary monopole, since fields of long lines in a far zone upon small distances between wires may be neglected.

In calculating a loss resistance R_{gA} of multi-folded antennas in the ground it is necessary to determine a loss resistance R_{gl} for lines of complex shape. Figure 4.7a shows the two-wire line of length $2L$, which in the middle is bent at an angle of $180°$. The losses of such line in a ground do not differ from the losses of the line shown in Figure 3.14b because the currents in the ground between projections of the wires 2 and 3 are practically absent. The currents between the wires 2 and 3 flow mainly along the connecting bridge AA', especially if the distance between these wires is less than the distance d between this bridge and the ground.

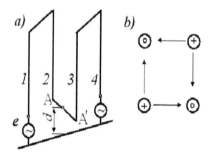

Figure 4.7 Losses in the ground for lines of complex shape (a) and with two wires in "conductor" (b).

The loss resistance of the long line, an each "conductor" of which consists of n wires, is smaller by a factor n^2 than the loss resistance of an ordinary line. This situation is illustrated by Figure 4.7b for the case when $n = 2$. The current of each wire of "conductor" flows into the n directions to n wires of the other "conductor", thereby forming a system of n^2 resistances, included in parallel. Each resistance is equal to the resistance R_{gl0} of a sector between single wires. Thus, the loss resistance in the ground of multi-folded radiator with a gap is equal to

$$R_{gA} = R_{ge} + \frac{1}{4}\sum_{m=0}^{n}\frac{R_{gl0}}{n_m^2}, \tag{4.10}$$

where n_m is the number of wires in each "conductor" of the line m.

Values R_{gA} for some antennas are given in Table 4.1. The table demonstrates that if the transverse dimension of an antenna is the same, increasing of the number of loops affects weakly the losses in the ground.

Table 4.1 Loss Resistance of Multi-folded Radiators with a Gap

Type of radiator	Loss resistance in a ground R_{gA}	Loss resistance in wires R_{wA}
Folded	$R_{ge} + 0.25R_{gl0}$	$\dfrac{R_{\sim}L}{\sin^2 2kL}\left(1 - \dfrac{\sin 4kL}{4kL}\right)$
Two-folded	$R_{ge} + 0.313R_{gl0}$	$\dfrac{2R_{\sim}L}{\sin^2 4kL}\left(1 - \dfrac{\sin 8kL}{8kL}\right)$
Four-folded	$R_{ge} + 0.328R_{gl0}$	$\dfrac{4R_{\sim}L}{\sin^2 8kL}\left(1 - \dfrac{\sin 16kL}{16kL}\right)$
Linear	R_{ge}	$\dfrac{R_{\sim}L}{2\sin^2 2kL}\left(1 - \dfrac{\sin 2kL}{2kL}\right)$

Further we consider losses in the antenna wires caused by a skin effect. As is known, the surface resistance of a round copper wire per one meter is

$$R_{\sim} = \sqrt{f} / 24a, \qquad (4.11)$$

where f is a frequency, in MHz, a is a wire radius, in mm. The resistance of the steel wire is 2.3 times as much.

The surface resistance causes the longitudinal attenuation of electromagnetic waves. Therefore, the propagation constant of a wave along the wire and the wave impedances of the long line and the monopole become complex quantities. One can find, substituting them into the expression for the input impedance, this impedance with allowance for the skin effect losses.

The imaginary additive to the propagation constant in the first place increases the magnitude of the active impedance R_A, since the latter is small in comparison with the reactive impedance X_A everywhere except the vicinity of resonances. An addition to a given above radiation resistance R_Σ is a sought value of loss resistance R_{wA} in the wires.

In Table 4.1 the values R_{wA} are given for several variants of multi-folded radiators with a gap, when attenuation in the wires is weak. It is believed that the radii of wires are small in comparison with the distances between them, i.e. the proximity effect and the corresponding redistribution of current over the wire cross-section are neglected. For comparison, the table demonstrates the loss resistance in the wires of a monopole.

The table shows that the losses in the wires cause the appearance of additional maxima on the curve for the active impedance when $kL = (2m + 1)\pi/2$ (m is a natural number), i.e. near the parallel resonances of the long lines. In these bands of the frequency range it is impossible to ignore the skin effect losses. At the low frequencies ($kL \ll 1$) the loss resistance increases proportionally to the number and the length of wires.

One can write that the input admittance of the N-folded radiator with the ground connection of the unexcited wire with allowance for the losses in the ground and in the wires is equal to

$$Y_A = \frac{1}{j120\ln(b/a)\tan NkL + R_{gl} + R_{wl}} + \frac{1}{4Z_{N/2}}, \tag{4.12}$$

where $Z_{N/2}$ is the input impedance of the $N/2$-folded antenna of two-wire "conductors" with a gap (in view of the losses), and R_{wl} and R_{gl} are the loss resistances of the two-wire long line in the wires and in the ground, relatively, at that

$$R_{wl} = \frac{NR_{-}L}{\cos^2 NkL}.$$

Calculation and experimental verification confirmed the rightness of the obtained results. Figures 4.8 and 4.9 show the input impedances of two-folded and four-folded radiators with a gap in different frequency bands. The experimental values of the resonant frequencies in the figures are shifted in comparison with the calculated results in the direction of lower frequencies. This small shift is caused by the fact that the calculation did not take into account the length of the horizontal connecting bridges between the wires.

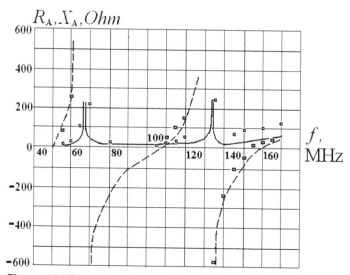

Figure 4.8 Input impedance of two-folded radiator with a gap.

Further, we pass to Q-factor. Q-factor (quality) is an important electrical characteristic of the antenna. It defines in particular the frequency band within which one may obtain the given level of matching an antenna with a cable without a change of a tuning. Q-factor characterizes the rate of changing the antenna input impedance as the result of the influence of various external factors and can be used to quantify the sustainability of the antenna tuning, if the sustainability is understood as the preservation of the results of the antenna tuning. From this standpoint, the higher the quality factor, the worse the stability, and vice versa.

The parameter Q is the magnitude similar to the quality factor of the resonant circuit. It is calculated at the point of the serial resonance of the antenna in accordance with the formula

$$Q_1 = \frac{\omega_i}{2R_{Ai}}\frac{dX_A}{d\omega}\bigg|_{\omega=\omega_i} = \frac{(kL)_i}{2R_{Ai}}\frac{dX_A}{d(kL)}\bigg|_{kL=(kL)_i}. \tag{4.13}$$

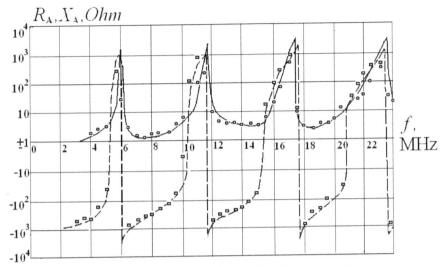

Figure 4.9 Input impedance of four-folded radiator with a gap.

Here R_{Ai} is the active component of the input impedance on the frequency f_i, in the vicinity of which R_{Ai} it is considered to be constant.

The expressions for calculating the electrical length $(kL)_i$ and Q-factor of four-folded radiator with a gap in the first four points of the serial resonances are given in Table 4.2 as an example. The table also indicates for the comparison the electrical length and the quality factor of a quarter-wavelength radiator and of a short linear radiator with a matching device. The table uses the following designations: W_m is the impedance of the linear radiator, in particular of the equivalent radiator with radius a; $W_l = 120 \ln(b/a)$ is the wave impedance of two-wire long line; $\Delta = \sqrt{W_l / (4W_m + 5W_l)}, p = \lambda/(4L)$.

Table 4.2 Expressions for Calculating Electrical Length and Quality of Four-Folded and Linear Radiators

Radiator	Number of resonance	Electrical length	Q-factor
Four-folded	1	$\frac{1}{4}\tan^{-1}\left(9.63\frac{W_m}{W_l} - 0.604\right)$	$\frac{1}{R_{A1}}\left(-0.94W_m + 0.38W_l + 18.2\frac{W_m^2}{W_l}\right)$
"	2	$\frac{1}{2}\tan^{-1}\left(8\frac{W_m}{W_l} - 0.5\right)$	$\frac{1}{R_{A2}}\left(0.56W_l + 6.3\frac{W_m^2}{W_l}\right)$
"	3	$\frac{1}{4}\tan^{-1}\left(1.66\frac{W_m}{W_l} - 0.1\right)$	$\frac{1}{R_{A3}}\left(0.5W_m + 1.14W_l + 1.62\frac{W_m^2}{W_l}\right)$
"	4	$(\pi - \delta)/2$	$\frac{1}{R_{A4}}[0.79(1+\Delta^2/4)W_m +$ $+ (1.03 + 3.34\Delta^2 + 0.196/\Delta^2)W_l]$
Quarter-wavelength	1	$\pi/2$	$W_m/93$
Short with a matching device	1	$\pi/(2p)$	$\pi W_m/[4pR_A\sin^2(\pi/2p)$

4.3 ANTENNA WITH MEANDERING LOAD

In this section we consider an example of using a multi-folded structure—a meandering load, which is applied in the capacity of a horizontal load of a wire antenna. The antenna with a meandering load is analyzed by means of the theory of coupled lines.

As is known, the horizontal load serves for increasing an effective height of the antenna by changing a current distribution along a radiating vertical wire. The equivalent length of the horizontal segment of the antenna is equal to

$$l_e = \frac{1}{k}\tan^{-1}\left(\frac{W_H}{W_V}\cot kL\right) \approx \frac{W_V l}{W_H}. \qquad (4.14)$$

If dimensions of the horizontal segment are small in comparison with a wave length, then its equivalent length is inversely proportional to its wave impedance W_H and directly proportional to its length l and also to wave impedance W_V of the vertical segment. Therefore, the horizontal load is usually fabricated in the form of several parallel wires, whose ends are connected with each other (Figure 4.10a).

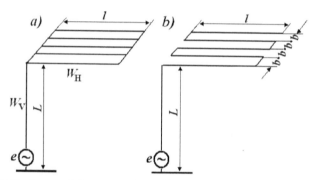

Figure 4.10   Inverted-L antenna (a) and antenna with meandering load (b).

Figure 4.10b shows the antenna circuit, in which the wires of the horizontal load are connected with each other in series [39]. In this circuit, the current path along the load is extended, and its equivalent length also increases. The antenna was called in accordance with the shape of the horizontal load.

In order to calculate the input impedance of the load we use, as already mentioned, the theory of coupled lines located above the ground. The current and the potential of each wire of asymmetric line, consisting of N parallel wires, are defined by the expressions (3.3), and the wave impedances between the wires are defined by the expressions (3.4).

For example, calculating the input impedance of a three-wire load presented in Figure 4.11 begins with writing the boundary conditions for the currents and the potentials:

$$\begin{array}{lll} u_1(0) = u_2(0), & i_1(0) + i_2(0) = 0, & i_3(0) = (0), \\ u_2(l) = u_3(l), & i_2(l) + i_3(l) = 0, & u_1(l) = e \end{array} \qquad (4.15)$$

It is considered that the radii of wires are small as against distances b between them. Since the height L of their suspension is commensurable with the wire length l, then the method of average potentials is used for calculating potential coefficients. Taking into account the mirror image, we obtain:

$$\alpha_{11} = \alpha_{22} = \alpha_{33} = \alpha_1/(2\pi\varepsilon), \tag{4.16}$$

where

$$\alpha_1 = \ln(l/a) - 0.307 - sh^{-1}(l/2L) + \sqrt{1 + (2L/l)^2} - 2L/l.$$

Similarly

$$\alpha_{12} = \alpha_{21} = \alpha_{23} = \alpha_{32} = \alpha_2/(2\pi\varepsilon), \ \alpha_{13} = \alpha_{31} = \alpha_3/(2\pi\varepsilon). \tag{4.17}$$

Here

$$\alpha_2 = sh^{-1}(l/b) - \sqrt{1 + (b/l)^2} + b/l - sh^{-1}\left(l/\sqrt{4L^2 + b^2}\right) + \sqrt{1 + \left(4L^2 + b^2/l^2\right)} - \sqrt{4L^2 + b^2}/l^2,$$

and α_3 is obtained from α_2, if to replace b by $2b$. Since the distance between adjacent wires of the load is small in comparison with L and l, it is possible put for simplifying calculations that

$$\alpha_2 = \alpha_3. \tag{4.18}$$

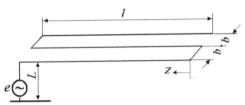

Figure 4.11 Meandering load of three wires.

This assumption does not cause significant error because $sh^{-1}x = \ln(x + \sqrt{x^2 + 1})$, i.e. the error is logarithmic. It is also justified by the fact that the length of connecting wires (of bridges) at the load ends can be taken into account in the used method only indirectly (by means of replacing l by $l + b$).

From (3.4) and (4.16)–(4.18) we obtain

$$\rho_{ns} = \begin{cases} \rho_1, n = s \\ \rho_2, n \neq s \end{cases}, \tag{4.19}$$

where $\rho_1 = 60\alpha_1$, $\rho_2 = 60\alpha_2$. Coefficients β_{ns} and α_{ns} are connected by relationship (3.5). From (3.4), (3.5) and (4.19) it is follows that

$$W_{ns} = \begin{cases} W_1 = 1/(c\beta_{11}), n = s \\ W_2 = -1/(c\beta_{12}), n \neq s \end{cases} \tag{4.20}$$

and for three-wire load

$$\frac{1}{W_1} = \frac{\rho_1 + \rho_2}{(\rho_1 - \rho_2)(\rho_1 + 2\rho_2)}, \quad \frac{1}{W_2} = \frac{\rho_2}{(\rho_1 - \rho_2)(\rho_1 + 2\rho_2)}. \tag{4.21}$$

Substituting (4.3) into the boundary conditions (4.15) and taking into account (4.19) and (4.21), we obtain:

$$U_2 = U_1, \ I_2 = -I_1, \ I_3 = 0, \quad U_3 = U_1 + j(\rho_2 - \rho_1)I_1 \tan kl,$$

$$I_1 = j\frac{2U_1 \tan kl}{\rho_1 + 2\rho_2 - \rho_1 \tan^2 kl}, \; U_1 = e\frac{\rho_1 + 2\rho_2 - \rho_1 \tan^2 kl}{\cos kl\left[\rho_1 + 2\rho_2 - (3\rho_1 - 2\rho_2)\tan^2 kl\right]}. \tag{4.22}$$

From here the input impedance of three-wire load

$$Z_3 = \frac{u_1(l)}{i_1(l)} = -j\cot kl\frac{\rho_1 + 2\rho_2 + (2\rho_2 - 3\rho_1)\tan^2 kl}{3 - \tan^2 kl}. \tag{4.23}$$

In order to test the effect of the adopted assumption (4.18) on the result of the calculating electrical characteristics of the antenna we shall consider that $\rho_3 = 60\alpha_3$. In this case the electrodynamics wave impedances are equal to

$$W_{11} = W_{33} = \frac{(\rho_1 - \rho_3)(\rho_1^2 + \rho_1\rho_3 - 2\rho_2^2)}{\rho_1^2 - \rho_2^2}, \; W_{22} = \frac{\rho_1^2 + \rho_1\rho_3 - 2\rho_2^2}{\rho_1 + \rho_3},$$

$$W_{13} = W_{31} = \frac{(\rho_1 - \rho_3)(\rho_1^2 + \rho_1\rho_3 - 2\rho_2^2)}{\rho_1\rho_3 - \rho_2^2}, \; W_{12} = W_{21} = W_{23} = W_{32} = \frac{\rho_1^2 + \rho_1\rho_3 - 2\rho_2^2}{\rho_2},$$

and expression for Z_3 will take the form

$$Z_3 = -j\cot kl\frac{A + B\tan^2 kl}{C + D\tan^2 kl}, \tag{4.24}$$

where

$$A = 2\left[\frac{1}{W_{11}} - \frac{3}{W_{12}} + \frac{1}{W_{22}} - \frac{1}{W_{13}}\right]^{-1}, \; B = A(2\rho_{12} - \rho_{11} - \rho_{13})\left(\frac{1}{W_{11}} - \frac{1}{W_{12}}\right) + 2(\rho_{12} - \rho_{11}),$$

$$C = 2 + A\left(\frac{1}{W_{11}} - \frac{1}{W_{12}} - \frac{1}{W_{13}}\right),$$

$$D = (2\rho_{12} - \rho_{11} - \rho_{13})\left[\frac{2}{W_{13}} + A\left(\frac{1}{W_{11}} - \frac{1}{W_{12}} - \frac{1}{W_{13}}\right)\left(\frac{1}{W_{11}} - \frac{1}{W_{12}}\right)\right].$$

As the calculation results show, impedances Z_3, calculated in accordance with (4.23) and (4.24) for real antennas differ from each other by no more than 1–2%.

The example of calculating the input impedance of a three-wire load shows that the use of the approximate expressions (4.19) and (4.20) has little effect on the calculation result, if the inequality $a \ll b \ll l$ is true. At the same time, the application of these approximations greatly simplifies the calculation and allows to use it for load of any number of wires.

Determinant Δ_N in accordance with (3.5) is written in the form

$$\Delta_N = \begin{vmatrix} \alpha_1 & \alpha_2 & \alpha_2 & \cdots & \alpha_2 \\ \alpha_2 & \alpha_1 & \alpha_2 & \cdots & \alpha_2 \\ \alpha_2 & \alpha_2 & \alpha_1 & \cdots & \alpha_2 \\ \cdots & \cdots & \cdots & \cdots & \cdots \\ \alpha_2 & \alpha_2 & \alpha_2 & \cdots & \alpha_1 \end{vmatrix}. \tag{4.25}$$

It is easy to check to see that for N, equal to 1, 2, 3,

$$\Delta_N = (\alpha_1 - \alpha_2)^{N-1} [\alpha_1 + (N-1)\alpha_2]. \tag{4.26}$$

The method of mathematic induction allows to prove that this expression is true for any positive integer N. For this it is necessary to show that the rightness of this expression for the determinant Δ_N means its rightness for the determinant Δ_{N+1}. We expand the determinant by the elements of the first line:

$$\Delta_{N+1} = \alpha_1 \Delta_N + \alpha_2 \sum_{r=2}^{N+1} (-1)^{r+1} M_{1r} = \alpha_1 \Delta_N - N\alpha_2 D_N. \tag{4.27}$$

Here M_{1r} is minor of the determinant Δ_{N+1}, $D_N = M_{12}$ is the determinant of the Nth order:

$$D_N = \begin{vmatrix} \alpha_2 & \alpha_2 & \alpha_2 & \dots & \alpha_2 \\ \alpha_2 & \alpha_1 & \alpha_2 & \dots & \alpha_2 \\ \alpha_2 & \alpha_2 & \alpha_1 & \dots & \alpha_2 \\ \dots & \dots & \alpha_2 & \dots & \dots \\ \alpha_2 & \alpha_2 & \alpha_2 & \dots & \alpha_1 \end{vmatrix}.$$

For it the expression similar to expression (4.27) is true:

$$D_N = \alpha_2 \Delta_{N-1} - (N-1)\alpha_2 D_{N-1}.$$

At the same time in accordance with (4.25)

$$\Delta_N = \alpha_1 \Delta_{N-1} - (N-1)\alpha_2 D_{N-1}.$$

From the last two equalities

$$D_N = \Delta_N + (\alpha_2 - \alpha_1)\Delta_{N-1}. \tag{4.28}$$

Substituting into (4.28) the value Δ_N from (4.26) and the value Δ_{N-1}, which in accordance with (4.26) is equal to $\Delta_{N-1} = (\alpha_1 - \alpha_2)^{N-2} [\alpha_1 + (N-2)\alpha_2]$, we obtained

$$D_N = \alpha_2(\alpha_1 - \alpha_2)^{N-1}, \tag{4.29}$$

From here in accordance with (4.27)

$$\Delta_{N+1} = \alpha_2(\alpha_1 - \alpha_2)^{N-1} [\alpha_1 + (N-2)\alpha_2] - N p_2^2 (\alpha_1 - \alpha_2)^{N-1} = (\alpha_1 - \alpha_2)^N (\alpha_1 - N\alpha_2),$$

as required.
Since

$$\Delta_{ns} = \begin{cases} \Delta_{N-1}, n = s \\ (-1)^{n+s} \alpha_2 (\alpha_1 - \alpha_2)^{N-2}, n \neq s \end{cases}, \tag{4.30}$$

i.e. accordingly (4.20) electrostatic wave impedances are equal to

$$W_1 = (\rho_1 - \rho_2) \frac{\rho_1 + (N-1)\rho_2}{\rho_1 + (N-2)\rho_2}, \quad W_2 = (\rho_1 - \rho_2) \frac{\rho_1 + (N-1)\rho_2}{\rho_2}. \tag{4.31}$$

Boundary conditions for currents and potentials of wires of the load differ for even and odd number of wires. For an odd number of wires $N = 2m - 1$ the boundary conditions are:

$$\begin{aligned} u_{2n-1}(0) &= u_{2n}(0), & i_{2n-1}(0) + i_{2n}(0) &= 0, & i_{2m-1}(0) &= 0, \\ u_{2n}(l) &= u_{2n+1}(l), & i_{2n}(l) + i_{2n+1}(l) &= 0, & u_1(l) &= e. \end{aligned} \tag{4.32}$$

In this case, as shown in [40], the input impedance of the load is

$$Z_{2m-1} = \frac{u_1(l)}{i_1(l)} = -j(\rho_1 - \rho_2) \times$$

$$\tan \beta \frac{2(A_m + A_{m-1} - 1) + H_{2m-1}\left[1 - A_m - A_{m-1} + (B_m + B_{m-1})/(2M\sin^2 \beta)\right]}{2(A_{m-1} - A_m) + H_{2m-1}\left[A_m - A_{m-1} + (B_{m-1} - B_m)/(2M\sin^2 \beta)\right]}, \quad (4.33)$$

where

$$\beta = kl, \quad A_m = m + 2\sum_{r=1}^{m-1} r\cos 2(m-r)\beta, \quad B_m = \cos 2m\beta,$$

$$M = \frac{\rho_2}{\rho_1 + 2(m-1)\rho_2}, H_{2m-1} = (1 - 4M\sin^2 \beta)/\sum_{s=0}^{m-1}(B_s - 2M_s A\sin^2 \beta).$$

Similarly, for an even number of wires $(N = 2m)$

$$Z_{2m} = -j(\rho_1 - \rho_2)\tan\beta \frac{4M\tan^2 \beta(C_m + C_{m-1} - \cos^2 \beta) + H_{2m}(D_m + D_{m-1})}{4M\tan^2 \beta(C_{m-1} - C_m) + H_{2m}(D_{m-1} - D_m)}, \quad (4.34)$$

where

$$C_m = m + \sum_{r=0}^{m-1}(2r+1)\cos 2(m-r)\beta, D_m = \frac{\cos 2(m+1)\beta}{\cos\beta}, H_{2m} = \left[1 - 4M\tan^2 \beta\sum_{s=0}^{m-1}C_s\right]/\sum_{s=0}^{m-1}D_s.$$

For the sake of convenience of calculating, numerators and denominators of expressions (4.33) and (4.34) can be presented as an expansion in powers of the value $\tan^2\beta$ [40].

In Table 4.3 expressions for calculating the load impedances with different numbers of wires are given. If $\tan\beta < 1$, the calculation accuracy in accordance with (4.34) and with expressions, presented in Table 4.3, is determined by the number of calculated terms. If $\beta \ll 1$, then, limited in (4.34) by the first two terms of the numerator and denominator, we find:

Table 4.3 Impedances of Loads

Number of wires	Expressions for calculating
2	$-0.5j\cot\beta\left[\rho_1 + \rho_2 + (\rho_2 - \rho_1)\tan^2 \beta\right]$
3	$-j\dfrac{\cot\beta}{3 - \tan^2 \beta}\left[\rho_1 + 2\rho_2 + (2\rho_2 - 3\rho_1)\tan^2 \beta\right]$
4	$-j\dfrac{\cot\beta}{4(1 - \tan^2 \beta)}\left[\rho_1 + 3\rho_2 - 2(3\rho_1 - \rho_2)\tan^2 \beta + (\rho_1 - \rho_2)\tan^4 \beta\right]$
5	$-j\dfrac{\cot\beta}{5 - 10\tan^2 \beta + \tan^4 \beta}\left[\rho_1 + 4\rho_2 - 10\rho_1\tan^2 \beta + (5\rho_1 - 4\rho_2)\tan^4 \beta\right]$
6	$-j\dfrac{\cot\beta}{2(1 - 3\tan^2 \beta)(3 - \tan^2 \beta)} \times$ $\left[\rho_1 + 5\rho_2 - 5(3\rho_1 + \rho_2)\tan^2 \beta + 3(5\rho_1 - 3\rho_2)\tan^4 \beta - (\rho_1 - \rho_2)\tan^6 \beta\right]$

$$Z_N = -j\frac{\rho_1 + (N+1)\rho_2}{N\beta}\left\{1 - \frac{N\beta^2[N\rho_1 - (N-1)\rho_2]}{3[\rho_1 + (N-1)\rho_2]}\right\}. \tag{4.35}$$

This expression is true also for the odd number of wires.

Knowing the load impedance, one can determine all electrical characteristics of the antenna. Figure 4.12a presents for two antennas with meandering load the calculated curves and experimental values of the reactance. Dimensions of the load (in meters) are: $L = 200$, $l = 300$, $a = 0.1$. The number of wires in the load is equal to four. The distance between the wires axes of the first antenna load (curve 1) is equal to $b = 5$, of the second antenna (curve 2) $b = 2$. This experiment was performed on a model. The coincidence of experimental values with calculation is good. For comparison, the input reactance of the inverted-L antenna with the same sizes is given in the figure by dotted lines (curves 3 and 4). As can be seen from the figures, the use of meandering load increases the electrical length of the antenna and shifts its resonances in the direction of low frequencies.

Figure 4.12b presents the results of calculation and experiment for the antenna with meandering load of six wires. Dimensions of the antenna are: $L = 50$, $l = 70$, $b = 1.54$, $a = 3.75 \cdot 10^{-3}$. Here the coincidence of results of the calculation and experiment also is good.

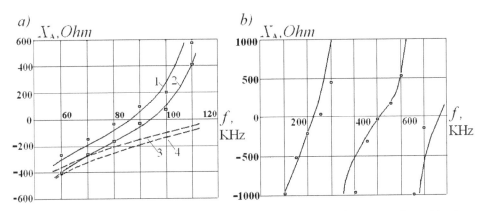

Figure 4.12   Reactance of antennas with meandering loads of four (a) and six (b) wires.

Figure 4.13a presents the calculated curves and experimental values of the input impedance for similar antenna (curves 1) with six wires in the load and with dimensions: $L = 50$, $l = 45$, $b = 1.54$, $a = 2. \, 5 \cdot 10^{-2}$. For comparison, the input impedance of the inverted L-antennas with the same height and with the length of the horizontal load 90 and 45 m (curves 2 and 3) are given. In Figure 4.13b the calculated efficiency of these antennas is given. As seen from the figures, in the range of 200–500 kHz efficiency of the antenna with meandering load is substantially higher than the efficiency of the inverted-L antenna of the same size and is comparable with the efficiency of such antenna having double length of the load.

The use of antenna with meandering load in MF range allows reducing a size of a plot, which the antenna occupies, an area of the grounding and a price of construction.

The first specimen of this antenna was put into operation in 1983 in the town Pavlovo on the radio center of the Baltic Shipping Company. In 1987, the radio center of commercial sea port in Ventspils (Latvia) was equipped with three antennas. The effectiveness of the new antenna has been tested by measuring the field's strength,

produced by new antennas in comparison with the inverted-L antenna. Antennas
operated with the same transmitter. Tests have shown that the field created by the
antenna with a meandering load is greater in 1.5–1.6 times. This is equivalent to
increasing the transmitter power in 2.3–2.6 times.

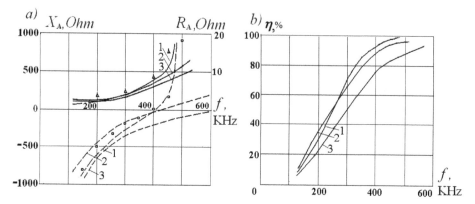

Figure 4.13 Input impedances (*a*) and efficiency of antenna with meandering load and
inverted-L antennas with different length of load (*b*).

The correct selection of elements of antenna device requires calculation of currents
and voltages arising therein. For example, the magnitudes of the voltages between the
wires ends and the ground determine the choice of insulators, on which the load is
suspended. Diameters of wires depend on the maximum currents, etc.

We examine in particular the load of the three wires using condition (4.18) for
simplifying calculation. As shown earlier, substituting the boundary conditions (4.15) in
the equations (3.3) for the currents and the potentials along the load wires, we obtain
equalities (4.22). By means of (4.22) and (4.31) in accordance with the first equation of
the set (3.3) one can obtain expressions for the currents along the load wires. The current
in the first wire is

$$i_1(z) = I \sin(2\beta + kz), \tag{4.36}$$

where

$$I = j\frac{U_1}{\cos^2 \beta \left(\rho_1 + 2\rho_2 - \rho_1 \tan^2 \beta\right)} = j\frac{e}{\cos^3 \beta \left[\rho_1 + 2\rho_2 - (3\rho_1 - 2\rho_2)\tan^2 \beta\right]}.$$

Similarly,

$$i_2(z) = -I \sin(2\beta - kz), \quad i_3(z) = I \sin kz. \tag{4.37}$$

The total current in the load wires is

$$i(z) = \sum_{n=1}^{3} i_n(z) = I(2\cos 2\beta + 1)\sin kz, \tag{4.38}$$

i.e. current is distributed depending on coordinate *z* in accordance with sinusoidal law.

Figure 4.14 shows the current distribution along the wires of a three-wire load,
constructed in accordance with (4.36) and (4.37), and the total current, constructed in
accordance with (4.38). If to straighten mentally the load wire, it will be seen that the
current of the wire and the total current are distributed in accordance with sinusoidal
law.

Potential does not submit to this law. The expressions (3.3), (4.22) and (4.31) give for the potential another result:

$$u_1(z) = U_1 \cos kz + j(\rho_1 - \rho_2)I_1 \sin kz = U_1 \cos kz - U \cot \alpha \sin kz, \quad (4.39)$$

where

$$U = \frac{2e(\rho_1 - \rho_2)\tan^2 \beta}{\cos \beta \left[\rho_1 + 2\rho_2 - (3\rho_1 - 2\rho_2)\tan^2 \beta\right]}.$$

Similarly,

$$u_2(z) = U_1 \cos kz + U \cot \beta \sin kz, \quad u_3(z) = U_1 \cos kz + U \cot kz. \quad (4.40)$$

Thus, the potential of the load wire is the sum of two summands. Both summands and its sum are shown in Figure 4.14.

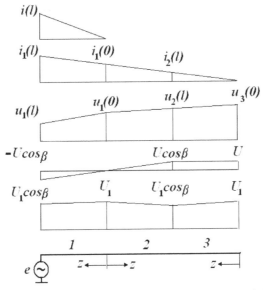

Figure 4.14 Currents and potentials along the wires of the load.

Sinusoidal law of the current distribution along the load wire remains valid for a greater number of wires. For example, if a load consists of four wires,

$$i_1(z) = I \sin(3\beta + kz), \; i_2(z) = -I \sin(3\beta - kz), \; i_3(z) = I \sin(\beta + kz), \; i_4(z) = -I \sin(\beta + kz), \quad (4.41)$$

where

$$I = j\frac{U_1}{\cos^3 \beta \left[\rho_1 + 3\rho_2 - (3\rho_1 + \rho_2)\tan^2 \beta\right]}$$

$$= j\frac{e}{\cos^4 \beta \left[\rho_1 + 3\rho_2 - 2(3\rho_1 - \rho_2)\tan^2 \beta + (\rho_1 - \rho_2)\tan^4 \beta\right]}.$$

The calculated results of the input current of the load (for different number N of wires) and the potential at the end of the load are given in Tables 4.4 and 4.5. The presented magnitudes of the current and the potential are maximal in all frequency range up to the frequency of serial resonance. Therefore, it is possible to choose the diameters of the wires and the type of insulators in accordance with these magnitudes.

Figure 4.15 gives as an example the calculated values of the maximum potentials and currents of the load for an antenna with the dimensions: $L = 50$, $l = 45$, $b = 1.54$, $a = 2.5 \cdot 10^{-2}$. The number of wires in the load is $N = 6$. The maximum potential is specified with respect to the voltage e at the load input and to emf at the antenna input. The maximum current is specified with respect to e and to the input current J_A of the antenna. It was taken into account that

$$e_A = e \frac{\cos k\left((L + l_e)\right)}{\cos k l_e}, J_A = i_1(l) \frac{\sin k\left(L + l_e\right)}{\sin k l_e}, \tag{4.42}$$

where $l_e = \dfrac{1}{k} \tan^{-1}(-jW / Z_N)$ is an equivalent length of the load with input impedance Z_N.

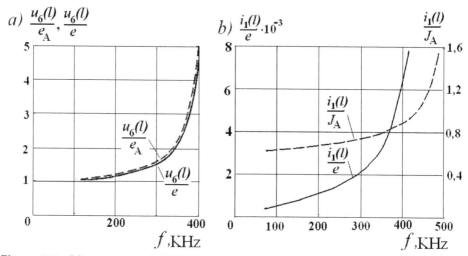

Figure 4.15 Maximum potential (*a*) and maximum current (*b*) as functions of frequency.

Table 4.4   Input Current of the Load

Number of wires	i_{max}
1	$j\dfrac{e \sin \beta}{\rho_1 \cos \beta}$
2	$j\dfrac{e \sin 2\beta}{\cos^2 \beta\left[\rho_1 + \rho_2 - (\rho_1 - \rho_2)\tan^2 \beta\right]}$
3	$j\dfrac{e \sin 3\beta}{\cos^3 \beta\left[\rho_1 + 2\rho_2 - (3\rho_1 - 2\rho_2)\tan^2 \beta\right]}$
4	$j\dfrac{e \sin 4\beta}{\cos^4 \beta\left[\rho_1 + 3\rho_2 - 2(3\rho_1 - \rho_2)\tan^2 \beta + (\rho_1 - \rho_2)\tan^4 \beta\right]}$
5	$j\dfrac{e \sin 5\beta}{\cos^5 \beta\left[\rho_1 + 4\rho_2 - 10\rho_1 \tan^2 \beta + (5\rho_1 - 4\rho_2)\tan^4 \beta\right]}$
6	$j\dfrac{e \sin 6\beta}{\cos^6 \beta\left[\rho_1 + 5\rho_2 - 5(3\rho_1 + \rho_2)\tan^2 \beta + 3(5\rho_1 - 3\rho_2)\tan^4 \beta - (\rho_1 - \rho_2)\tan^6 \beta\right]}$

Table 4.5   Potential on the End of Load Wire

Number of wires	u_{max}
1	$u_1(0) = \dfrac{e}{\cos \chi}$
2	$u_2(l) = \dfrac{e\left[\rho_1 + \rho_2 + (\rho_1 - \rho_2)\tan^2 \beta\right]}{\left[\rho_1 + \rho_2 - (\rho_1 - \rho_2)\tan^2 \beta\right]}$
3	$u_3(0) = \dfrac{e\left[\rho_1 + 2\rho_2 + (\rho_1 - 2\rho_2)\tan^2 \beta\right]}{\cos \beta \left[\rho_1 + 2\rho_2 - (3\rho_1 - 2\rho_2)\tan^2 \beta\right]}$
4	$u_4(l) = \dfrac{e\left[\rho_1 + 3\rho_2 + 2(\rho_1 - 3\rho_2)\tan^2 \beta + (\rho_1 - \rho_2)\tan^4 \beta\right]}{\left[\rho_1 + 3\rho_2 - 2(3\rho_1 - 2\rho_2)\tan^2 \beta + (\rho_1 - \rho_2)\tan^4 \beta\right]}$
5	$u_5(0) = \dfrac{e\left[\rho_1 + 4\rho_2 + 2(\rho_1 - 6\rho_2)\tan^2 \beta + \rho_1 \tan^4 \beta\right]}{\cos \beta \left[\rho_1 + 4\rho_2 - 10\rho_1 \tan^2 \beta + (5\rho_1 - 4\rho_2)\tan^4 \beta\right]}$
6	$u_6(l) = \dfrac{e\left[\rho_1 + 5\rho_2 + (3\rho_1 - 23\rho_2)\tan^2 \beta + 3(\rho_1 + \rho_2)\tan^4 \beta + (\rho_1 - \rho_2)\tan^6 \beta\right]}{\left[\rho_1 + 5\rho_2 - 5(3\rho_1 + \rho_2)\tan^2 \beta + 3(5\rho_1 - 3\rho_2)\tan^4 \beta - (\rho_1 - \rho_2)\tan^6 \beta\right]}$

4.4   MULTI-LEVEL ANTENNA WITH ADJUSTABLE DIRECTIONAL PATTERN IN A VERTICAL PLANE

Considered in the previous section, an antenna with meandering load is a concrete example of using structures of parallel wires for improving electrical characteristics of an antenna. Multilevel antenna is the other example of such kind.

Vertical linear radiators, for example whip antennas, are widely used for a short-wave radio communication of mobile objects. They take up little space and do not interfere with the overview. Their disadvantage is that the maximum radiation is directed horizontally if only the electrical length of the antenna is small. With the growth of an electrical length the horizontal signal decreases. If the antenna height L is greater than 0.7λ, a main lobe of a radiation pattern in a vertical plane separates from a ground and the radiation in the horizontal direction drops sharply. Changing of the height, for example by a telescopic construction, allows to improve the antenna characteristics. But such mechanical tuning consumes a lot of time and complicates the design of the antenna. Folded structures allow to create an antenna, in which the length of the radiating segment is changed without changing the geometric dimensions of the device [41].

In the simplest embodiment, the antenna is designed for operation in two frequency bands (similar to the telescopic antenna, which may have two geometrical heights during operation). The antenna is made as two-tiered and consists of two radiators, which are located one above the other and connected together. The upper radiator is linear, the lower radiator is folded (Figure 4.16). The exciting emf e_1 and e_2 are connected in both wires of the folded radiator and may vary in amplitude and phase by means of tuning circuit.

If the antenna operates in the first frequency band, emfs are included in phase, creating in both wires of folded radiator the currents in one direction. These currents become by the current of the linear radiator. As a result the current is created along the entire antenna, and the height of the radiating segment is equal to the total height of the antenna.

If the antenna operates in the second frequency band, the current is created only in the wires of the folded radiator. With this aim excited emfs are included in the anti-phase, and their magnitudes are chosen so that the potential at the point of joining of the linear radiator to the folded radiator is zero. This eliminates the immediate excitation of the linear radiator. The upper segment of the antenna, consisting of one wire, can be excited by electromagnetic fields of the currents flowing along the wires of the folded radiator. However, if the length of the upper segment is far from the resonance (is not a multiple of $\lambda/2$), the current along this segment is small, i.e., the height of the radiating section of the antenna is equal to the height of the folded radiator.

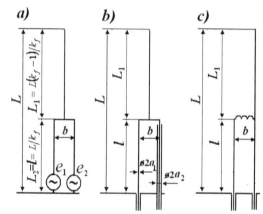

Figure 4.16 Two-tiered antenna: common circuit (*a*), with coaxial cable (*b*), with reactive load (*c*).

Changing the height of the radiator allows to provide the operation in two frequency bands. If the frequency ratio in each band is equal to k_f, the total frequency ratio is equal to k_f^2. The height of the folded radiator is chosen to be equal to

$$l = L/k_f. \tag{4.43}$$

The considered circuit of the antenna can be generalized for use in N of the frequency bands (instead two bands). For this, the number of tiers of the antenna should be increased to N. The two upper tiers are similar with the described embodiment. The lower ends of each wire of the folded radiator connect with the upper points of the folded radiator of the next (third) tier etc. Overall frequency ratio of N-tiered antenna is equal to k_f^N. The heights of the lower tier and the rest tiers are given by expressions

$$L_n = \begin{cases} L/k_f^{n-1}, n = N, \\ L\left(K_f - 1\right)/k_f^n, n \neq N, \end{cases} \tag{4.44}$$

where n is the tier number, counting from the top.

Multi-tiered antenna creates a new prospect for the development of a broadband antenna. In this direction the most significant results were obtained earlier by means of connecting concentrated capacitive loads and optimization of these loads [19]. Calculations show (see Section 5.5) that the capacitive loads allow extending the antenna range in the direction of higher frequencies with a sufficiently high level of matching, ensuring the frequency ratio of the order of 10. But at the same time the directional pattern in the vertical plane deteriorates sharply in high frequency part of a range. With allowance for obtaining acceptable directional patterns in the vertical plane the frequency ratio does not exceed three. Using of the multi-tiered structure and the capacitive loads in the wires of each tier allows to ensure in a wide range the high level of matching and the required directional pattern.

Let us return to the two-tiered variant of the antenna, more precisely to its excitation in the anti-phase mode. The antenna will radiate, if to provide asymmetry in the folded radiator, i.e. it is necessary to obtain different amplitudes of the currents in the left and the right branches of the radiator. With this aim one of the wires must be accomplished in the form of a coaxial cable (see Figure 4.16b), i.e. generator must be included not in the lower, but in the upper point of the wire. Another way of creating an asymmetry is connection of a reactive load in one of the wires (see Figure 4.16c).

In the presence of asymmetry not only anti-phase currents, but also in-phase currents will flow in the branches of the folded radiator. In-phase currents are caused by the presence of the ground (see the last paragraph of Section 3.2 and Figure 3.8). These in-phase currents create radiation. However, for the sake of simplicity we shall conventionally call by anti-phase mode the mode of antenna operation, when the potential at the point of joining of the linear radiator to the folded radiator is zero.

In order to analyze the two-tiered antenna we shall apply the theory of electrically coupled long lines, described in Section 3.1. This theory allows to find the currents and the potentials along each wire of the line and emf of generators providing the required operation mode. In this case, the equivalent line (Figure 4.17) is considered in the general form—with two generators in one of the branches and two complex loads. The set of equations (3.3) for the three wires in this case takes the form:

$$i_1 = I_1 \cos kz_1 + j[U_1 / W_{11} - U_2 / W_{12}]\sin kz_1, \quad u_1 = U_1 \cos kz_1 + j(\rho_{11}I_1 + \rho_{12}I_2)\sin kz_1,$$

$$i_2 = I_2 \cos kz_1 + j[U_2 / W_{22} - U_1 / W_{12}]\sin kz_1, u_2 = U_2 \cos kz_1 + j(\rho_{12}I_1 + \rho_{22}I_2)\sin kz_1,$$

$$i_3 = I_3 \cos kz_3 + j(U_3 / W_{33})\sin kz_3, \qquad u_3 = U_3 \cos kz_3 + j\rho_{33}I_3 \sin kz_3. \qquad (4.45)$$

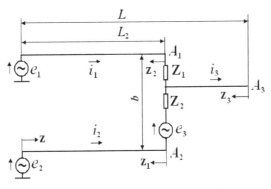

Figure 4.17 Equivalent asymmetric line.

The boundary conditions for the currents and the potentials are

$$i_3\big|_{z_3=0}=0,\ i_1+i_2\big|_{z_1=0}=i_3\big|_{z_3=L-l},\ u_1\big|_{z_1=l}=e_1,\ u_2\big|_{z_1=l}=e_2,$$

$$u_1-Z_1i_1\big|_{z_1=0}=u_2-Z_2i_2\big|_{z_1=0}+e_3=u_3\big|_{z_3=L-l}.\qquad(4.46)$$

Equalities (4.45) and the boundary conditions (4.46) are the set of equations with six unknown magnitudes U_i, I_i ($i = 1, 2, 3$). Substituting (4.45) into (4.46), we obtain:

$$I_3 = 0,\ I_2 = -I_1 + j(U_3/W_{33})\sin k(L-l),\ U_1 = Z_1I_1 + U_3\cos k(L-l),$$

$$U_2 = -e_3 - Z_2 I_1 + U_3\cos k(L-l)[1 + j(Z_{22}/W_{33})\tan k(L-l)],$$

$$e_1 = I_1\cos kl[Z_1 + j(\rho_{11}-\rho_{12})\tan kL] + U_3\cos kl\cos k(L-l)[1-(\rho_{12}/W_{33})\tan kl\tan k(L-l)],$$

$$e_2 = -e_3\cos kl - I_1\cos kl[Z_2 + j(\rho_{22}-\rho_{12})\tan kL] +$$
$$U_3\cos kl\cos k(L-l)\left[1-\frac{\rho_{22}-jZ_2}{W_{33}}\tan kl\tan k(L-l)\right].\qquad(4.47)$$

The rest two formulas permit to express the magnitudes I_1 and U_3 through emfs of the generators. But the corresponding expressions are cumbersome, are hence not used for further analysis and thus not presented here. The previous four equalities after substituting into (4.45) allow expressing the current distribution along each wire as a function of magnitudes I_1 and U_3:

$$i_1(z) = I_1\cos k(L-z)[1+j(Z_1/W_{11}+Z_2/W_{12})\tan k(l-z)]+j(e_3/W_{12})\sin k(l-z)+$$
$$U_3[j(1/W_{11}-1/W_{12})\cos k(L-l)+Z_2\sin k(L-l)/(W_{12}W_{33})]\sin k(L-z),$$
$$i_2(z) = -I_1\cos k(L-z)[1+j(Z_1/W_{12}+Z_2/W_{22})\tan k(l-z)]-j(e_3/W_{22})\sin k(l-z)+$$
$$U_3[j(1/W_{22}-1/W_{12})\cos k(L-l)-Z_2\sin k(L-l)/(W_{22}W_{33})+\qquad(4.48)$$
$$j\sin k(L-l)\cot k(L-z)/W_{33}]\sin k(l-z),$$
$$i_3(z) = j(U_3/W_{33})\sin k(L-z).$$

Further, we shall consider the specific embodiments of antennas as partial cases of general equivalent circuit. The circuit of two-tiered antenna with the coaxial cable is shown in Figure 4.16b. Here a few of elements of the overall circuit is absent, i.e.

$$Z_1 = Z_2 = e_2 = 0.$$

In the anti-phase mode in accordance with the boundary condition $u_3\big|_{z_3=L-l}=0$ we find that $U_3 = 0$. Then from (4.47) we obtain emf of generators

$$e_1 = j(\rho_{11}-\rho_{12})I_1\sin kl,\ e_3 = -j(\rho_{22}-\rho_{12})I_1\tan kl = -j(\rho_{22}-\rho_{12})\sec kl/(\rho_{11}-\rho_{12}).\ (4.49)$$

It is necessary to emphasize that the relationship between emf of two generators is an obligatory condition for providing anti-phase mode in the antenna. The currents along the antenna wires in this mode according to (4.48) are

$$i_1(z) = I_1\cos k(l-z)+\frac{\rho_{22}-\rho_{12}}{W_{12}}I_1\tan kl\sin k(l-z),$$

$$i_2(z) = -I_1 \cos k(l-z) + \frac{\rho_{22} - \rho_{12}}{W_{22}} I_1 \tan kl \sin k(l-z), \quad i_3(z) = 0. \tag{4.50}$$

The expressions (4.50) confirm that the currents along the first and the second wires contain in-phase and anti-phase components. The total antenna current (sum of currents) varies along the antenna similarly to the current along the linear radiator of the length l—in full accordance with the results presented in Chapter 3:

$$i_1(z) + i_2(z) = -\frac{(\rho_{22} - \rho_{12})(\rho_{11} - \rho_{12})}{\rho_{11}\rho_{22} - \rho_{12}^2} I_1 \tan kl \sin k(l-z). \tag{4.51}$$

Here it is taken into account that (see Section 3.1)

$$\frac{1}{W_{22}} = \frac{\rho_{11}}{\rho_{11}\rho_{22} - \rho_{12}^2}, \frac{1}{W_{12}} = \frac{\rho_{12}}{\rho_{11}\rho_{22} - \rho_{12}^2}.$$

The current distribution along the antenna wires in the anti-phase mode is shown in Figure 4.18a. The boundary between the components of current in each wire is given by a dotted line, and the total current of the wire and the total current of the antenna are given by solid curves. Impedance on the output of each generator, exciting asymmetric long line, is equal to

$$jX_{A1} = \frac{e_1}{i_1(0)} = j(\rho_{11} - \rho_{12})\tan kl\left[1 + \frac{\rho_{22} - \rho_{12}}{W_{12}}\tan^2 kl\right]^{-1},$$

$$jX_{A3} = \frac{e_3}{i_2(l)} = j(\rho_{22} - \rho_{12})\tan kl. \tag{4.52}$$

These expressions allow determining approximately the reactive component of the loading impedance of each generator (similar to the formula for the input impedance of an equivalent two-wire long line, which allows determining approximately the reactive component of the input impedance of a linear antenna).

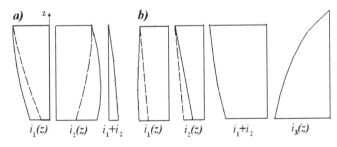

Figure 4.18 Currents in the two-tiered antenna with a coaxial cable in anti-phase (*a*) and in-phase (*b*) modes.

From the viewpoint of radiation, as seen from (4.50) and Figure 4.18a, the antenna in anti-phase mode consists of two parallel radiators of the height l with in-phase currents in the base:

$$i_1^{(i)} = \frac{\rho_{22} - \rho_{12}}{W_{12}} I_1 \tan kl \sin kl, i_2^{(i)} = -\frac{\rho_{22} - \rho_{12}}{W_{22}} I_1 \tan kl \sin kl.$$

The radiation resistance of each radiator consists of the self-resistance and the mutual resistance multiplied by the ratio of the currents. In particular, for the first radiator we write:

$$R_{A1} = R_{11} + R_{12} \frac{i_2^{(i)}(0)}{i_1^{(i)}(0)}. \tag{4.53}$$

where R_{11} is a self-radiation resistance, R_{12} is the mutual radiation resistance, and $R_{12} \approx R_{11}$, since the radiator heights are the same and the distance between the radiators is small in comparison with the wave length. Thus,

$$R_{A1} = R_{11}(1 - W_{12}/W_{22}), \quad R_{A2} = R_{11}(1 - W_{22}/W_{12}). \tag{4.54}$$

In the anti-phase mode the electric field strength in the far region and the directional pattern coincide with the similar characteristics of the conventional linear radiator of a height l. Expressions (4.50)–(4.54) are sufficiently simple and allow determining the influence of antenna geometric dimensions upon the current magnitude in each wire and upon the electrical characteristics of the radiator. More precisely, the input impedance of the antenna can be calculated by using an algorithm of calculation, based on the integral equation for the current, the Moment Method and the systems of piecewise sinusoidal basis functions.

Let us move on to an analysis of the in-phase mode. For the implementation of this mode one must ensure equality of potentials in both branches of the folded radiator, i.e.,

$$u_1(z) = u_2(z). \tag{4.55}$$

Applying this condition to the set of equations (4.45), we find:

$$e_1 = jI_1 \frac{\rho_{11}\rho_{22} - \rho_{12}^2}{\rho_{22} - \rho_{12}} \sin kl - jI_1 W_{33} \frac{\rho_{11} + \rho_{22} - 2\rho_{12}}{\rho_{22} - \rho_{12}} \cot k(L-l), \quad e_3 = e_1 \sec l. \tag{4.56}$$

Currents along the wires consist in this case of the in-phase components only:

$$i_1(z) + i_2(z) = I_1 W_{33} \frac{\rho_{11} + \rho_{22} - 2\rho_{12}}{\rho_{11}\rho_{22} - \rho_{12}^2} \cot k(L-l) \sin k(l-z) + I_1 \frac{\rho_{11} - \rho_{12}}{\rho_{22} - \rho_{12}} \tan kl \sin k(l-z)$$

$$+ I_1 \left(1 + \frac{\rho_{11} - \rho_{12}}{\rho_{22} - \rho_{12}}\right) \cos k(l-z); \quad i_3(z) = I_1 \left(1 + \frac{\rho_{11} - \rho_{12}}{\rho_{22} - \rho_{12}}\right) \frac{\sin k(L-z)}{\sin k(L-l)}. \tag{4.57}$$

The current distribution along the wires is shown in Figure 4.18b. Impedances on the output of each generator, exciting asymmetric long line, are

$$jX_{A1} = \frac{e_1}{i_1(0)} = j \frac{\rho_{11}\rho_{22} - \rho_{12}^2}{\rho_{22} - \rho_{12}} \tan kl \frac{1 - W_{33} \frac{\rho_{11} + \rho_{22} - 2\rho_{12}}{\rho_{11}\rho_{22} - \rho_{12}^2} \cot kl \cot k(L-l)}{1 + \frac{W_{33}}{W_{11}} \frac{\rho_{11} + \rho_{22} - 2\rho_{12}}{\rho_{22} - \rho_{12}} \tan kl \cot k(L-l) - \frac{\rho_{12}}{\rho_{22} - \rho_{12}} \tan^2 kl},$$

$$jX_{A3} = \frac{e_3}{i_2(l)} = j \frac{\rho_{11}\rho_{22} - \rho_{12}^2}{\rho_{11} - \rho_{12}} \tan kl \left[1 - W_{33} \frac{\rho_{11} + \rho_{22} - 2\rho_{12}}{\rho_{11}\rho_{22} - \rho_{12}^2} \cot kl \cot k(L-l)\right]. \tag{4.58}$$

The radiation resistance is calculated according to formulas similar to (4.53). But R_{11} is the resistance of the radiator of height L, the current along which is determined by

(4.57). Since the derivative of the current has discontinuity on the border of segments, the calculation should use the technique described in Section 1.5, i.e. should take into account the break of the current derivative. Correspondingly it is necessary to replace the known expression (1.31) by the equality of type (1.68)

$$E_z(J) = -j\frac{15}{k}\left\{ \frac{2\exp(-jkR_0)}{R_0}\frac{dJ(0)}{dz} - \left[\frac{\exp(-jkR_{11})}{R_1} + \frac{\exp(jkR_{12})}{R_1}\right]\frac{dJ(l_1)}{dz} + \right.$$

$$\left. \sum_{m=2}^{M}\left[\frac{\exp(-jkR_{m1})}{R_{m1}} + \frac{\exp(-jkR_{m2})}{R_{m2}}\right]\left[\frac{dJ(l_m+0)}{dz} - \frac{dJ(l_m-0)}{dz}\right]\right\}, \qquad (4.59)$$

Here R_{m1} and R_{m2} are the distances from observation point to the segments' borders in the upper and the lower arms of the radiator, M is the number of borders,

$$\frac{dJ(l_m+0)}{dz} \text{ and } \frac{dJ(l_m-0)}{dz}$$

are the values of the current derivatives from the left and the right of point $z = l_m$.

It should be noted that for the same diameters of the antenna wires the formulas become far simpler.

As an example of the two-tiered antenna with a coaxial cable we consider the antenna with dimensions (in meters): $L = 1.0$, $l = 0.39$, $b = 0.037$, $a_1 = 0.002$, $a_2 = 0.025$. Here, b is the distance between the axes of the wires of the folded radiators, a_1 and a_2 are the radii of the wires (see Figure 4.16b). Calculation of the antenna characteristics is made by means of the Moment method.

Figure 4.19 shows the calculated curves for the directional patterns in the vertical plane—in the in-phase (a) and anti-phase (b) modes. The model of the antenna was made in full size. The results of the experimental verification are given for frequencies 150 and 300 MHz. As can be seen from the figures, the coincidence of the calculation and the experiment is quite satisfactory. The high level of radiation in the direction perpendicular to the axis of the radiator (along the ground) is provided in the double frequency range. However, in the anti-phase mode when the length of the third wire (of the upper segment of the antenna) is a multiple of half the wavelength, i.e. at frequencies 245 and 490 MHz, the main lobe of the directional pattern is located at a large angle

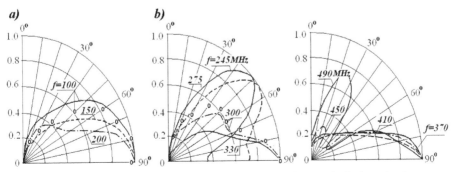

Figure 4.19 Directional patterns of antenna in the vertical plane in the in-phase (a) and anti-phase (b) modes.

to the horizontal. Here the current along the third wire is too large. The dimensions of the antenna must be chosen such that the resonance frequencies were lying outside the operating range.

Figure 4.20 shows the current distribution along the antenna wires in the anti-phase mode, including the current distribution along the left wire and the connecting bridge between the wires of the folded radiator, and also the current distribution along the right wire and the upper (third) wire. These current distributions of currents are the graphic illustration of the processes in the anti-phase mode of the two-tiered antenna. The results are shown at four frequencies, including frequencies 245 and 490 MHz with increased current of the third wire.

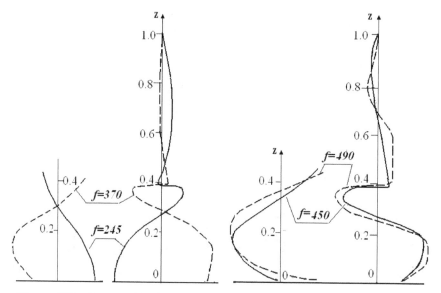

Figure 4.20 Current distribution along the wires of two-tiered antenna with coaxial cable at four frequencies (in anti-phase mode).

The circuit of a two-tiered antenna with a reactive load is shown in Figure 4.16c. Here a few of elements of the overall circuit is absent also, i.e. $Z_2 = e_3 = 0$. Let us assume that $\rho_{11} = \rho_{22}$. In the anti-phase mode the boundary condition $u_3|_{z_3 = L-l} = 0$ should be executed. This means that the emfs of the generators are connected by the relationship

$$e_2 = -e_1 + Z_1 I_1 \cos kl. \tag{4.60}$$

Currents along the antenna wires in this mode

$$i_1(z) + i_2(z) = jZ_1 I_1 W_{33} \frac{1}{\rho_{11} + \rho_{12}} \sin k(l-z), \ I_3(z) = 0. \tag{4.61}$$

Reactances of generators' load are

$$jX_{A1} = \frac{Z_1 + j(\rho_{11} - \rho_{12})\tan kl}{1 + j(Z_1/W_{11})\tan kl}, jX_{A2} = \frac{j(\rho_{11} - \rho_{12})\tan kl}{1 + j(Z_1/W_{12})\tan kl}. \tag{4.62}$$

Resistances are calculated in accordance with (4.53), and

$$\frac{i_2^{(i)}(0)}{i_1^{(i)}(0)} = \frac{W_{11}}{W_{12}}.$$

The in-phase mode has the salient features. Since the load is connected in the upper section of the folded radiator, it is impossible to produce the equality of voltages in both its wires. Let us assume

$$u_1(z) - u_2(z) = U \cos k \, (l - z).$$

For executing this condition it is necessary that

$$e_2 = e_1 - Z_1 I_1 \cos kl. \tag{4.63}$$

The currents along the wires are calculated in accordance with the expressions

$$i_1(z) + i_2(z) = 2I_1 \cos k(l-z) + \frac{4I_1 W_{33}}{\rho_{11} + \rho_{12}} \cot k(L-l) \sin k(l-z) + j\frac{Z_1 I_1}{\rho_{11} + \rho_{12}} \sin k(l-z),$$

$$i_3(z) = 2I_1 \frac{\sin k(L-z)}{\sin k(L-l)}. \tag{4.64}$$

Reactive impedances of generators' load are

$$jX_{A1} = \frac{Z_1 + j(\rho_{11} + \rho_{12}) \tan kl - 2jW_{33} \cot k(L-l)}{1 + j(Z_1 / W_{11}) \tan kl + \dfrac{2W_{33}}{\rho_{11} + \rho_{12}} \tan kl \cot k(L-l)},$$

$$jX_{A2} = \frac{j(\rho_{11} + \rho_{12}) \tan kl - 2jW_{33} \cot k(L-l)}{1 - j\dfrac{Z_1}{W_{12}} \tan kl + \dfrac{2W_{33}}{\rho_{11} + \rho_{12}} \tan kl \cot k(L-l)}. \tag{4.65}$$

The current distribution along the wires of the antenna with the reactive load in the in-phase and anti-phase modes is shown in Figure 4.21. The in-phase currents are designated by symbol (i), the anti-phase currents are designated by symbol (a).

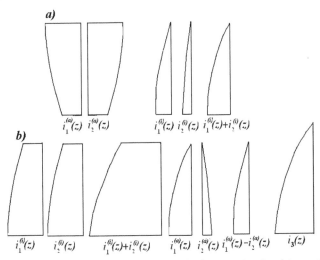

Figure 4.21 Currents in the antenna with the reactive load in anti-phase (*a*) and in-phase (*b*) modes.

Summarizing the results presented in this section, one should make the following conclusion. The principle of changing electrical height of the radiator without changing its geometric dimensions, realized in the circuit of a two-tiered antenna, is very promising, quite efficient and requires careful study to implement it in real structures.

5

Antennas with Concentrated Loads

5.1 IMPEDANCE LONG LINE

Distribution of a current along the radiator defines electrical characteristics of the radiator, i.e. its input impedance, directional pattern, etc. In order to change this distribution, one can use extraneous fields (exciters) or loads—both distributed and concentrated loads. Even only one load can significantly change the current distribution, and hence the electrical characteristics of the antenna.

If a great number of loads are placed along the antenna at small electrical distances from each other, we can consider that they are included uniformly and continuously along the entire antenna length, and this means that the radiator with a finite number of loads turned into the radiator with distributed load, i.e. it turned into the impedance radiator. The boundary conditions on the surface of this radiator along the z-axis of the cylindrical coordinates system, between the points $z = -L$ and $z = L$ are given by (2.51).

The impedance radiator unlike the metal radiator has additional degrees of freedom. By means of loads one can solve the inverse problem of the thin antennas theory—to create an antenna with required electrical characteristics. A particular case of the problem is creation of the radiator, which ensures in a wide frequency range a high matching level and the field maximum in the plane, perpendicular to the radiator axis. This problem has a great practical importance.

A typical linear radiator (thin, without loads) fails to meet these requirements. The reactive component of its input impedance is great everywhere, except in the vicinity of the serial resonances. That results in the antenna mismatch with a cable. If the radiator arm is larger than 0.7λ, the radiation in the plane, perpendicular to antenna axis, decreases, since the current distribution along a thin linear monopole without loads (Figure 5.1a) is close to the sinusoidal, and at high frequencies anti-phase segments are formed on the current curve (Figure 5.1b, curve 1).

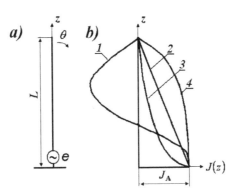

Figure 5.1 A linear monopole (*a*) and the laws of current distribution along it (*b*).

If to connect concentrated loads along the radiator length, one can, depending on their magnitudes and points of connection, obtain the other, not sinusoidal distribution of the current. The experimental results show that a radiator with linear or exponential in-phase current distribution exhibits good performance (high matching level, required shape of the vertical directivity pattern) in a wide frequency range. In particular, such distribution is created with the help of capacitive loads [42, 43]. These results confirm the known fact that the radiation maximum in the direction, perpendicular to the dipole axis, is attained, if the current is in-phase along the entire length of the antenna. Moreover, a long radiator with an in-phase current has high radiation resistance, which allows increasing the matching level.

Thus, the experiment confirms that the use of concentrated loads for the creation of antenna with the desired characteristics is promising. But before moving on to this problem we must consider an antenna with constant and changing surface impedance and also an equivalent impedance long line.

The main results of applying the method of induced emf and solving the integral equation for analysis of impedance antennas are given in Sections 1.5 and 2.5. An ordinary two-wire long line is an equivalent of a metal radiator. An impedance two-wire long line is an analogous equivalent of an impedance radiator. But unlike the metallic radiator the tangential component of the electric field on the surface of the impedance radiator is not equal to zero, and this causes an additional voltage decrease on each element of wire length. According to (2.51) and (2.52), this voltage decrease is equal to

$$dU = \frac{Z(z)}{2\pi a} J(z)dz, \tag{5.1}$$

Infinitesimal small element dz of the impedance line, which is equivalent to a symmetric radiator, comprises, beside an inductance $d\Lambda = \Lambda_1 dz$ and a capacitance $dC = C_1 dz$, the complementary impedance $\dfrac{Z(z)}{\pi a}dz$ (coefficient 2 takes into account that the radiator consists of two wires). Here Λ_1 and C_1 are the inductance and the capacitance per unit length of the radiator.

The impedance long line, which is equivalent of the impedance radiator, is shown in Figure 5.2.

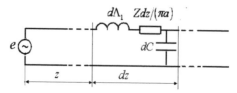

Figure 5.2 Equivalent long line in the case of constant surface impedance.

If the surface impedance is constant along the long line, the telegraph equations for such line are

$$-\frac{dU(z)}{dz} = J(z)\left(j\omega\Lambda_1 + \frac{Z}{2\pi a}\right), \quad -\frac{dJ(z)}{dz} = j\omega C_1 U(z), \tag{5.2}$$

from which

$$\frac{d^2U(z)}{dz^2} + k_1^2 U(z) = 0, \quad \frac{d^2J(z)}{dz^2} + k_1^2 J(z) = 0, \tag{5.3}$$

where

$$k_1^2 = k^2 - j\frac{Z\omega C_1}{\pi a}. \tag{5.4}$$

We consider that the capacitance between the radiator arms per unit of their length is equal to half of the self-capacitance C_0 of an infinitely long wire per unit of its length. Since the radius of the radiator is much less than its length, the surface of zero potential may be placed on the distance $2L$ from the radiator axis, i.e. capacitance per unit length of the line, which is equivalent of the symmetrical radiator, is equal to

$$C_1 = C_0 / 2 = \frac{\pi\varepsilon}{\ln(2L/a)}, \tag{5.5}$$

where from

$$k_1^2 = k^2 - j\frac{\omega\varepsilon Z}{aLn(2L/a)}. \tag{5.6}$$

Solving equations (5.3), we find the current $J(z)$ and the input impedance Z_l of the open at the end impedance long line:

$$J(z) = J(0)\frac{\sin k_1(L-z)}{\sin k_1 L}, \quad Z_l = -jW \cot k_1 L, \tag{5.7}$$

where $J(0)$ is the generator current, W is the wave impedance of the long line:

$$W = \frac{k_1}{\omega C_1} = 120\frac{k_1}{k}\ln\frac{2L}{a}. \tag{5.8}$$

Since the input current $J(0)$ of the line is $J(0) = e/Z_l$, the current $J(z)$ along the line is equal to

$$J(z) = j\frac{ke}{120k_1 \ln(2L/a)\cos k_1 L}\sin k_1(L - |z|). \tag{5.9}$$

That coincides with (2.55), i.e. with a first approximation for the antenna current.

If the surface impedance is changing along the antenna (more precisely, if this magnitude is piecewise constant), the equivalent impedance long line is a non-uniform line, i.e. it is a stepped line. It consists of N uniform sections of length l_m with the wave impedance W_m, current J_m and voltage u_m. We shall designate the surface impedance of

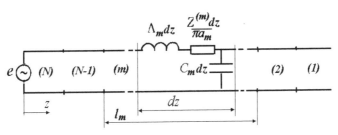

Figure 5.3 Equivalent long line in the case of changing surface impedance.

the segment m as $Z^{(m)}$ (Figure 5.3). A comparison of the segment m of the impedance radiator and of the impedance line allows coming to a formula, obtained in Section 2.5 for the propagation constant k_m and to an expression for the wave impedance W_m, similar to (5.8):

$$W_m = 120 \frac{k_m}{k} \ln \frac{2L}{a_m}. \tag{5.10}$$

From the theory of the long lines it is known that

$$u_m = U_m \cos(k_m z_m + \varphi_m), \quad J_m = jJ_m \sin(k_m z_m + \varphi_m), \tag{5.11}$$

and $I_m = U_m/W_m$. Since the voltage and the current along the stepped long line, which is equivalent to the radiator, are continuous:

$$u_m\Big|_{z_m=0} = u_{m-1}\Big|_{z_{m-1}=l_{m-1}}, \quad J_m\Big|_{z_m=0} = J_m\Big|_{z_{m-1}=l_{m-1}},$$

then

$$I_m = I_{m-1} \frac{\sin\left(k_{m-1} l_{m-1} + \varphi_{m-1}\right)}{\sin \varphi_m}, \quad U_m = U_{m-1} \frac{\cos\left(k_{m-1} l_{m-1} + \varphi_{m-1}\right)}{\cos \varphi_m}. \tag{5.12}$$

Dividing the first of expressions (5.12) onto the second expression and taking into account (5.10) and (5.11), we find:

$$\tan \varphi_m = \frac{k_m}{k_{m-1}} \tan\left(k_{m-1} l_{m-1} + \varphi_{m-1}\right). \tag{5.13}$$

Equalities (5.13) enable us to express the amplitude and phase of the current in any section through the parameters of sections and one of the currents:

$$\varphi_m = \tan^{-1}\left\{\frac{k_m}{k_{m-1}} \tan\left[k_{m-1} l_{m-1} + \tan^{-1}\left\{\frac{k_{m-1}}{k_{m-2}} \tan\left(k_{m-2} l_{m-2} + \dots + \tan^{-1}\frac{k_2}{k_1}\dots\right)\right\}\right]\right\},$$

$$I_m = I_n \prod_{p=m-1}^{N} \frac{\sin \varphi_p}{\sin\left(k_{p-1} l_{p-1} + \varphi_{p-1}\right)}. \tag{5.14}$$

The last expression is true for the Nth section, if we accept that $\prod_{p=N+1}^{N} = 1$.

Since the current of the generator is

$$J(0) = J_N \sin(k_N l_N + \varphi_N),$$

then

$$I_m = A_m J(0), \tag{5.15}$$

where

$$A_m = \prod_{p=m+1}^{N} \sin\varphi \Big/ \prod_{p=m}^{N} \sin(k_p l_p + \varphi_p).$$

Expressions (5.11) together with (5.14) and (5.15) present the approximate laws for the distribution of the voltage and the current along the radiator with piecewise constant surface impedance.

The required value of the surface impedance and a predetermined law of its change along the radiator can be realized by means of the concentrated loads, connected in the antenna wire. Consider the monopole with the height L and N loads Z_m (Figure 5.4a). Let loads be located uniformly at distance b along the antenna. If the distance b is small ($kb \ll 1$), the current distribution along the antenna undergoes practically no change, if to replace the concentrated loads by continuous surface impedance $Z(z)$, distributed across a length of each segment. Assume that the surface impedance of the antenna segment m is constant and equal to $Z^{(m)}$. Then

$$Z_m = bZ^{(m)}/(2\pi a_m), \qquad (5.16)$$

where a_m is the radius of this segment.

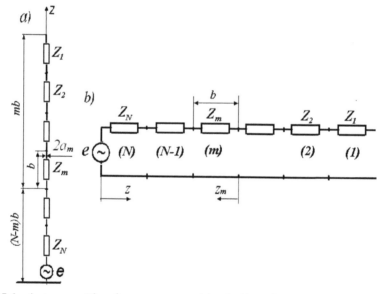

Figure 5.4   Antenna with a few concentrated loads (*a*) and an equivalent long line (*b*).

5.2   METHOD OF AN IMPEDANCE LONG LINE

As already mentioned, the current distribution along the antenna with a piecewise constant surface impedance is, in the first approximation, identical to the current distribution along an open at the end stepped long line (Figure 5.4b), i.e. a transmission line with stepwise variation of propagation constant (see Figure 5.3). In this case, one can write the expression for the generalized wave propagation constant γ_m at the segment m of the long line in accordance with the equation for current $J_m(z)$ along the segment m of a radiator in the form:

$$-\gamma_m^2 = k^2 - j2k\chi Z^{(m)}/(a_m Z_0). \qquad (5.17)$$

Here, $\chi = 0.5/\ln(2L/a)$ is a small parameter, a is middle radius. In the general case, γ_m is a complex magnitude. In a particular case, when this magnitude is purely imaginary ($\gamma_m = jk_m$), the current distribution at the segment m of the line has sinusoidal character.

If the law of changing the propagation constant, which allows ensuring the required current distribution, is found, expressions (5.17) and (5.16) can be used to calculate, first, the surface impedance $Z^{(m)}$ and, second, concentrated loads Z_m, respectively. The current of the segment m of a stepped line at arbitrary γ_m is

$$J(z_m) = I_m \sinh(\gamma_m z_m + \varphi_m), \ 0 \le z_m \le b, \qquad (5.18)$$

where I_m and φ_m are the amplitude and the phase of the current on the segment m, respectively, and z_m is the coordinate, measured from the segment end, i.e. $z_m = (N - m + 1)b = z$.

Suppose we want to obtain the given current distribution along the long line,

$$J(z) = J_A f(z), \ 0 \le z \le L, \qquad (5.19)$$

where J_A is the input current of the line (current of generator), $f(z)$ is the real and positive distribution function, which corresponds to the in-phase current. We equate currents $J(z)$ and $J(z_m)$ at the beginning and the end of the each segment. In this case, if the segment length is small, the current distribution along the line is close to the required. In accordance with (5.18) and (5.19), at $z_m = b$ and $z_m = 0$

$$I_m \sinh\left(\gamma_m b + \varphi_m\right) = J_A f\left[(N-m)b\right], \quad I_m \sinh \varphi_m = J_A f\left[(N-m+1)b\right].$$

If to divide the left and right parts of the first equation onto the respective parts of the second equation and to confine by the first terms of expansion of hyperbolic functions with small arguments into series (considering that b is a small magnitude), we get

$$\tanh \varphi_m = \gamma_m b / \left\{ \frac{f\left[(N-m)b\right]}{f\left[(N-m+1)b\right]} - 1 \right\}. \qquad (5.20)$$

For the segment $(m + 1)$, similarly to (5.20),

$$\tanh \varphi_{m+1} = \gamma_{m+1} b / \left\{ \frac{f\left[(N-m-1)b\right]}{f\left[(N-m)b\right]} - 1 \right\}. \qquad (5.21)$$

The voltage and the current are continuous along the stepped line, hence

$$\tanh \varphi_{m+1} = (\gamma_{m+1}/\gamma_m) \tanh (\gamma_m b + \varphi_m). \qquad (5.22)$$

Equations (5.20) and (5.21) present a set of equations that allow to relate γ_m and γ_{m+1} with each other. The solution of this set shows that magnitude γ_m is independent of γ_{m+1}:

$$\gamma_m = \frac{1}{b}\sqrt{1 - \frac{2f\left[(N-m)b\right] - f\left[(N-m-1)b\right]}{f\left[(N-m+1)b\right]}}. \qquad (5.23)$$

As is clear from (5.19), function $f(z)$ characterizes the law, in accordance with which the amplitude of the current changes along the radiator. In the case of in-phase distribution of the current along the antenna, its directional pattern in the vertical plane has a form

$$F(\theta) = \sin\theta \int_{-L}^{L} f(z)\exp(jkz\cos\theta)dz. \tag{5.24}$$

Calculations show that in this case in contrast to the sinusoidal distribution, the radiation maximum with growing frequency does not deviate from the perpendicular to the radiator axis. Increasing L/λ makes the main lobe narrower and increases the maximal directivity.

At linear distribution of the in-phase current amplitude (see Figure 5.1b, curve 2),

$$J_2(z) = J_A(1-z/L),$$

where $z = (N - m + 1)b - z_m$, i.e.

$$f_2(z) = (L-z)/L = [(m-1)b + z_m]/(Nb). \tag{5.25}$$

The linear distribution is a particular case of the exponential one (see Figure 2.1b, curves 3 and 4):

$$J_{3,4}(z) = J_A\frac{\exp(-\alpha z)-\exp(-\alpha L)}{1-\exp(-\alpha L)},$$

that is

$$f_{3,4}(z) = \frac{\exp(-\alpha z)-\exp(-\alpha L)}{1-\exp(-\alpha L)} = \frac{\sinh\{(\alpha/2)[(m-1)b+z_m]\}}{\sinh(\alpha Nb/2)}, \tag{5.26}$$

where α is the logarithmic decrement. If α is positive, the curve of a current is concave, i.e. the current quickly decreases from the maximum value near the generator to zero near the free end of the antenna. If α is negative, the curve of a current is convex, i.e. the current is more uniformly distributed along the dipole. The steepness of a curve depends on the value of α. It is easy to show that if α tends to zero, the expression for $J_{3,4}(z)$ turns into $J_2(z)$.

The antenna input impedance in the first approximation is equal to the input impedance of the stepped long line:

$$Z_l = -jW_N \coth(\gamma_N b + \varphi_N).$$

Here, as is seen from (5.8), $W_N = \gamma_N W/k$, where W is the wave impedance of a metal monopole of the same dimensions without loads.

If $m = N$, we find for the linear current distribution, using equalities (5.21) and (5.22) and taking into account that $f(0) = f_2(0) = 1$:

$$Z_l = -j(W/kb)[f_2(-b)-1]. \tag{5.27}$$

As seen from this expression, reducing the reactive component of the input impedance requires a slow variation of function $f(z)$ near the antenna base, so that the difference in square brackets should be a small magnitude—of the order of kb. Otherwise, the reactive component of input impedance will be great.

For the exponential distribution, replacing $f_2(-b)$ with $f_3(-b)$, we obtain from (5.27)

$$Z_{A3} = -j(W/kl)f_x(\alpha L/2) \tag{5.28}$$

where $f_x(x) = x(1 + \coth x)$. The graph of function $f_x(x)$ is given in Figure 5.5a. In particular for the linear distribution

$$Z_{A2} = -j\frac{W}{kb}\left(\frac{N+1}{N}-1\right) = jW/(kL).$$

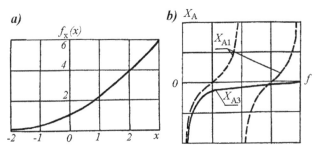

Figure 5.5 Graph of function $f_x(x)$ (*a*) and input impedances of uniform and non-uniform line (*b*).

Figure 5.5*b* compares the input impedance X_{A1} of a uniform line with sinusoidal current distribution and the input impedance X_{A3} of a non-uniform line with exponential current distribution depending on the frequency. Here, the magnitude α is assumed constant. In the first case the input impedance has a form of a cotangent, in the second case the curve smoothly approaches to zero with growing frequency, and that allows ensuring good matching in a wide range. As seen from Figure 5.5*a* and expression (5.28) for Z_{A3}, if α decreases, the input impedance of a long line at a given frequency diminishes. It means that a decrease of α results in a decrease of the reactive component of the antenna input impedance. Simultaneously the effective height grows, since the area, bounded by the curve of the current, increases; hence, the radiation resistance grows, too. Thus, at exponential distribution, it is expedient to decrease α in particular, into the region of negative values.

According to (5.16) and (5.17),

$$-\gamma_m^2 = k^2 - jk\chi Z_m / (30b),\tag{5.29}$$

In order to create the in-phase current distribution, magnitude γ_m should be purely real or purely imaginary along the entire antenna, and magnitude γ_m^2, correspondingly, only positive or only negative:

$$\text{sign } \gamma_m^2 = \text{const } (m).$$

In order to the in-phase distribution has been realized in a wide frequency range, magnitude γ_m should be real:

$$\gamma_m^2 > 0.\tag{5.30}$$

Indeed, if the values of γ_m (and also φ_m) are purely imaginary, the hyperbolic sine in the formula (5.18) would become the trigonometric sine. With frequency growth, an argument of a sine will increase and will exceed π, and the sine will change sign. If γ_m is real, then, as seen from (5.28), if function $f(z)$ decreases monotonically, the sign of φ_m coincides with sign of γ_m. As it follows from (5.23), in order that γ_m was real, following condition must be carried out

$$f[(N-m)b] \le \frac{1}{2}\{f[(N-m+1)b] + f[(N-m-1)b]\},\tag{5.31}$$

i.e., function $f(z)$ cannot be convex.

Two variants of carrying-out of condition (5.30) in a wide frequency range follow from (5.29). The first variant takes place at

$$k^2 \ll jk\ \chi Z_m/(30\ b),\tag{5.32}$$

i.e.

$$\gamma_m^2 = jk\chi Z_m / (30b). \tag{5.33}$$

If one takes into account that parameter χ is, strictly speaking, a complex magnitude ($\chi = \chi_1 - j\chi_2$), then the admittance of load is according with (2.19),

$$Y_m = 1/Z_m = j\omega C_m + 1/R_m, \tag{5.34}$$

where

$$C_m = 4\pi\varepsilon\chi_1 /(b\gamma_m^2), \quad R_m = b\gamma_m^2 /(4\pi\varepsilon\omega\chi_2).$$

As follows from (5.34), in order that magnitude γ_m^2 has been positive, each load should be executed as a parallel connection of a resistor and a capacitor (Figure 5.6a). The resistance of the resistor should vary in inverse proportion to frequency, and the capacitance of the capacitor should remain constant. When creating an actual antenna, it is expedient to choose the value of R_m for the middle frequency of band.

In order to achieve required current distribution $f(z)$ along the antenna, the magnitude γ_m should correspond to (5.23). Its substitution into (5.34) gives

$$C_m = 4\pi\varepsilon\chi_1 b\left\{1 - \frac{2f[(N-m)b] - f[(N-m-1)b]}{f[(N-m+1)b]}\right\}^{-1}, \quad R_m = \frac{\chi_1}{\chi_2\omega C_m}. \tag{5.35}$$

By comparing (5.31) and (5.35), one can easily verify that, if inequality (5.31) holds, the values of C_m are non-negative. For exponential and linear distribution,

$$C_{m3} = \frac{8\pi\varepsilon\chi_1}{\alpha^2 b\{1 + \coth[\alpha(m-1)b/2]\}}, \quad C_{m2} = \frac{4\pi\varepsilon}{\alpha}\chi_1(m-1), \quad R_m = \frac{\chi_1}{\chi_2\omega C_m}. \tag{5.36}$$

One can see from (5.36) that in the particular case, if one must obtain a law of current distribution, close to the linear, capacitances of loads should decrease towards to the free end of the antenna in proportion to the distance from it:

$$C_{m2} = C_{N2}(m-1)/(N-1), \tag{5.37}$$

where C_{N2} is the capacitance of the capacitor near the antenna base. The resistances of resistors should grow towards to the free end of the antenna:

$$R_{m2} = R_{N2}(N-1)/(m-1). \tag{5.38}$$

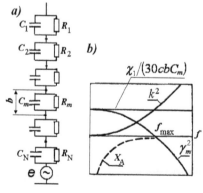

Figure 5.6 An antenna circuit with capacitors and resistors (a) and the frequency dependence of propagation constant (b).

Thus, to create the in-phase current distribution, which ensures high electrical characteristics of an antenna in a wide frequency range, each load should represent parallel connection of the resistor and the capacitor. For the first time the expediency of using a complex load for creation of a linear current distribution was demonstrated in [44]. But later on, a greatest attention was given to antennas with capacitive loads. The calculation results show that if resistors are connected in parallel with capacitors, the linear law of current distribution along the radiator is carried out more precisely, and the operating frequency range increases. However, connection of resistors results in decreasing the antenna efficiency, so the issue of their application should be solved in each particular case.

Figure 5.6*b* shows a graph of γ_m^2 depending on frequency for an antenna with capacitive loads:

$$\gamma_m^2 = -k^2 + \chi_1 4\pi\varepsilon /(bC_m).$$ (5.39)

In order for the propagation constant to be real at a given frequency f, capacitances of capacitors should not exceed

$$C_m \le \frac{\chi_1}{30k^2bc} = \frac{2.54 \cdot 10^5 \chi_1}{f^2 b}.$$ (5.40)

Here, c is the speed of light, capacitance C is measured in farads, if frequency f is given in Hertz's. In the case of linear distribution, capacitance C_{N2} of the capacitor near the antenna base is greater than other capacitances and should be chosen in accordance with (5.40). Similarly, under other distributions, this expression determines the maximum capacity.

As follows from (5.39), the propagation constant at low frequencies is real along the entire antenna. As the frequency increases, the magnitudes γ_m become purely imaginary (first of all, on segments, adjoining to the generator), i.e. the current distribution along these segments of the radiator becomes sinusoidal, and the main lobe of the vertical directional pattern deviates from the perpendicular to the dipole axis. This effect limits the antenna frequency range from above. From below, the range is limited by frequencies, where the reactive component of input impedance is still great. In order that magnitude γ_m does not become purely imaginary with increasing frequency, the capacitances of capacitors should decrease with growth of frequency (e.g., vary in inverse proportion to square of the frequency).

Regarding the second variant of realization of condition (5.40) in a wide frequency range, with its help one can make similar conclusions. This variant takes place, if the second summand of the right part of (5.29) is proportional to k^2:

$$-\gamma_m^2 = k^2[1 - j\chi Z_m /(30\ kb)],$$ (5.41)

i.e. load Z_m represents negative inductance Λ_m:

$$Z_m = -j\omega |\Lambda_m|,$$ (5.42)

where $|\Lambda_M| = 30b(1 + \gamma_m^2 /k^2)/(\chi c)$. At small γ_m/k the inductance is independent of frequency f.

In this case, the value of $\gamma_m^2 = k^2[\chi |\Lambda_m| c /(30b) - 1]$ is positive, if

$$|\Lambda_m| > 30b /(\chi c).$$ (5.43)

The negative inductance is an element of a circuit, which has purely reactive and negative impedance, proportional to f. This element is equivalent to a frequency-dependent capacitance:

$$-j\omega|\Lambda_m| = 1/(j\omega C_m), \qquad (5.44)$$

where

$$C_m = 1/(\omega^2|\Lambda_m|) = C_{m0}f_0^2/f^2, \qquad (5.45)$$

C_{m0} is the magnitude of capacitance C_m at frequency $f = f_0$. C_{m0} is independent on f.

Thus, in order to retain the in-phase current distribution in a wide frequency range, the capacitances of concentrated loads, which connected in an antenna wire, should vary in inverse proportion to the square of the frequency. As one can easily verify, inequality (5.30) will be true at all frequencies, and constraint (5.32) will be removed, if the negative inductances are connected in series with loads, determined by expression (5.34).

The proposed method allows making a number of practical conclusions. In order that concentrated loads may efficiently influence the current distribution, the distance between them should be small in comparison with the wave length. For creating the wideband (wide-range) radiator only capacitors must be used as reactive elements, since inclusion of reactive two-terminal networks of a more complex type, whose structure includes inductance coils, results in narrowing of the operating range. Capacitors enable to create along an antenna in a wide frequency range an electromagnetic wave with real propagation constant, which corresponds to the exponential change of the current amplitude with positive decrement (concave curve of the current). Obtaining a convex curve of the current with the help of simple concentrated elements (resistors, capacitors, inductance coils) is impossible. Among distributions with positive α, the antenna with distribution, which is close to linear one and is created by capacitances decreasing to the free end of the antenna in proportion to the distance from it, has a higher matching level and narrower main lobe of the directional pattern.

Method of the impedance long line and results of its use were first described in [45].

5.3 METHOD OF A LONG LINE WITH LOADS

Together with the method of impedance long line there is another approximate method for calculating magnitudes of the loads, which provide the given current distribution [46]. By means of this method, one can find on the basis of the given current distribution the law in accordance with which the equivalent length of the long line grows along it. That allows to calculate the load of each segment, corresponding to this law. Use of this method and the method of impedance long line gives analogous results.

Figure 5.4a shows an asymmetrical radiator of height L with N loads, which are uniformly located along it at a distance b from each other. A current distribution along the radiator in a first approximation is similar to the current distribution along the open at the end long line, with the impedances Z_m connected in series. The current distribution along each segment, located between adjacent loads, has a sinusoidal character:

$$J(s_m) = J_m \sin k(s_m + l_{e,m-1}), 0 \le s_m \le b, \qquad (5.46)$$

where $s_m = (N - m + 1)b - z$ is the coordinate, measured from the end of segment m, J_m is the current amplitude in the segment m, $l_{e,m-1}$ is the equivalent length of all preceding segments with $(m - 1)$ loads and the total length $(m - 1)b$. The values $l_{e,m}$ and $l_{e,m-1}$ are mutually related by the expression:

$$-jW \cot kl_{em} = Z_m - jW \cot k\,(b + l_{e,m-1}). \tag{5.47}$$

Here W is the wave impedance of the line. Note that the value $l_{e,m}$, if $m = N$, is equal to the equivalent length L_e of the radiator.

Expression (5.47) permits to find the magnitude Z_m:

$$Z_m = -jW[\cot kl_{em} - \cot k(b + l_{e,m-1})]. \tag{5.48}$$

If the distance between the loads is small ($kb<<1$), then

$$Z_m = -jW \frac{\sin k(b+l_{e,m-1}-l_{em})}{\sin kl_{em}\,\sin k(b+l_{e,m-1})} \approx -jW \frac{k(b+l_{e,m-1}-l_{em})}{\sin^2 kl_{em}}. \tag{5.49}$$

With the help of expressions (5.48) and (5.49) one can calculate the magnitudes of loads. For that it is necessary to know the function, in accordance with which the equivalent length grows along the line. The choice of this function depends on the current distribution along the radiator, which must be obtained in the first stage of solving the synthesis problem. In the general case

$$J(s_m) = J_A f(z), 0 \le s_m \le b, (N-m)b \le z \le (N-m+1)b, \tag{5.50}$$

where J_A is the current amplitude in the antenna base, and $f(z)$ is the function of the current distribution. Henceforth we shall assume that the function $f(z)$ is real and positive, i.e. we shall consider only the in-phase distributions.

Suppose we want to obtain along the antenna a given current distribution $J(z)$. For this we assume that the current $J(s_n)$ at the beginning and the end of each line segment coincides with the current $J(z)$. If the lengths of the segments are small, the current distribution along the line is close to the required one. In the general case we have in accordance with (2.32) and (2.36)

$$J_{n+1} \sin k\,(b + l_{en}) = J_A f[(N-n-1)b]\,at\;s_{n+1} = b,$$

$$J_{n+1} \sin kl_{en} = J_A f[(N-n)b]\,at\;s_{n+1} = 0.$$

Divide the left and right sides of the first equality into the corresponding sides of the second one. Considering that the magnitude b is small and retaining only the first terms of the expansion of trigonometric functions into a series, we obtain

$$1 + kb \cot kl_{en} = f[(N-n-1)b]/f[(N-n)b],$$

i.e.

$$l_{en} = \frac{1}{k}\tan^{-1}\frac{kb}{f[(N-n-1)b]/f[(N-n)b]-1}. \tag{5.51}$$

As is seen from (5.51), the equivalent length l_{em} is frequency dependent. Knowing l_{en}, one may in accordance with the expression (5.48) find the values of loads. They also have a frequency-dependent nature:

$$Z_n = j\frac{W}{kb}\left\{\frac{f[(N-n-1)b]}{f[(N-n)b]} - 2 + (1+k^2b^2)\frac{f[(N-n+1)b]}{f[(N-n)b]}\right\}. \tag{5.52}$$

An input impedance of the open at the end transmission line is a first approximation to the reactive impedance of the antenna. In general case it is equal to

$$Z_A = -jW \cot kL_e = -j\frac{W}{kb}[f(-b)-1]. \tag{5.53}$$

This expression shows that the function $f(z)$ should change slowly near the antenna base and the difference $f(-b) - 1$ should be small, of the order of kb. Otherwise, the reactive impedance of the antenna will be great.

Consider, for example, the exponential distribution (5.26) of the current amplitude along the radiator. The linear distribution (5.25) is its particular case. The equivalent lengths of the long lines for the exponential and linear distribution in accordance with (5.51), (5.26) and (5.25) are equal to

$$l_{em3,4} = \frac{1}{k}\tan^{-1}\frac{kb[1-\exp(-m\alpha b)]}{\exp(\alpha b)-1}, l_{em2} = \frac{1}{k}\tan^{-1}(mkb). \tag{5.54}$$

As is seen from (5.54), if $\alpha > 0$, the equivalent length of the antenna arm may not exceed a quarter of the wave length. The input impedance of a long line with the exponential and linear current distribution may be written in the form

$$Z_{A3,4} = -j\frac{W[\exp(\alpha b)-1]}{kb[1-\exp(-\alpha L)]}, Z_{A2} = -jW/(kL). \tag{5.55}$$

Using expressions (5.52) and (5.27), we find the magnitudes of the loads, which provide the exponential law of distribution for the amplitude of the current along the radiator:

$$Z_m = -j\frac{W}{kb}\frac{2(ch\alpha b - 1) + k^2 b^2 \exp(-\alpha b)}{1-\exp(-m\alpha b)}\{1-\exp[-(m-1)\alpha b]\}. \tag{5.56}$$

If the product αb is not small, then, neglecting the second summand of the numerator, we obtain

$$Z_m = 1/(j\omega C_m), \tag{5.57}$$

where

$$C_m = \frac{b[1-\exp(-m\alpha b)]}{2Wc(\cosh\alpha b - 1)}, \tag{5.58}$$

and as is seen from this formula, the sign of C_m coincides with the sign of α.

Thus, in order to obtain an exponential distribution of current amplitude and a sufficiently great magnitude of decrement, one must use the capacitive loads. Capacitors allow creating only the concave current distribution ($\alpha > 0$). In order to obtain a convex distribution ($\alpha < 0$), capacitances must be negative. If $\alpha b \ll 1$, then, confining by the first terms of the functions expansion into a series, we find from (5.56):

$$Zm = 1/(j\omega C_m) + j\omega\Lambda_m, \tag{5.59}$$

where

$$C_m = \frac{m}{cW\alpha}, \Lambda_m = -\frac{Wb(m-1)}{cm}. \tag{5.60}$$

In order to obtain the exponential distribution with a small decrement α, the negative inductances Λ_m should be included in series with capacitors. They can be neglected, if the first term of (5.59) is much larger than the second one, i.e. $\alpha \gg k^2 b(m-1)$. If $\alpha = 0$,

$$Z_m = j\omega\Lambda_m = -jkbW(m-1)/m, \tag{5.61}$$

i.e. in the case, when the loads are fabricated in the form of negative inductances, proportional to $(m - 1)/m$, we obtain a purely linear distribution.

As it follows from the analysis made in this section, the method of the long line with loads and the method of the impedance long line lead to similar results. Comparison of these results allows applying these methods to specific problems, using specific details of the current distribution along the radiators. Results obtained by means of these methods, can be used for solving the problem of optimization of antennas with loads by the mathematical programming method.

5.4   METHOD OF MATHEMATICAL PROGRAMMING

Use of the mathematical programming method [47] plays a major role in solving the inverse problems. The mathematical programming method allows determining optimal parameters of an antenna, in particular its geometric dimensions, magnitudes of the connected in the antenna concentrated and distributed loads, etc. It allows obtaining radiators with the given characteristics or with characteristics so close to the given characteristics as much as possible.

This remark is due to the fact that the variation interval of radiator parameters is bounded, i.e., not every value of antenna electrical characteristic can be realized practically. Different characteristics are optimal for different values of parameters. Moreover, an antenna should exhibit certain properties not at a single fixed frequency, but in the entire operation range. Therefore, the selected parameters are the result of a compromise, which is reached with the help of the mathematical programming method.

The problem of mathematical programming in the general case is stated as follows: one has to find vector $\vec{x}$ of parameters that minimizes some objective function $F(\vec{x})$ under imposed constraints $\varphi_i(\vec{x}) \geq 0$. Depending on the type of functions $F(\vec{x})$ and $\varphi_i(\vec{x})$, mathematical programming is divided into linear, convex and nonlinear ones. In the case at hand, the problem is solved by nonlinear programming methods, since the type of function $F(\vec{x})$ is unknown.

The objective function (or the general functional) is a sum of several partial functional $F_j(\vec{x})$ with weighting coefficients p_j and penalty function F_{ip}:

$$F(\vec{x}) = \sum_j p_j F_j(\vec{x}) + \sum_i F_{ip}. \tag{5.62}$$

The partial functional is an error function for one or the other characteristic. The weighting function allows to take into account an importance of this characteristic and a sensitivity of corresponding functional to results of changing the vector. A penalty function is zero, if the parameters lie within the given intervals, and has a great magnitude, even if only one of the parameters falls outside the interval limits.

Present an example of antenna parameters. Controlled parameters x for an antenna with concentrated loads are magnitudes of the loads, coordinates z_n of the points of their placement and the wave impedance W of the cable. The loads are the simple elements or the sets of simple elements (capacitors with capacitances C_n, coils with inductances Λ_n and resistors with resistances R_n). Values z_n, W, C_n, Λ_n and R_n are to be real, positive and frequency-independent, and z_n are to be smaller than antenna length L. These requirements, naturally, limit the interval of parameters change.

Different ways of an error function formation are known. For example, the quasi-Tchebyscheff criterion gives the good results:

$$F_j(\vec{x}) = \frac{1}{N_f}\left[\frac{f_{j0}}{f_{j\min}(\vec{x})}-1\right]\left\{\sum_{n_f}\left[\frac{(f_{j0}/f_j\langle\vec{x}\rangle)-1}{(f_{j0}/f_{j\min}\langle\vec{x}\rangle)-1}\right]^s\right\}^{1/s}. \qquad (5.63)$$

Here, N_f is a number of points of the independent argument (e.g. a number of frequencies in a given range), n_f is a point number (e.g. a frequency number), $f_j(\vec{x})$ is one of the electrical characteristics of an antenna, $f_{j\min}(\vec{x})$ is its minimal value in the considered interval, f_{j0} is a hypothetical value of the characteristic, which must be reached, S is the index of power, allowing to control the method sensitivity.

A root-mean-square criterion is other error function

$$F_j(\vec{x}) = \frac{1}{N_f N_l}\sum_{n_f=1}^{N_f}\sum_{n_l=1}^{N_l}[f_j(\vec{x})-f_{j0}]^2. \qquad (5.64)$$

Here, N_f and N_l are numbers of points of the independent argument (e.g. a number of frequencies in given range and a number of points on the wire), n_f is a frequency number, n_l is a point number, $f_j(\vec{x})$ is one of the electrical characteristic of an antenna (e.g. a current or a voltage), f_{j0} is a hypothetical value of the characteristic, which must be reached.

The choice of function $f_j(\vec{x})$ depends on the stated problem. For example, for creation of a wide-band radiator one must use as functions $f_j(\vec{x})$ the travelling wave ratio (TWR) in the cable and the pattern factor (PF), which is equal to the average level of radiation at predetermined angles:

$$\text{TWR} = \frac{2a}{a^2+b^2+1+\sqrt{(a^2+b^2+1)^2-4a^2}}, \text{PF} = \frac{1}{K}\sum_{k=1}^{K}F(\theta_k). \qquad (5.65)$$

Here $a = R_A/W$, $b = X_A/W$ are respectively the active and reactive components of the antenna impedance, relative to a wave impedance of a cable, K is a number of angles θ_k within the limits of angular sector from θ_1 to θ_K (e.g. from 90° to 60°), and $F(\theta_k)$ is a magnitude of normalized directional pattern in the vertical plane for an angle θ_k. If resistors with the resistances R_n are used as the loading elements, it is necessary to supplement the set of $f_j(\vec{x})$ by the function of antenna efficiency

$$\eta_A = 1 - \frac{1}{J_A^2 R_A}\sum_{n=1}^{N}|J_n|^2 R_n, \qquad (5.66)$$

where N is the number of loads, J_n and J_A are the currents in the load n and in the antenna base, respectively.

If it is necessary to obtain a given current distribution $J(z)$, it is expedient to use as functions $f_j(\vec{x})$ (the electrical characteristics of an antenna) either real and imaginary current components

$$f_1 = \text{Re } J(z, f), f_2 = \text{Im } J(z, f), \qquad (5.67)$$

or the amplitude and the phase of the current:

$$f_3 = |J(z,f)|, f_4 = \tan^{-1}\left[\text{Im}(z,f)/\text{Re}(z,f)\right]. \qquad (5.68)$$

In the cases, when analytical expression for objective function $F(\vec{x})$ is absent, one can find the minimum of this function by a numerical method, based on searching the gradient. The gradient method is an iterative procedure, in which we go step by step from one set of parameters $\vec{x}_m$ to another set $\vec{x}_{m+1}$ in the direction of the maximal decrease of the function (the steepest descent method):

$$\vec{x}_{m+1} = \vec{x}_m - \alpha_m gradF(\vec{x}_m). \tag{5.69}$$

Here m is the iteration number, α_m is the scale coefficient, determined as a result of a linear searching the minimum of the functional in the direction of anti-gradient.

The minimum of the functional and the values of parameters, which are correspond to this minimum, are determined for each iteration. Each iteration is, in essence, a search for the minimum of a function of one variable – α. The method with increasing the step (e.g. with doubling it) and subsequent interpolation function in the considered interval by a polynomial of the given power is the most rational. It is convenient to apply the cubic interpolation, since the number of interpolation nodes is great enough (four), and the root of the derivative (the value of α, causing the derivative to vanish) is found analytically. If the first step results in an increase, rather than decrease, of the objective function, the step should be reduced by a factor of 10^p, where $p = 1, 2...$, whereupon the linear search goes on again with doubling of a step.

A modification of the steepest descent method is the method of the conjugate gradients. In this case the iteration 1, $(Q - 1)$, $(2Q + 1)$ and so on are calculated according to anti-gradient (here Q is a number of parameters) and the rest steps correspond to the expression

$$\vec{x}_{m+1} = \vec{x}_m - \alpha_m \vec{G}_{m'} \tag{5.70}$$

where

$$\vec{G}_m = gradF(\vec{x}_m) + \left| gradF(\vec{x}_m)/ gradF(\vec{x}_{m-1}) \right|^2 \vec{G}_{m-1}.$$

The calculation is over, when the decrease of the objective function from iteration to iteration becomes smaller than a preset value, or the magnitude of iterations exceeds certain limit ($m \geq M$).

The mathematical programming method (synthesis) presupposes frequentative computations of the antenna electrical characteristics at different initial parameters (analysis). Performing such calculations requires incorporation of a special program into the synthesis software. This program allows to determine at given loads and exciting emf's all electrical characteristics of an antenna, i.e. calculating functions $f_j(\vec{x})$ for known vector $\vec{x}$ of initial parameters.

The most laborious in the calculation is computation of the self- and the mutual impedances between the antenna segments (between so-called short dipoles). Therefore, in order to speed up the calculations, it is expedient to fixate, for example, points of placing concentrated loads, in order to the coordinates of short dipoles and their mutual impedances do not might change from iteration to iteration. If there are enough many loads, i.e. the distances between them are small in comparison with the wave length, this restriction will have no effect on the synthesis results.

As the initial values of the antenna parameters, one must use the magnitudes, found by the approximate method, according to the physical content of the problem. The results of calculations show that the computational process in this case is accelerated, and most importantly, the error probability decreases, since the process of optimization at the arbitrary choice of the initial parameters may lead to a local, rather than true extremum of the objective function. Examples of the approximate physical methods are presented in the following sections.

The synthesis program, based on the mathematical programming method, permits to bring the problem solution to an end. Other methods of solving often stop and do not reach the goal. For example, earlier the synthesis of the antennas with given electrical characteristics was broken up into two stages: at the first stage the distribution of current was computed. The parameters of the antenna, providing such distribution, must be determined at the second stage. The first question has been investigated sufficiently. It covers a wide class of the tasks (the task of creating a wide-band antenna is one of possible variants). Far less attention has been paid to the second stage of synthesis.

In principle, if the required current distribution along a wire antenna is known, one can split the wire of an antenna onto short dipoles and define currents at the centers of these dipoles. The amplitudes of piecewise-sinusoidal basis functions are equal to the magnitudes of the currents at the corresponding antenna points. It is easy to calculate the magnitudes of loads, which one must connect at these points to obtain the desired currents.

But the impedances of loads, calculated by this method, consist of active and reactive components, which are changed with frequency. The calculated active component of load impedance may be obtained negative, and this is an evidence of impossibility to create such distribution with the help of passive elements. As to the reactive component, it is necessary to solve still the problem of its implementation in the given frequency range with the help of a set of simple elements. Therefore, it is necessary to solve the problem of creating an antenna with the chosen type of loads in order to ensure in the desired range not the given current distribution, but the current distribution close to the desired distribution as much as possible. This problem, as the problem of creating a wide-band radiator, is solved by the mathematical programming method.

The method of mathematical programming offers wide scope for solution of various problems of synthesis.

5.5 SYNTHESIS OF WIDE-BAND RADIATOR

As already mentioned, the described approximate methods of the analysis of antennas with concentrated loads (method of the impedance transmission line and method of the transmission line with loads) allow making a number of practical conclusions. The distance between loads should be small in comparison with the wavelength. For creating the wide-band dipole only capacitors must be used as reactive elements. Capacitors enable to create along an antenna in a wide frequency range in-phase current distribution in the form of concave curve. The antenna with distribution, which is close to linear one and is created by capacitances, decreasing to the free end of the antenna in proportion to the distance from it, has a higher matching level and narrower main lobe of the directional pattern.

On the one hand, the mentioned methods have a sufficiently general character. They allow to derive analytical expressions for impedances of loads, which ensure different laws of the current distribution along the radiator. On the other hand they have an approximate character, i.e. verification of obtained results by means of rigorous calculation and experiment is necessary.

Consider as an example a monopole of height 6 m and radius $6.67 \cdot 10^{-3}$ m with ten capacitors, which are located along it at distance 0.6 m from each other. The capacitance of the capacitor near the antenna base is adopted equal to 17.7 pF. In this case the propagation constant γ_n is real along the entire antenna up to frequency 40 MHz. The capacitances of other capacitors are calculated in accordance with (5.37). Calculating the electrical characteristics of the antennas was performed by means of a general-purpose program based on the Moment Method and built in accordance with the procedure described in Section 2.7.

Figure 5.7a shows the frequency dependence of active R_A and reactive X_A components of the input impedance as well as TWR in a cable with wave impedance 75 Ohm for the described asymmetrical antenna. Here, for the sake of comparison, the experimental data obtained on a full-scale model-up are given. Figure 5.7b presents the calculated directional patterns in the vertical plane. Placement of inductance in the antenna base and selection of cable allow to obtain a higher level of matching. Curves TWR_{200} and TWR_{200}^c demonstrate the level of matching with the cable, whose wave impedance is equal to 200 Ohm, without compensation of antenna reactance and with its compensation by means of the inductance 76.4 nH.

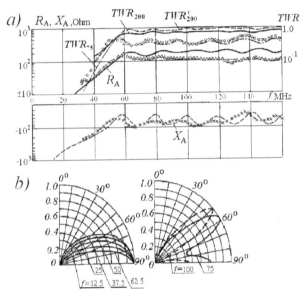

Figure 5.7 Input characteristics (a) and directional patterns (b)
of the radiator with constant capacitive loads.

Calculation and experiment confirm that characteristics of radiators with loads are much better than their characteristics without loads. Under the identical requirements to electrical characteristics and under the identical dimensions, the frequency ratio is changed from 1.3 to 1.5 MHz for thin whip antennas, from 1.5 to 3—for radiators with capacitors, and from 3 to 4—for radiators with capacitors and resistors. In the case of the radiators with loads the upper limit of the frequency ratio is defined by the deviation of the main lobe of directional pattern from the perpendicular to the radiator axis. As to the matching level, it remains high in a wider range (with the frequency ratio about 10).

As is shown in Section 5.2, in order to retain the in-phase current distribution in a wide frequency range, the capacitances of concentrated loads, located along the antenna wire, should vary in inverse proportion to square of the frequency. Calculation confirms that the antennas with frequency-dependent capacitances have a wider frequency range than the antennas with constant capacitances. Figure 5.8a gives TWR for three variants of the monopoles in the cable with the wave impedance 75 Ohm. The calculations are performed for the antenna of the height 12 m and the radius 0.03 m, with ten capacitors located at the distance 1.2 m from each other (the upper and lower capacitors are placed at the distances 0.6 m from the ends of the monopole).

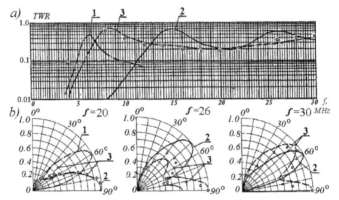

Figure 5.8   Input characteristics (*a*) and directional patterns (*b*) of radiators without loads (1), with constant (2) and frequency-dependent (3) capacitive loads.

Curve 1 in Figure 5.8 corresponds to the radiator without loads (to the whip antenna), curve 2—to the radiator with loads, whose capacitances are frequency independent. Here, the capacitance C_{N0} of the lower capacitor was chosen equal to 177 pF. In this case the propagation constant γ_n is real along the entire antenna up to frequency 10 MHz. Capacitances C_{n0} of others capacitors are decreased to the free end of the antenna in proportion to the distance from it. This allows to obtain the law of current distribution along the radiator, which is close to the linear one. Curve 3 is plotted for the radiator with frequency-dependent capacitive loads. Their capacitances are changed in accordance with (5.45), where $f_0 = 20$ MHz.

Table 5.1 shows lower f_1 and upper f_2 frequencies of the operating range of each antenna. At the frequency f_1 TWR becomes greater than 0.2, at the frequency f_2 the field strength along the perpendicular to the antenna axis becomes less than 0.7 of the maximum (as a rule, it corresponds to the second maximum on curve of TWR). TWR of the whip antenna with growing frequency quickly decreases below the level of 0.2, and the value of frequency, corresponding to this point, is taken as f_2. Besides, Table 5.1 reports the range width $\Delta f = f_2 - f_1$ and frequency ratio $k_f = f_2/f_1$.

As is seen from Figure 5.8*a* and Table 5.1, the level of matching of the variant 3 at low frequencies approaches to the level of matching of a whip antenna, and the upper frequency of the variant 3 in comparison with the variant 2 is displaced to the right, since the main lobe of directional pattern deviates from the perpendicular to the antenna axis at a higher frequency. In addition, the minimum TWR increases in the middle of the operating range.

Table 5.1   Frequency Ratio of Radiators.

Version of antenna	Frequency, MHz		Range width Δ_f, MHz	Frequency ratio k_f
	f_1	f_2		
1	5.2	7.7	2.5	1.5
2	12.3	26.0	13.7	2.1
3	6.3	34.0	27.7	5.4

Figure 5.8*a* presents also the results of experimental verification of TWR for the variant 3. The measurements were performed on the model of the antenna on a scale 1:10. The frequency range was split into short intervals, and the capacitance of the capacitor used in each interval was equal to the capacitance, calculated for the middle of the interval. The agreement of calculated and experimental data was good. Figure 5.8*b* demonstrates for the same antennas variants the calculated curves and the measured directional patterns in the vertical plane.

The task of implementation of frequency-dependent capacitances is not easy, but is promising. Yet, the use of tunable capacitors seems by more realistic. In this case expression (5.45) determines the optimal dependence of the capacitance on the frequency. The continuous changing of its magnitude can be replaced with stepwise switching, which is easier implemented.

The method of impedance long line allows finding the potential capabilities of antennas with loads. Furthermore, the results, obtained by means of this method, can be used for solving the optimization problem of an antenna with loads by the mathematical programming method, described in Section 5.4.

The results of the synthesis of wide-range antennas with loads were given in [48]. The program of synthesis was used for selection of the optimal capacitive loads, allowing to obtain maximal TWR and PF in the predetermined range of frequencies $f_1 - f_2$. Calculations showed that it is enough 4-5 iterations for the synthesis of antenna. Number of optimized electrical characteristics has little effect on the synthesis time. For example, duration of optimizing TWR and PF is almost the same as duration of optimizing only TWR. Selection of criterion is almost not affected the results of synthesis and the calculation time. For example root-mean-square criterion has no advantage over quasi-Tchebyscheff criterion. So, only the last one was used in subsequent calculations. As a hypothetical value of the characteristic, which must be reached (e.g., TWR), it is expedient to select the maximum, because its decrease leads to deterioration of the results. Increasing the index S in quasi-Tchebyscheff criterion leads to faster convergence of the process. This index was adopted in the calculations equal to $S = 6$. The all weighting coefficients p_j were taken by identical.

Figure 5.9 gives the basic dimensions of two antennas with loads. The first antenna is the monopole of height 11.31 m with four capacitors, irregularly spaced along the

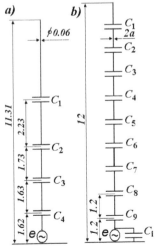

Figure 5.9 Synthesized antennas with four (*a*) and nine (*b*) capacitors.

radiator. The second antenna is the monopole of height 12 m with nine capacitors spaced equidistantly. The capacitance $C_i = 15$ pF, equal to the capacitance of a typical ceramic insulator, was connected at the base of the second antenna in parallel with its input.

The results of the synthesis of the considered antennas are presented in Tables 5.2 and 5.3. The basic characteristics of the radiators are given in Table 5.2. Optimal capacitive loads are given in Table 5.3. In Table 5.2 the following designations are used: N is the number of capacitors, N_f is the number of frequencies, used in the range for calculations, M is the required number of iterations.

Table 5.2 Main Characteristics of Antennas

Variant	L, m	a, m	N	$f_1 - f_2$, MHz	N_f	M	TWR min	PF min
1	11.31	0.03	4	11.5–16.5	11	4	0.310	0.860
2	"	"	"	"	"	5	0.360	0.812
3	12	0.03	9	7.5–15	16	4	0.123	0.819
4	"	"	"	15–30	"	4	0.273	0.610
5	"	"	"	30–60	"	5	0.360	0.562
6	"	0.15	"	7.5–15	"	4	0.205	0.813
7	"	"	"	15–30	"	5	0.414	0.680
8	"	"	"	30–60	"	4	0.380	0.605
9	"	0.03	"	8.5–13	10	3	0.217	0.870
10	"	"	"	13–22	"	5	0.216	0.790
11	"	"	"	22–60	20	8	0.204	0.437
12	"	0.15	"	8.5–13	10	5	0.314	0.829
13	"	"	"	13–22	"	4	0.278	0.859
14	"	"	"	22–60	20	5	0.322	0.565

Table 5.3 Optimal Capacitive Loads

Variant	Optimal capacitive loads, pF								
	C_1	C_2	C_3	C_4	C_5	C_6	C_7	C_8	C_9
1	44.3	33.2	91.2	432	-	-	-	-	-
2	84	164	143	182	-	-	-	-	-
3	37.3	81.1	127	181	58	369	516	691	883
4	8.7	20.6	36.5	51.1	58.7	53.3	50.6	88.1	156
5	2.0	3.9	5.4	6.1	9.5	12.2	15.7	21.1	18.3
6	51.2	134	219	340	477	633	804	981	1150
7	10.7	28.4	57.0	76.7	86.4	86.8	53.5	216	409
8	4.5	19.0	11.4	15.8	26.4	29.8	32.4	23.0	35.7
9	33.9	72.8	115	164	223	296	385	492	608
10	8.4	18.4	30.3	40.4	47.1	53.3	82.8	151	248
11	1.7	5.6	12.2	11.7	17.6	41.0	30.5	56.4	240
12	21.1	0.2	39.2	231	519	909	1380	1900	2450
13	20.3	76.3	122	78.5	0.1	107	351	761	1340
14	2.9	26.9	0.3	42.1	22.6	59.1	35.8	55.0	73.1

The frequency dependences of TWR and PF of the antenna with the height 11.31 m (the first variant) for the different number m of the iteration are shown in Figure 5.10a,b. The input characteristics and the directional patterns of the synthesized antenna are presented in Figure 5.10c,d. As is seen from the figure, the curve of TWR has a maximum in the area of series resonance, and the magnitudes of PF decrease gradually with growth of f. The capacitances increase to the antenna base, but not monotonically.

The second variant is distinguished from the first variant by the fact that the wave impedance of a cable is used as a regulated parameter in addition to the magnitudes of the capacitances. The optimal wave impedance is equal to $W=238$ Ohm. In this case the antenna has the optimal characteristics in the area of the parallel resonance.

The other variants of Tables 2.2 and 2.3 concern the antenna of a height 12 m with an insulator at the base. Frequency ratio for the antennas 3–8 is adopted equal to two. As is seen from Table 5.2, the increase of antenna radius from 0.03 m to 0.15 m at frequencies up to 30 MHz, results in growing minimal TWR approximately by a factor 1.5. The variation of radius has a weaker effect upon the minimal PF.

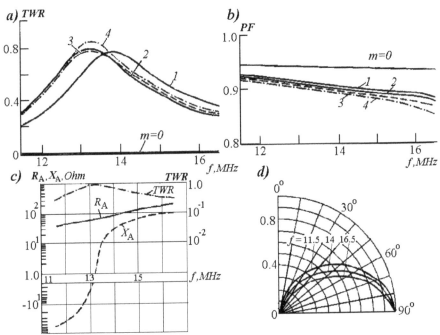

Figure 5.10 Change of TWR (a) and PF (b) depending on m, and input impedance (c) and the vertical directional pattern (d) of synthesized antenna of the height 11.31 m.

Figure 5.11 shows the electrical characteristics of variants 3–5 (the radius of the antenna is 0.03 m) as well as of the whip antenna with the same geometrical dimensions and the same capacitance C_i of the insulator. The characteristics of the antennas with radius 0.15 m (variants 3–8) are similar. As seen from the Figure 5.11, the curve of TWR can have two maximums at high frequencies. The curve of PF does not decrease monotonically with frequency, but has extremums too. In addition to calculated curves, Figure 5.11 shows (by dots and other symbols) the results of experimental verification, carried out on the models on a scale 1:5. The calculation and experiment agree well.

As it follows from Table 5.2 (variants 3–8), if frequency ratios in various sub-ranges are identical, the level of antenna matching with a cable can be different in the different sub ranges. This level substantially rises, if the frequency grows. In order to obtain more uniform and, on the whole, better characteristics over the entire frequency range (at unaltered number of sub-ranges), it is expedient to split the total range onto such fractions that the frequency ratio of sub-ranges increases with increasing frequency. The results of solving this problem are presented in Table 5.2 as variants 9–14.

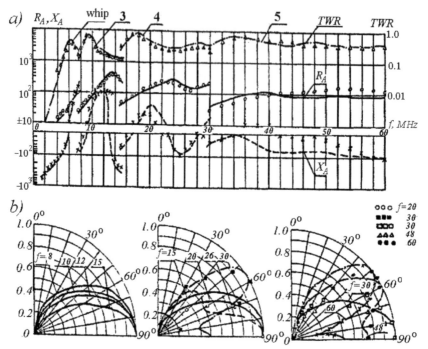

Figure 5.11 Input characteristics of the antennas of the height 12 m and the radius 0.03 m (*a*) and their vertical directional patterns (*b*).

The electrical characteristics of variants 12–14 as well as of the monopole of radius 0.15 m without loads (with capacitance C_i of the insulator at the base) are given in Figure 5.12. Data of Table 5.2 confirm a general increase of TWR level in comparison with variants 6–8. In each sub-range, increasing the antenna radius causes the rise of minimal TWR (approximately by a factor 1.5), together with the rise of minimal PF at high frequencies.

The results of optimization of 12-meter antennas with capacitance $C_i = 15$ pF at the base are used to plot in Figure 5.13 the curves for the minimal TWR depending on relative antenna length L/λ_{max} (λ_{max} is the maximum wavelength of the range) at various frequency ratios k_f and different antenna radii a. These curves determine the maximum attainable characteristics, which can be obtained with the help of antennas with constant capacitive loads.

The calculation results show that, if this is necessary, the operating range of the antenna can be expanded in the direction of high frequencies at a sufficiently high level of TWR, but the vertical directional patterns deteriorate significantly in the additional (high-frequency) fraction of the range. In this connection, the frequency ratio of an antenna with capacitive loads does not exceed 3 (at PF ≥ 0.5 and TWR ≥ 0.2). As it is shown in the Section 4.4, the application of a multi-tiered structure allows in a wide range to ensure desired directional pattern. Joint use of both principles, i.e. of the multi-tiered structure and the capacitive loads in the wires of each tier allows in a wide range to ensure high level of matching and desired directional pattern.

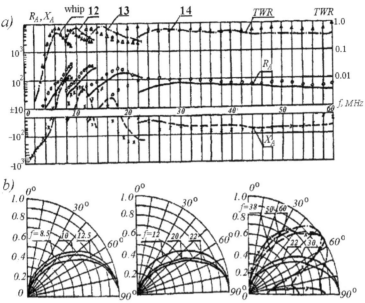

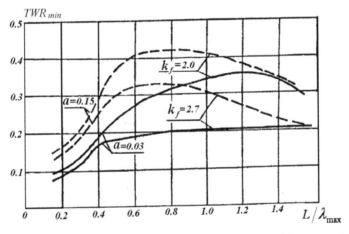

Figure 5.12 Input characteristics of the antennas of the height 12 m and the radius 0.15 m (*a*) and their vertical directional patterns (*b*).

Figure 5.13 The maximum level of matching for the antenna with constant capacitances.

Thus, the method of mathematical programming is an efficient method of optimization of antennas with capacitive loads. Its software may be used for optimization of antennas with loads of other kinds. It can also be applied to solution of other synthesis problems, for example, in order to find loads ensuring the required current distribution along the radiator.

5.6 SYNTHESIS OF CURRENT DISTRIBUTION: REDUCING SUPERSTRUCTURES IMPACT

It is necessary to emphasize that similarly the previous problem of ensuring the required current distribution along the radiator does not mean a rigorous coincidence of an obtained current distribution with the given distribution, but creating the distribution, closest to the required as far as possible.

Examples of antenna synthesis with the given current distribution, realized in a certain frequency range, were presented in [46]. The calculation was performed for the described in Section 5.5 monopole of height 6 m with ten capacitive loads. Figure 5.14a shows the equivalent lengths measured along the monopole from its free end to the points, where the capacitors must be installed. In Figure 5.14b the capacitances of these capacitors are given. The tasks were considered: creating a linear distribution of current

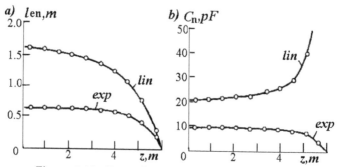

Figure 5.14 Equivalent length of antenna (a) and capacitances, calculated by approximated method (b).

(curves, along which the equivalent lengths and the magnitudes of the capacitances are presented, are designated by label "lin") and creating an exponential distribution of current with the logarithmic decrement $\alpha = 2$ (corresponding curves are designated by label "exp"). The calculations are performed by the approximated method of a long line with loads in accordance with the expressions (5.54) and (5.58) at frequency $f = 40$ MHz.

These results were used for strict calculating the amplitude and the phase along antennas with the loads. They are given in Figure 5.15a for a linear distribution and in Figure 5.15b for an exponential distribution. As can be seen from the figures, at $f = 40$ MHz the amplitude distribution is close to the required one, the phase curves have a slight slope. When the frequency changes (at $f = 30$ and $f = 50$ MHz), the amplitude and the phase distribution of the current are not conserved.

In order to provide the required current distribution in the continuous range from 40 to 80 MHz, the synthesis of the antenna was made by the method of mathematical programming. The results are shown in Figure 5.16 for the linear (a) and exponential (b) distributions respectively. The amplitude and phase of the current are obtained as result of optimization of electrical characteristics of the antenna. The error function was formed, using the root-mean-square criterion. The values, calculated by a method of the long line with loads at the middle frequency $f = 60$ MHz, were taken as a zero approximation. In the calculation it was adopted that the number of frequencies in the given range is equal to 9, and the number of the points on the wire is equal to 11.

The results were improved significantly. In each figure the four curves for the current amplitude are drawn: curve, labeled by f_0, corresponds to the required distribution, and curves labeled by $f = 40$, 60 and 80 corresponds to the result of synthesis at frequencies 40, 60 and 80 MHz. As is seen from the figures, the obtained distribution is, on the whole, close to the given distribution, but is not identical to it. However, this difference is not caused by the inexact method. Primarily the reason of this difference is the limited potential opportunities of antennas. Thus, in addition to the successful solution of the problem the method permits to determine the potential opportunities of the antennas.

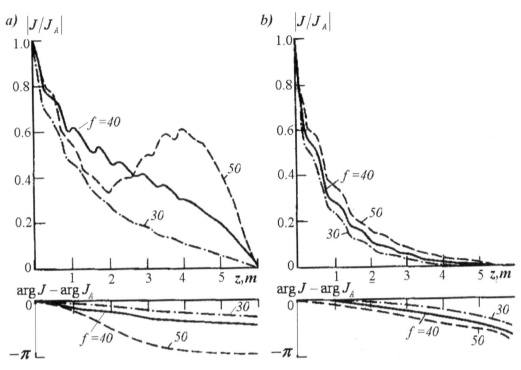

Figure 5.15   Currents in the antenna with approximate loads designed for creating the linear (*a*) and exponential (*b*) distribution of the amplitude at f = 40 MHz.

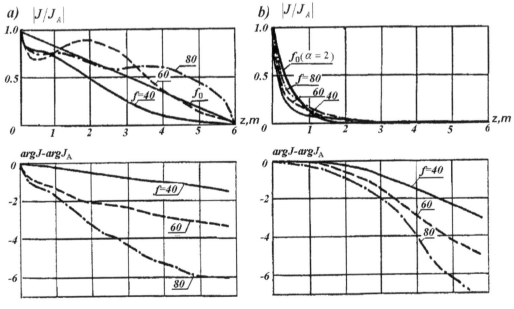

Figure 5.16   Linear (*a*) and exponential (*b*) distributions of the current in the antenna with loads.

The use of loads also gives freedom in choosing the antenna length (taking into account the possibilities of manufacture and installation), since they permit securing the desired characteristics in the required frequency range at given antenna length. The freedom in choosing the radiator length enables weakening the effect of the adjacent metal bodies, e.g. of the superstructures, on the directional pattern of an antenna or an antenna array. Figure 5.17 shows the calculation results for the directional pattern of a monopole, situated near a metal superstructure in a shape of a circular metal cylinder of finite length. The directional patterns in the horizontal plane are calculated at two frequencies of *HF* range.

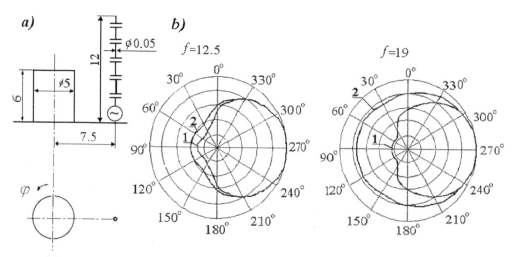

Figure 5.17 An antenna near a superstructure (*a*) and its horizontal pattern (*b*).

Two options are considered: 1—the monopole without loads of the height 6 m and the diameter 0.016 m, 2—the monopole of the height 12 m and the diameter 0.06 m with 9 capacitive loads, selected with the aim to ensure the optimal electrical characteristics on the frequencies from 8 to 22 MHz. The relative placement of the superstructure and the monopole as well as the superstructure dimensions are shown in Figure 5.17*a*. The circular cylinder during calculation was replaced with a wire structure from equidistant conductors, located along generatrices of the cylinder and the radii of its end surface. As is seen from Figure 5.17*b*, the radiation of an ordinary monopole in the direction of superstructure decreases sharply, and the use of the monopole with loads allows to lessen this effect.

Figure 5.18 demonstrates similar results for the uniform linear array, situated near the superstructure. The mentioned above two variants of monopoles are adopted as radiators of the array. The relative placement of the superstructure and radiators as well as the superstructure dimensions are given in figure, the phase shift between the currents of the radiators is adopted zero. The calculation results show that in the upper part of the frequency range the influence of the superstructure on the directional pattern of array, consisting of the monopoles without loads, is slighter than its influence on the directional pattern of the separate monopole. This is, apparently, related to the fact that the superstructure does not hinder the propagation of electromagnetic waves from the side radiators. Nevertheless, the use of monopoles with loads in this case also allows to reduce the impact of the superstructure and to increase the signal in its direction.

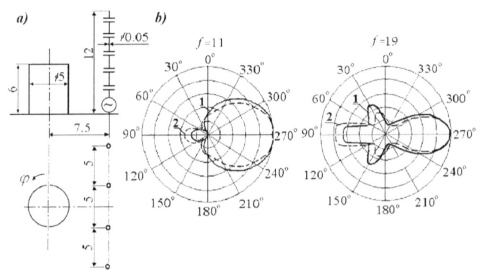

Figure 5.18 A linear array near a superstructure (*a*) and its pattern in the horizontal plane (*b*).

6

Synthesis of
Directional Radiators

6.1 THE SHAPE OF A CURVILINEAR RADIATOR WITH
MAXIMUM DIRECTIVITY

In the previous chapter the synthesis problem of a straight radiator with concentrated loads was regarded. Together with loads the radiator shape substantially affects the antenna's characteristics, in particular its directivity.

One must refine the considered problem. It is the optimization of the shape of the thin radiator with the aim of obtaining maximal signal in the predetermined direction. Unlike the previous chapter the problem is solved for a single frequency (or for a single electrical length of the radiator). A thin curvilinear radiator of an arbitrary geometry is situated in a lossless medium in the single vertical plane, e.g. in the plane zOy of rectangular coordinate system (Figure 6.1a). For a certainty directivity is calculated in the y-direction. Selecting the radiator shape is limited by the necessity to exclude super directivity in order to decrease the reactance of the antenna. The reason for such restriction is negative properties of the super directive antennas, which impede the realization of small-sized antennas of such kind.

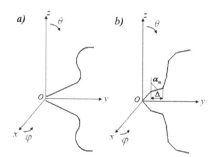

Figure 6.1 Symmetrical radiator of an arbitrary geometry in the shape of a curve (a) and a broken (b) line.

A rigorous analysis of the problem of the super directivity is given in [49]. This analysis studies an antenna in a shape of a sphere (Figure 6.2), in which only surface currents exist. It is assumed that the field has a circular symmetry relative the axis z, as well as it is symmetrical with respect to the equatorial plane. This simplification means a radiation concentration in the E-plane and a circular directional pattern in the H-plane. Simplification of the solved problem does not interfere to make the most general conclusions about the features of small antennas with high directivity.

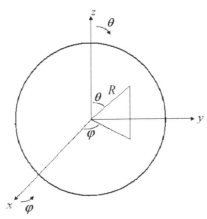

Figure 6.2 An antenna in the shape of a sphere.

Under these conditions the electromagnetic field in a spherical coordinates system has only three components, which are not equal to zero: E_R, E_θ and H_φ. Maxwell's equations take the form:

$$\frac{1}{R\sin\theta}\frac{\partial(H_\varphi\sin\theta)}{\partial\theta}=j\omega\varepsilon_0 E_R,\; -\frac{1}{R}\frac{\partial(RH_\varphi)}{\partial R}=j\bar\omega\varepsilon_0 E_\theta,\; \frac{1}{R}\left[\frac{\partial(RE_\theta)}{\partial R}-\frac{\partial E_R}{\partial\theta}\right]=-j\omega\mu_0 H_\varphi.\quad (6.1)$$

Substituting E_R and E_θ from the first two equations in the third equation and introducing a potential function $U : H_\varphi = \dfrac{1}{R}\dfrac{\partial U}{\partial\theta}$, we obtain an equation

$$\frac{\partial^2 U}{\partial R^2}+\frac{1}{R^2\sin\theta}\frac{\partial}{\partial\theta}\left(\sin\theta\frac{\partial U}{\partial\theta}\right)+k^2 U=0 \qquad (6.2)$$

Applying the method of separating variables (the eigenfunction method), in accordance with which

$$U=\mathfrak{R}(R)\Theta(\theta), \qquad (6.3)$$

we come as a result to the following two equations:

$$\frac{d^2\mathfrak{R}}{dR^2}+\left[k^2+\frac{n(n+1)}{R^2}\right]R=0,\quad \frac{1}{\sin\theta}\frac{d}{d\theta}\left(\sin\theta\frac{d\Theta}{d\theta}\right)+n(n+1)\Theta=0. \qquad (6.4)$$

They have particular solutions, respectively, in the form of Hankel functions of the second kind of order integer plus one half $\mathfrak{R}(R)=C_n\sqrt{R}H^{(2)}_{n+1/2}(kR)$ and of the Legendre polynomials $\Theta(\theta)=P_n(\cos\theta)$. Thus

i.e.

$$U = \sqrt{R} \sum_{n=0}^{\infty} C_n H_{n+1/2}^{(2)}(kR) P_n(\cos\theta), \tag{6.5}$$

In his turn

$$H_\varphi = \frac{1}{\sqrt{R}} \sum_{n=1}^{\infty} C_n H_{n+1/2}^{(2)}(kR) \frac{dP_n(\cos\theta)}{d\theta}. \tag{6.6}$$

$$E_\theta = \frac{1}{\omega\varepsilon_0 R^{3/2}} \sum_{n=1}^{\infty} C_n \frac{d}{dR}\left[R H_{n+1/2}^{(2)}(kR) \right] \frac{dP_n(\cos\theta)}{d\theta}. \tag{6.7}$$

In order to satisfy the condition of the field symmetry with respect to the equatorial plane, the indices n in these expressions must take on only the odd values.

If to replace at large distances from the radiator, i.e. at large values of kR, the Hankel functions with their asymptotic magnitudes, we may write for the far field

$$H_\varphi = \sqrt{\frac{2}{\pi}} \frac{e^{-jkR}}{R} \sum_{n=1}^{N} (-1)^{\frac{n+1}{2}} C_n \frac{dP_n(\cos\theta)}{d\theta}, \quad E_\theta = \sqrt{\mu_0/\varepsilon_0} H_\varphi, \tag{6.8}$$

where $N \leq kR/2$. Substitution of the fields into the expression for the directivity

$$D = \frac{4\pi R^2 \left[\vec{E}, \vec{H}\right]_{n\,max}}{\oint_S \left[\vec{E}, \vec{H}\right]_n dS} \tag{6.9}$$

and calculation of the coefficients corresponding to the case of the maximal directivity allows to find

where

$$D_{max} = 2 \sum_{n=1}^{N} \left[\bar{P}_n^{(1)}(0) \right]^2, \tag{6.10}$$

$$\bar{P}_n^{(1)}(\cos\theta) = \sqrt{\frac{2n+1}{2n(n+1)}} P_n^{(1)}(\cos\theta), \quad P_n^{(1)}(\cos\theta) = \frac{dP_n(\cos\theta)}{d\theta}.$$

From (6.10) it follows that the directivity increases with the increasing number of the series terms and does not depend on the antenna dimensions. The total current on the sphere surface ($R = a$) is equal to

$$J_\theta = H_\varphi \big|_{R=0} = 2\pi a J_\theta \sin\theta =$$

$$2\pi\sqrt{a} E_{max} \sin\theta \sum_{n=1}^{N} \sqrt{\frac{2n+1}{2n(n+1)}} \frac{(-1)^{n/2} \bar{P}_n^{(1)}(0)}{\sum_{n=1}^{N}\left[\bar{P}_n^{(1)}(0)\right]^2} H_{n+1/2}^{(2)}(ka) P_n^{(1)}(\cos\theta). \tag{6.11}$$

A graph of the function $P_n^{(1)}(\cos\theta)$ for some values of n is given in Figure 6.3. The values of the functions $H_{n+1/2}^{(2)}(ka)$ for the same n are shown in Table 6.1. As it is seen from Figure 6.3, the individual components of the series (6.11) are the alternate short segments of the currents with the different phases, wherein the number of these segments is equal to the index n. Obviously, that the more number of segments with currents of different phases, the less the radiated field for the same total current. Also obviously that the more often the currents alternate, the greater currents, creating the field of sufficient magnitude.

That is confirmed by Table 6.1. At the same value of ka it is seen that the greater n, the greater the absolute value of the current of each segment. And the current turns out almost purely reactive.

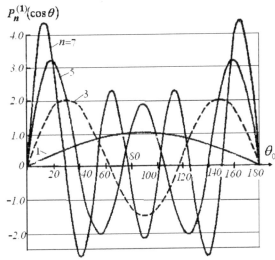

Figure 6.3 The graph of function $P_n^{(1)}(\cos\theta)$.

Performed analysis shows that, in principle, a high directivity can be obtained by means of the relatively small-sized antenna. But this antenna must necessarily be variably-phase antenna. With growing directivity the number of alternations increases and anti-phase segments become shorter. Great reactive currents are inherent to this antenna. These currents must be distributed on individual segments with high precision. Great reactive currents are associated with high quality (Q) and low stability of a system and lead to low efficiency and a narrow bandwidth. Such systems require the use of powerful transmitters. In consequence of these negative characteristics, designing and making super directive antennas of small dimensions become inexpedient.

Table 6.1 Values of Hankel Function $H_{n+1/2}^{(2)}(ka)$.

n	$ka = 1$	$ka = 2$	$ka = 3$	$ka = 4$
1	0.2403+j0.4311	0.3474+j0.2797	0.2758–j0.0502	0.0926–j0.1836
3	0.0072+j0.8764	0.0527+j1.1843	0.1213+j0.4054	0.1829+j0.1744
5	0.0001+j797.44	0.0021+j14.834	0.0131+j1.7929	0.0413+j0.5288
7	0.0000+j1.12·10^5	0.0001+j416.15	0.0006+j2.4352	0.0040+j3.1818
9	0.0000+j2.83·10^7	0.0000+j35183	0.0000+j609.72	0.0001+j42.676

Analogous results are given in [50, 51].

We return to the optimization of a radiator with a flat curvilinear profile [52]. In accordance with (6.9)

$$D = \frac{4\pi\left|\vec{F}(\theta_0,\varphi_0)\right|^2}{\int\limits_0^{2\pi}\int\limits_0^{\pi}\left|\vec{F}(\theta,\varphi)\right|^2\sin\theta\,d\theta\,d\varphi}, \qquad (6.12)$$

where $\vec{F}(\theta, \varphi)$ is a vector directional pattern of the radiator, and θ_0 and φ_0 are angular coordinates defining the direction of maximal radiation. For the radiator located in the plane zOy components of the current in spherical coordinates system are

$$J_x(l) = 0, \quad J_y(l) = J(l)\sin\alpha, \quad J_z(l) = J(l)\cos\alpha. \tag{6.13}$$

Here l is the coordinate along the radiator, α is a value of an angle θ in the radiator point with the coordinate l. Components of the radiator pattern with arm length equal to L are

$$F_y(\theta,\varphi) = \int_{-L}^{L} J(l)\sin\alpha \exp\left[jk\left(y\sin\theta\sin\varphi + z\cos\theta\right)\right]dl, \tag{6.14}$$

$$F_z(\theta,\varphi) = \int_{-L}^{L} J(l)\cos\alpha \exp\left[jk\left(y\sin\theta\sin\varphi + z\cos\theta\right)\right]dl. $$

Placing the radiator in a vertical plane allows to simplify the procedure of the calculation and optimization of the directivity. One can consider that the antenna directivity is equal to the product

$$D = D_\varphi D_{\theta'} \tag{6.15}$$

where

$$D_\varphi = 1, D_\theta = \frac{2|F_\theta(\theta_0,\varphi_0)|^2}{\int_0^\pi |F_\theta(\theta,\varphi_0)|^2 \sin\theta d\theta}.$$

In order to exclude the effect of the super directivity, the directivity is divided into reactivity coefficient:

$$D_1 = \frac{D}{\Gamma} = \frac{|F_\theta(\theta_0,\varphi_0)|^2}{2\pi\int_0^L |J(l)|^2 dl}. \tag{6.16}$$

Minimization of the functional allows to find the shape of the curve, provided the maximal directivity for a given arm length L. The mentioned minimization is accomplished by two ways. The first way was used in [52]. The radiator arm was replaced by a broken line consisting of M straight segments with equal length $\Delta = L/M$ (see Figure 6.1b). The directional pattern of this radiator is calculated by the formula

$$F_\theta(\theta,\varphi) = \sum_{m=1}^{M}\left\{\exp\left[jk\Delta\sum_{i=1}^{m-1}\cos(\alpha_i - \theta)\right]\cdot\cos(\alpha_m - \theta)\times\int_0^\Delta J(l_m)\exp\left[jkl_m\cos(\alpha_m - \theta)\right]dl_m\right\}. \tag{6.17}$$

Here the integral is replaced by the sum of the integrals along the segments, and l_m is the coordinate along segment m. The first exponent determines the phase of the current at the segment beginning, and the second exponent—along the segment. Each integral over the segment contributes its share to the radiation. It is assumed that the current distribution along the entire wire is sinusoidal, i.e. on the segment m we obtain

$$J(l_m) = J_A \frac{\sin\left[k\Delta\left(M - m + 1 - l_m / \Delta\right)\right]}{\sin kl}.$$ (6.18)

where J_A is the current in the feed point of the radiator.

Calculation of the integrals allows obtaining a rather cumbersome formula for $F_\theta (\theta, \varphi)$. Actually it is a function of one variable—α. The selection of this function is performed by trial-and-error method. Optimal geometry for the arm of length 0.75λ, found when $M = 10$, is shown in Figure 6.4 (curve 1). The calculated magnitude of the directivity is equal to 6.2.

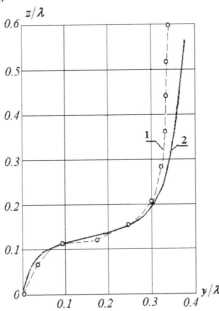

Figure 6.4 Optimum configuration of the arm with length 0.75λ, calculated by the first (curve 1) and the second (curve 2) way.

The second version of the calculation [53] did not confirm this result. Apparently, a significant error was caused by using sinusoidal distribution not along each segment, but along the entire broken line. In the second calculation, the equation for the current along the antenna has been written in matrix form. A curvilinear radiator was represented as a set of short dipoles connected in series. The segments forming the short dipoles are not considered to be straight, but selected in the form of the cubic splines. That permits to decrease the number of segments and to approximate more accurately the shape of the curve. The piecewise constant (pulsed) functions are used as the basis functions for the current, and this choice apparently reduces the calculation accuracy as compared with using piecewise sinusoidal functions. The shape of the antenna axis is determined depending on the directivity. The calculation was performed for the curvilinear antenna with arm length 0.75λ, divided into three segments with lengths of 0.0714λ, 0.4286λ and 0.25λ. The antenna radius is 2 mm. Optimum configuration of the arm is shown in Figure 6.4 (curve 2). The calculated magnitude of the directivity is equal to 4.9. The directional patterns in the vertical plane and the curve of the current distribution along the radiator are shown in Figure 6.5a and 6.5b. Curve 1 in Figure 6.5a corresponds to the directional pattern of the straight radiator with arm length 0.75λ, curve 2 corresponds to that of the curvilinear radiator with the same arm length.

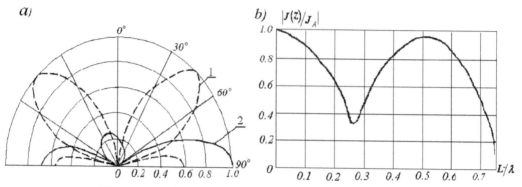

Figure 6.5 The directional patterns in the vertical plane (*a*)
and the curve of the current distribution along the radiator (*b*).

Let us compare this result with the directivity of the symmetrical straight vertical radiator. The directivity of the straight dipole with the arm length 0.75λ is close to 1. Maximal directivity of this dipole is obtained when the arm length is equal to 0.62λ. This directivity is equal to 3.2. Thus, the optimal directivity of curvilinear radiator with the arm length 0.75λ is significantly higher than the directivity of straight dipole with the same arm length and one and half times higher than the maximal directivity of the straight dipole. This result was obtained at the expense of reducing the vertical projection of the segments with negative current (of the segments, in which current flows in the opposite direction).

Indeed, comparing the versions of the straight and curvilinear dipoles, one can easily verify that the vertical projection of the curvilinear dipole with arm length 0.75λ is significantly decreased. From this point of view it is useful to compare the results of this section with the results of the Chapter 5, devoted shortening the electrical length of the radiator by incorporating capacitors. It should be emphasized that the reactance of capacitors has frequency-dependent character. This allows the creation of in-phase currents in antenna in a wide frequency band and thereby ensures the high characteristics, similar to a certain extent with characteristics in the area of series resonance of radiator. Curvilinear radiator of constant length and shape does not have such a frequency-dependent character, i.e. allows to get a positive result in one frequency only.

These critical remarks should not detract from the usefulness of the considered work. The idea of the optimization of antennas characteristics is important itself, irrespective of the parameters, selected for searching their optimal values. In accordance with the terminology, presented in Section 5.4, a set of changeable parameters is called a vector. A few options were considered as such vector: the magnitudes of concentrated loads, made in the form of simple elements (this option is described in Chapter 5), the shape of a curvilinear thin antenna (option presented in this Section), the dimensions of the straight rods—elements of an Yagi-Uda antenna (see Section 6.3). These variants are known, because they brought to a useful result. In addition, it is necessary to remember that optimizing the shape of a thin curvilinear radiator was accomplished many years ago, during the endless and futile debates about the possibility of practical using abstract theoretical distributions of the current along the wire.

At present, the shape selection can be executed at a higher level, in a frequency band instead of one frequency, with a sinusoidal current distribution along the each segment, with increased number of these segments. It is expedient also to consider the antenna wire in three-dimensional space.

6.2   CALCULATING DIRECTIVITY OF ANTENNA ARRAY ON THE BASIS OF THE MAIN DIRECTIONAL PATTERNS

As is well known, directivity D is one from basic electrical performances of any antenna. An antenna gain G is equal to

$$G = D\eta, \tag{6.19}$$

where η is an efficiency. Knowledge of these magnitudes allows planning the improvement of the antenna characteristics. The value of G can be defined by direct measurements. As regards magnitudes D and η, it is very difficult to measure them or to interpret the measurements' results. For example, for an evaluation D it is necessary to know the three-dimensional antenna patterns. But as a rule, these patterns are measured only in two main planes: horizontal and vertical. The calculation difficulties increase with decreasing the cross-section of the main lobe, i.e. with increasing directivity, caused for example by increased numbers of radiators in the antenna array.

Calculating the directivity of the narrow-beam antenna is described in [54]. It is based on the method of calculating the intermediate values in the directional pattern by means of using the measurements' results, which was proposed in [55].

Sometimes it is regarded that the magnitude of the antenna pattern in arbitrary direction is equal to

$$F(\theta,\ \varphi) = F_1(\theta)F_2(\theta), \tag{6.20}$$

where $F_1(\theta)$ is a pattern in a vertical plane xOz, $F_2(\varphi)$ is a pattern in a horizontal plane xOy, θ and φ are the angles in the spherical coordinates system, and x, y and z form a rectangular coordinates system (Figure 6.6). The calculations show, that the expression (6.20) is true only in a narrow area, limited by a main lobe of a directional pattern.

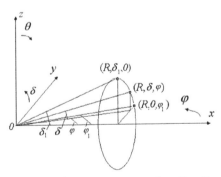

Figure 6.6   The coordinates system and a directional pattern.

The method proposed in [55] is based on revealing a curve, which is a locus of points with an identical signal. Here, the angle $\delta = \pi/2 - \theta$ is used instead of an angle θ. Respectively magnitudes of a directional pattern are equal to

$$f(\delta,\ \varphi) = f(\delta_1,\ 0) = f(0,\ \varphi_1), \tag{6.21}$$

where δ_1 and φ_1 are values of coordinates δ and φ in the points of intersections of a mentioned curve with planes xOz ($\varphi = 0$) and xOy ($\delta = 0$) respectively (see Figure 6.6).

Assume that the direction of the maximum radiation coincides with x-axis, and the pattern is symmetric about the planes xOz and xOy (Figure 6.7). For example, in-phase array, located in a plane yOz, has such directional pattern. In this case curves with identical directivity in the first approximation will be have the form of circles or ellipses:

$$f(\delta_1, 0) = f_1(\delta_1), \quad f(0, \varphi_1) = f_2(\varphi_1). \tag{6.22}$$

If to introduce for convenience a new angular coordinate β, measured from x-axis (see Figure 6.7), then as it is easy to show the new and the old coordinates will be connected among themselves by relation:

$$\beta = \cos^{-1}(\cos \delta \cos \varphi). \tag{6.23}$$

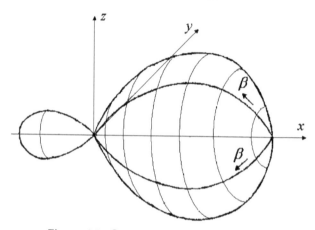

Figure 6.7 Symmetric directional pattern.

If the curves with identical directivity represent circles, i.e. if the main lobe of the three-dimensional pattern has circular symmetry, then

$$\delta_1 = \varphi_1 = \beta = \cos^{-1}(\cos\delta \cos\varphi). \tag{6.24}$$

In more common case these curves are ellipses. Let a_1 be length of its vertical axis, i.e. the arc length between the upper and the lower points of the pattern, corresponding to the given magnitude of a signal (to the given magnitude of the directional pattern). In the same way a_2 is the length of its horizontal axis, i.e. the arc length between the left and right points of the pattern corresponding to the given magnitude of a signal. Relation of lengths of vertical a_1 and horizontal a_2 axes is equal to $a = a_1/a_2$. Then at $a > 1$

$$\delta_1 = \cos^{-1}(\cos \delta\cos a\varphi), \quad \varphi_1 = \frac{1}{a}\cos^{-1}(\cos \delta\cos a\varphi), \tag{6.25}$$

and at $a > 1$

$$\delta_1 = a\cos^{-1}\left(\cos \frac{\delta}{a}\cos \varphi\right), \quad \varphi_1 = \cos^{-1}\left(\cos \frac{\delta}{a}\cos \varphi\right). \tag{6.26}$$

As is known, maximal directivity of antenna with the pattern, symmetrical about planes xOz and xOy, is equal to

$$D = \frac{\pi}{\displaystyle\int_0^\pi \int_0^{\pi/2} f(\delta,\varphi)\cos \delta d\delta d\varphi}, \tag{6.27}$$

whence

$$D = \cfrac{\pi}{\int\limits_0^{\pi/2}\int\limits_0^{\pi/2} f(\delta,\varphi)\cos\delta d\delta d\varphi + \int\limits_0^{\pi/2}\int\limits_0^{\pi/2} f(\delta,\psi)\cos\delta d\delta d\psi}. \qquad (6.28)$$

The first addend of a denominator corresponds to a forward half-space, the second addend—to a back half-space. Here in the second integral the change of variable $\varphi = \pi - \psi$ is performed. At $a < 1$ in accordance with (6.21), (6.22) and (6.25)

$$f(\delta,\varphi) = f_1(\delta_1) = f_1\left[\cos^{-1}(\cos\delta\cos a\varphi)\right]. \qquad (6.29)$$

At $a > 1$ according to (6.21), (6.22) and (6.26) one can obtain a similar expression. The expression (6.28) subject to (6.29) allows calculating the antenna directivity, if its directional patterns are given in the main planes.

In Figure 6.8 the experimental directional patterns of a planar uniform antenna array with in-phase excitation are given at the frequency 3.4 MHz. As one can see from the figure, the factor a is equal to 1 in intervals 160–180° and 135–145°, to 0,63 in an interval 145–160° and to 0,69 in an interval from 90° to 120° along an azimuthal angle. It means that the main lobe of the three-dimensional pattern has circular symmetry, i.e. the locus of points with identical signal, located on the main lobe, is a circumference. Such circular symmetry on some side lobes is absent, and that should be taken into account for increasing calculation accuracy. In an interval 120–135° the factor a is greater than 1. Calculation according to the described method at $a = 1$ gives a directivity value, equal to 18,6 dB. Calculation with allowance for $a > 1$ in the interval 120–135°gives an outcome 18.2 dB. The measured antenna gain is equal to 18 dB. Thus, one must admit a good conformance of calculated and experimental results.

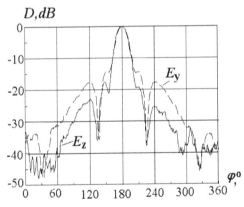

Figure 6.8 Experimental directional patterns of antenna.

For antennas with one narrow major lobe and small minor lobes, the theory (see, for example, [13]) recommends the expression

$$D = 41253/(\theta_1\theta_2), \qquad (6.30)$$

where θ_1 and θ_2 are half-power beam widths of the pattern (in degrees) in two mutually perpendicular planes. For planar arrays a better approximation is (see in [13])

$$D = 32400/(\theta_1\theta_2), \qquad (6.31)$$

The calculation in accordance with these formulas for the patterns, presented on Figure 6.8, gives accordingly 20.2 and 19.2 dB, i.e. a much greater error.

The program of directivity calculation used two procedures: the procedure of antenna pattern estimation at intermediate angles and the procedure of integrals calculation by summation of numerical masses. These methods can be used for the solution of other problems too, for example, for an estimation of increasing antenna directivity at the expense of decreasing side lobes.

If it is required to calculate, as far as the directivity will be changed due to diminution of a side lobes to level f_0, one must first determine the initial value of the directivity in accordance with (6.27) and next to reduce the side lobes, which exceed f_0 (for example in a vertical plane in the range of angles from φ_{11} to φ_{11} and in a horizontal plane in the range of angles from δ_{11} to δ_{11}) to level f_0 and to calculate the new directivity (at $a < 1$)

i.e.

$$D_1 = \pi \left\{ \int_0^\pi \int_0^{\pi/2} f(\varphi, \delta) \cos \delta d\varphi d\delta + \int_{\varphi_{11}}^{\varphi_{12}} \int_{\delta_{11}}^{\delta_{12}} \left[f_0 - f_1(\delta_1) \right] \cos \delta d\varphi d\delta \right\}^{-1}, \qquad (6.32)$$

$$\frac{D}{D_1} = 1 + D\left[\frac{f_0}{\pi}(\varphi_{12} - \varphi_{11})(\sin \delta_{12} - \sin \delta_{11}) - J \right], \qquad (6.33)$$

where

$$J = \frac{1}{\pi} \int_{\varphi_{11}}^{\varphi_{12}} \int_{\delta_{11}}^{\delta_{12}} f_1(\delta_1) \cos \delta d\delta d\varphi .$$

The program of the magnitude D calculation was performed in Matlab and presented in [54].

6.3   OPTIMIZATION OF THE DIRECTOR ANTENNAS

In this section we consider the problem of optimization of end-fire antenna array (of Yagi-Uda antenna). It is one of the most well-known directional antennas, in particular of *VHF* antennas. The standard version of this antenna, consisting of an active dipole, one or two reflectors and several equidistantly located directors of identical length is used for many years. Results of generalizing experimental data about these antennas and recommendations on selecting their geometric dimensions were given in [56]. The experimental data are showing that there is opportunity for significant increasing directivity, for expanding operating bandwidth and for lowering level of side lobes, if we select correctly the lengths of dipoles and the distances between them. But experimental optimization of such antennas is accomplished at the expense of long-term work and big money, since this optimization requires the creation of a reliable model with variable geometry and performance of numerous measurements of high quality.

Therefore, Yagi antenna, as already mentioned in Section 6.1, was among the first antennas, for which the optimization program has been specifically designed. Work on the program was brought to the real results.

The task of optimization of antenna arrays with passive dipoles as a rigorous task of mathematical programming was formulated in [57]. In [58] the task of creating end-fire array with maximal directivity was divided on two tasks: the choice of directors' lengths at the beginning and the choice of distances between directors after that. Most completely the problem was solved in [59], where independent selection of lengths and distances was used, and modern methods of mathematical programming were employed.

The antenna circuit is shown in Figure 6.9a, the general view of the standard antenna with vertical polarization is given in Figure 6.9b. In this case, the optimization problem has been formulated in the following way: it is required to determine the dipoles' lengths $2L_i$ (here i is the dipole number, L_i is the length of its arm) and their coordinates d_i along the axis of the antenna, providing the maximal directivity in the direction of the said axis for a given number N of the dipoles and restrictions on the total antenna length, on the dipoles' lengths, and on the distances between adjacent dipoles.

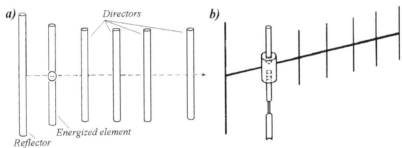

Figure 6.9   Circuit of Yagi antenna (a) and general view of a standard antenna with vertical polarization (b).

Comparison of this problem with definition of the problem of the mathematical programming in the general case (see Section 5.4) shows that the objective function is presented here in the form of a partial functional, i.e. in the form of the error function for a single characteristic (directivity). Vector $\tilde{x}$ of parameters is presented by magnitudes L_i and d_i. The objective function is written in the form

$$F(\tilde{x}) = \frac{1}{D(\tilde{x})} + \alpha F_p(\tilde{x} - \tilde{y}), \tag{6.34}$$

where $D(\tilde{x})$ is the directivity magnitude, $F_p(\tilde{x} - \tilde{y})$ is the penalty function, $\tilde{y}$ is the vector of allowable values of optimized parameters, and α is penalty factor.

The error function in this case is constructed using the criterion of Powell [60], which does not require the calculation of derivatives during changing direction of the vector $\tilde{x}$. New direction of this vector must be selected on the each iteration in accordance with the direction of maximal increasing directivity. When analyzing the antenna characteristics, it is considered that the antenna consists of the thin cylindrical wires, and the distribution of the currents along these wires obeys to the system of Hallen's equations:

$$\int_{-L_i}^{L_i} J_i(z_i) \sum_{j=1}^{N} G_{ij}\left(\left|z_j - z_i\right|\right) dz_i = -\frac{j}{Z_0}\left(C_i \cos kz_i + \frac{e_i}{2}\sin k|z_i|\right), i = 1, 2, \ldots N . \tag{6.35}$$

Here $J_i(z_i)$ and e_i are the functions of the current distribution and the extraneous emf on the dipole i (emf of the passive dipole is zero), $G_{ij} = \exp(-jkR_{ij})/(4\pi R_{ij})$ is Green's function, $R_{ij} = \sqrt{d_{ij}^2 + z_j - z_i)^2}$ is the distance between the points of the dipoles i and j, $Z_0 = 120\pi$ is the wave impedance of the free space. Comparing (6.35) and (2.5), it is easy to see that here Hallen's integral equations with approximation kernel is used.

The magnitude of the current is sought in the form of an expansion in spatial harmonics

$$J_i(z_i) = \sum_{m=1}^{M} I_{im} f_{im}(z_i),$$ (6.36)

where I_{im} is the complex amplitude of the harmonic m of the dipole i, and M is the number of harmonics, $f_{im}(z)$ are the power functions $f_{im}(z_i) = (1-|z_i|/L_i)^m$, proposed by Popovich [61], and trigonometric harmonics of King [62]:

$$f_{im}(z_i) = \begin{cases} \sin\left[k(L_u - |z_i|)\right], i = 1, \\ \cos\dfrac{kz_i}{m-1} - \cos\dfrac{kL_i}{m-1}, m \ge 2. \end{cases}$$

The functions $f_{im}(z)$ are used as spatial harmonics (the first harmonic is taken into consideration only on the excited dipole). The problem is reduced to the solution of the system of linear algebraic equations for the currents along the antenna segments. These equations are written in matrix form. Knowing the currents of the dipoles, one may find the field in the far region and the directivity in the direction of an antenna axis.

The results of the optimization of the director antennas are presented in [57] and [59]. It is the antennas with three, four and fifteen dipoles. Canonical array, described in [56], is taken as the first approximation for the array with fifteen dipoles. Selection of its geometrical dimensions in accordance with the procedure described above allows to double its maximal directivity. Initial (1) and optimum (2) directional patterns of the antenna are shown in Figure 6.10.

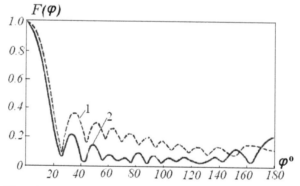

Figure 6.10   Initial (1) and optimum (2) directional patterns of the antenna with fifteen dipoles.

The initial dimensions of the antenna with four dipoles were taken in accordance with the dimensions of the standard receiving television antenna. Optimization of the antenna dimensions lead to directivity increase by 95%. Its directional patterns in the H-plane are shown in Figure 6.11. Here 1 is initial directional pattern, 2 and 3 are optimum directional patterns for three and four current harmonics respectively. Figure 6.11 shows that the use of the fourth harmonic improves the calculated directional pattern. However, a further increasing the number of harmonics does not affect the result.

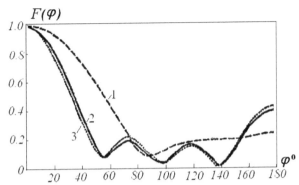

Figure 6.11   Directional patterns of the antenna with four dipoles: 1—initial pattern, 2—optimum pattern for three harmonics, 3—optimum pattern for four harmonics.

In the calculations of the antenna with three dipoles (Figure 6.12), two problems were included into objective function: increasing the directivity and the expansion of the operating bandwidth. Optimization allowed to increase the directivity in one and half times and to expand the bandwidth by 23%.

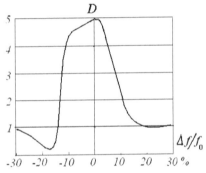

Figure 6.12   Frequency characteristic of antenna with three dipoles.

Comparison of the results, obtained in [58], with the results of optimization for the antenna with six dipoles, described in [59], shows the advantage of the method presented in [59]. The directivity of the last antenna is increased by 4% and the length of the antenna was decreased by 0.04λ.

During the ongoing research of the end-fire arrays, other characteristics of arrays were calculated together with directivity, in particular the input impedance of the excited dipole, the level of side lobes, the distribution of amplitudes and phases of the currents along the array. These characteristics can be measured on a real model, i.e. calculation can be compared with experimental results. The frequency dependence of these quantities are genuine characteristics of the antenna array, while, for example, the quality factor can be estimated only indirectly.

Results of the synthesis of the director arrays are presented in Table 6.2. Here N is the number of the dipoles, D_{max} is the maximum directivity, SLL is the level of side lobes, Z_{A1}, Z_A, and Z_{A2} are the input impedances at the lower, the main and the upper frequency of the range relatively, $\Delta f/f_0$ is the ratio of the bandwidth to the main frequency. The bandwidth is defined as the half-directivity bandwidth. Maximum lengths of the first five arrays are bounded by the magnitude 1.6λ, and lengths of the other arrays—by the magnitude 1.8λ.

Table 6.2 Characteristics of the End-Fire Arrays

N	D_{max}	SLL	Z_{A1}	Z_A	Z_{A2}	$\Delta f/f_0, \%$	Length
4	14.48	0.43	$16.2+j27.5$	$6.0+j91.2$	$9.4+j145.2$	$-5.3...+2.8$	0.76λ
5	18.77	0.36	$34.9-j74.5$	$4.1-j25.5$	$4.4-j3.2$	$-6.3...+1.8$	1.27λ
6	21.69	0.31	$26.2-j78.6$	$3.0-j36.1$	$2.1-j19.4$	$-7.3...+1.3$	1.6λ
7	24.43	0.31	$13.6-j16.6$	$3.6+j33.3$	$3.8+j65.0$	$-5.5...+2.0$	1.6λ
8	24.04	0.35	$28.7-j53.9$	$2.6-j3.5$	$2.3+j18.2$	$-7.0...+1.5$	1.6λ
6	24.39	0.32	$29.7-j9.7$	$6.9+j41.2$	$6.8+j73.4$	$-5.5...+1.5$	1.66λ
7	25.31	0.27	$6.9-j211.7$	$0.8-j180.6$	$1.9-j172.5$	$-7.5...+1.0$	1.8λ
8	25.19	0.28	$4.9-j214.2$	$0.7-j182.6$	$0.1-j173.7$	$-7.5...+1.5$	1.8λ

Calculations show that with increasing frequency the directivity changes faster than with decreasing, and the real part of input impedance is quite the opposite. When the number of dipoles is fixed, increasing antenna length causes the directivity to increase, until the length does not exceed a certain critical value (see the examples of arrays). When the length of an antenna is fixed, increasing the number of passive dipoles leads to the increase of the directivity, but not always. The frequency bandwidth of antennas is 7–9%, which is sufficient for most applications. Low input impedance makes difficult the matching of antennas with the standard cable. It is the main their drawback.

Table 6.3 Amplitudes and Phases of Dipoles Currentf

N	Number of dipoles							
	7				8			
	Array length							
	1.6λ		1.8λ		1.6λ		1.8λ	
	J_A	φ_A	J_A	φ_A	J_A	φ_A	J_A	φ_A
1	0.339	-41.7	38.25	-34.4	2.520	92.4	35.63	-34.6
2	0.534	124.0	53.18	126.4	1.926	-106.3	48.54	126.5
3	0.748	-69.1	65.79	-63.9	1.866	-101.7	62.41	-64.4
4	1.011	98.1	95.50	105.9	3.923	62.6	83.39	105.0
5	0.312	118.7	210.2	-85.6	1.502	99.8	33.58	-93.4
6	1.792	-83.7	332.2	89.7	12.55	-127.6	176.8	-84.8
7	1.128	92.2	31.02	-101.8	13.76	22.6	328.6	89.8
8	-	-	-	-	4.610	-132.6	25.5	-104.8

6.4 V-DIPOLE WITH CAPACITIVE LOADS

As already mentioned, one of the important tasks of antenna engineering is creating a radiator, which ensures in a wide frequency range a field maximum in the plane perpendicular to the radiator axis. An ordinary linear radiator fails to meet this requirement: if the radiator arm is larger than 0.7λ, the radiation in the plane, perpendicular to the antenna axis, decreases. In this case one can use V-antenna formed by two converging inclined wires. If arm length L is greater than 0.7λ, an ordinary V-antenna has preferential radiation along the bisector of the angular aperture. However, the side lobes of the directional pattern increase with growing frequency, and the main lobe of this pattern diminishes in the antenna plane. If the arm length is greater than about 1.25λ, the main lobe splits, and the radiation along the bisector sharply decreases.

Mounting of capacitive loads in the antenna wire allows to expand the frequency range, in which there are the directed radiation along the bisector of the angular aperture, and to increase the useful signal in this direction.

Consider a symmetric V-dipole with arm length L and arbitrary angular aperture $\alpha = \pi - 2\theta$ (Figure 6.13). The far field along the bisector of the angular aperture, which is created by an elementary segment $d\varsigma$ of the upper antenna arm with current $J(\varsigma)$, is

$$E_\theta(\varsigma)d\varsigma = E_\theta(0)[J(\varsigma)/J(0)]\exp(jk\varsigma \sin \theta_0)d\varsigma, \qquad (6.37)$$

where $E_\theta(0)d\varsigma$ is the field created by a segment of the upper arm located near point O with current $J(0)$, $k\varsigma \sin \theta_0$ is the path-length difference, ς is the coordinate counted off along the radiator axis. In order to the far fields of different segments of the upper arm coincide with each other in phase, the current distribution along this arm should conform to the expression

$$J(\varsigma) = J(0)f(\varsigma)\exp(-jk\varsigma \sin \theta_0), \qquad (6.38)$$

where $f(\varsigma)$ is a real and positive function.

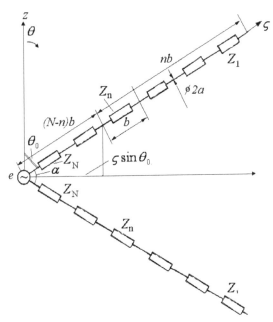

Figure 6.13   V-dipole with loads.

Let N loads Z_n be connected in each dipole arm, and they be connected uniformly along the antenna wire at a distance b from each other. If the load spacing is small ($kb \ll 1$), then, as well as for the linear dipole, the replacement of concentrated loads by distributed surface impedance $Z(\varsigma)$ practically does not change the current distribution along the antenna. We assume that the surface impedance of each antenna segment with load Z_n is constant and equal to $Z^{(n)}$.

As mentioned in Section 5.2, the current distribution along the antenna with piecewise constant surface impedance coincides in the first approximation with current distribution along an equivalent impedance line, i.e. along the line with stepwise variation of propagation constant (see Figure 5.4). Here, the wave propagation constant γ_n along the segment n is related to surface impedance $Z^{(n)}$ in accordance with (5.17). If the variation law of propagation constant is known, one can use magnitude γ_n for calculating with the help of (5.29) concentrated loads Z_n, which are needed for the embodiment of this law.

The current along the segment n of a stepped line is

$$J(\varsigma_n) = I_n \sinh(\gamma_n \varsigma_n + \varphi_n), \ 0 \le \varsigma_n \le b, \tag{6.39}$$

where I_n and φ_n are the current amplitude and phase at the segment n, respectively, and ς_n is the coordinate, counted off from the segment end, i.e. $\varsigma_n = (N - n + 1)b - \varsigma$. We equate current $J(\varsigma_n)$ at the beginning and the end of each segment to current $J(\varsigma)$, ensuring the phase coincidence of far fields from all dipole segments. The current inside each segment does not coincide with current $J(\varsigma)$. However, if the segments lengths are small, the current distribution along the line is close to $J(\varsigma)$.

According to (6.38) and (6.39), at $\varsigma_n = b$ and $\varsigma_n = 0$:

$$I_n \sinh(\gamma_n b + \varphi_n) = J(0)f[(N - n)b]\exp[-jk(N - n)b \sin\theta_0],$$

$$I_n \sinh\varphi_n = J(0)f[(N - n + 1)b]\exp[-jk(N - n + 1)b \sin\theta_0].$$

If to divide the left and right sides of the first equation onto the corresponding sides of the second equation, then considering that b is a small value and retaining only the first terms of series expansions for trigonometric functions of small arguments, we get

$$\tanh\varphi_n = \gamma_n b \Big/ \left\{ \frac{f[(N-b)]}{f[(N-n+1)b]}(1 + jkb \sin\theta_0) - 1 \right\}. \tag{6.40}$$

Similarly, to (6.40), for the segment $(n + 1)$

$$\tanh\varphi_{n+1} = \gamma_{n+1} b \Big/ \left\{ \frac{f[(N-n-1)b]}{f[(N-n)b]}(1 + jkb \sin\theta_0) - 1 \right\}. \tag{6.41}$$

Voltage and current are continuous along a stepped line. Therefore, (5.22) is true. Together with (6.40) and (6.41), it forms a set of equations that allows to relate γ_n and γ_{n+1}. From its solution it follows that magnitude γ_n is independent of γ_{n+1}:

$$\gamma_n = \frac{1}{b}\sqrt{1 - \frac{2f[(N-n)b - f[(N-n-1)b]]}{f[(N-n+1)b]} - 2jkb \sin\theta_0 \frac{f[(N-n)b] - f[(N-n-1)b]}{f[(N-n+1)b]}}. \tag{6.42}$$

This expression generalizes (5.23) for a linear dipole and transforms into it at $\theta_0 = 0$.

The possibility of implementation of propagation constant γ_n is determined by the possibility of manufacturing concentrated loads. According to (5.29), at low frequencies, when inequality (5.32) and equality (5.33), which follows from (5.32), are true, the load value is

$$Z_n = -j30(\gamma_n b)^2 / (kb\chi). \tag{6.43}$$

By substituting (6.42) into (6.43), we get

$$Z_n = R_n + 1/(j\omega C_n), \tag{6.44}$$

where

$$R_n = \frac{60}{\chi} \sin \theta_0 \frac{f[(N-n-1)b]-f[(N-n)b]}{f[(N-n+1)b]},$$

$$C_n = 4\pi\varepsilon_0 b\chi / \left\{ 1 - \frac{2f[(N-n)b]-f[(N-n-1)b]}{f[(N-n+1)b]} \right\}.$$

As seen from (6.44), each load should be a series connection of a resistor and a capacitor, where the resistance of the resistor is positive, if function $f(\varsigma)$ decreases monotonically with growing ς, and the capacitance of the capacitor is positive, if function $f(\varsigma)$ is concave. Here, the resistance depends on the angular aperture of the antenna and the form of function $f(\varsigma)$, whereas the capacitance depends only on the latter.

For a linear radiator with loads ensuring the maximal radiation in the plane, perpendicular to its axis, each load should, when condition (5.29) holds, represent a capacitor. Capacitors ensure real wave propagation constant γ_n and an in-phase current distribution along an antenna. For a V-dipole, the resistor is to be connected in series with the capacitor, and that will result in a phase delay of the current wave along an antenna wire. Such phase delay is necessary for a V-dipole, since it compensates the path-length difference from individual dipole segments to an observation point and ensures coincidence of phases of the far fields, created by segments along the bisector of the angular aperture.

The use of resistors in a transmitting antenna is inexpedient. This means that the loads of a V-dipole should not differ from the loads of a linear radiator, which ensure an in-phase current distribution along a wire.

At high frequencies, when condition (5.29) does not perform, the in-phase current distribution along a linear radiator takes place, if the load represents a negative inductance (a capacitance, which is inversely proportional to square of frequency). Similarly, the load for a V-dipole should be a series connection of a capacitor with a frequency-dependent capacitance and a resistor. In order for propagation constant to be real and the current distribution along an antenna may be in-phase, the capacitances should not exceed the value determined by inequality (5.40).

As an example, we shall consider V-dipole with arm length $L = 1.5$ m and radius 0.025 m. Fifteen capacitors are connected in each arm with spacing 0.1 m between each other (the first and last capacitors are placed at the distant 0.05 m from the end and center of the antenna). The capacitances of the capacitor nearest to the generator is equal to 33.5 pF in order that the propagation constant remains real at frequencies up to 100 MHz. The capacitances of other capacitors decrease to the antenna end according to the linear law. As shown in Section 5.2, in this case one can ensure the distribution of current along an antenna, close to linear distribution, and high level of matching with a cable.

Figure 6.14a shows the directivity of a V-dipole with capacitive loads (curve 1) and without loads (curve 2) along the bisector of angular aperture (the angle between dipole arms is equal to 90°). For the sake of comparison, Figure 6.14b shows similar curves for a linear dipole (curve 3 with loads, curve 4 without loads). The values of loads and the antenna arm length are the same. The calculations are performed in a frequency range from 100 to 500 MHz.

As can be seen from the figure, the directivity of a linear dipole without loads in the direction, perpendicular to its axis, quickly decreases at $L \approx (0.6-0.7)\lambda$. For a linear dipole with loads, this threshold value is found at $L \approx (1-1.2)\lambda$. V-dipoles, especially with loads, allow achieving a high directivity along the bisector of the angular aperture in a substantially wider frequency range: at frequencies from 350 ($L = 1.75\lambda$) to 500 MHz ($L = 2.5\lambda$). The loads increase the directivity of a V-dipole by a factor between 1.4 and 2.8.

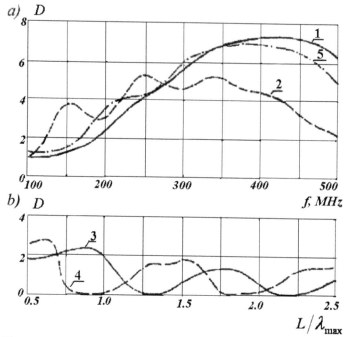

Figure 6.14 (*a*) Directivity of a V-dipole and (*b*) of a linear dipole.

Figures 6.15 and 6.16 give the directional patterns of the described V-dipole with loads (curve 1) and without loads (curve 2) in the plane of antenna and the plane perpendicular to it. The calculated curves are compared with the results of the experiment (circles).

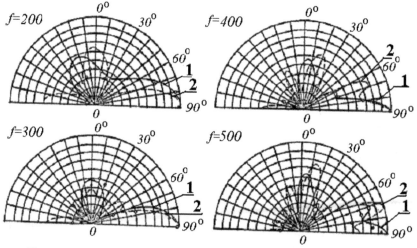

Figure 6.15 Directional patterns of a V-dipole in the antenna plane.

As shown in Section 5.4, electrical characteristics of an antenna with loads can be improved by optimal selection of the latter with the help of the mathematical programming method. Here, in order to calculate the zero approximation, it is expedient to use the method of a stepped (piecewise uniform) impedance transmission line. In the considered example, the mathematical programming method allows raising the directivity level of V-dipole with loads in the lower part of a frequency range (Figure 6.14*a*, curve 5).

V-dipole with capacitive loads can be used as a directional antenna of *VHF* range.

Additional possibilities associated with the use of V-antenna with loads and curvilinear shape of the wires. As is shown in Chapter 5, the connection of capacitive loads in the wires allows to change the current distribution along the radiator and to obtain owing to optimal choice of loads the electrical characteristics closest to the required characteristics as much as possible. In Section 6.1 it was demonstrated that by changing the shape of the radiator one can increase its directivity at a given frequency. Actually, both methods provide an additional degree of freedom when choosing parameters of the antenna with the aim of optimizing its characteristics. As is showed in the current section, the use of capacitive loads in the wires of V-antenna allows to increase directivity and to expand the frequency range. Use of a curvilinear shape of wires gives additional possibilities.

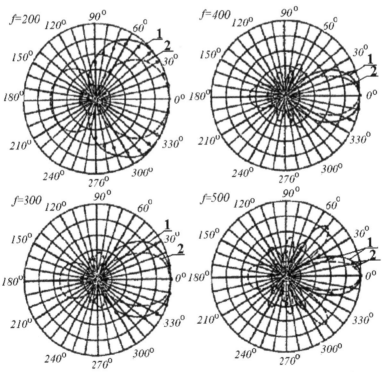

Figure 6.16 Directional patterns of a V-dipole in the plane perpendicular to the antenna plane.

Circuit of a V-radiator with rectilinear arms of length L and an angular aperture $\alpha = \pi - 2\theta_0$ is given in Figure 6.13. Capacitive loads are uniformly located in the radiator wires. Figure 6.17 presents the promising circuit of the curvilinear arm of the V-radiator with loads.

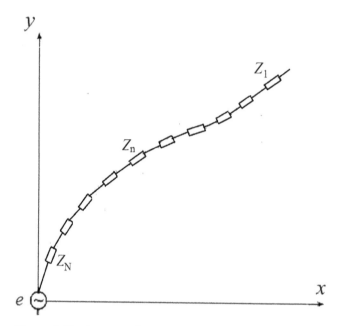

Figure 6.17 An arm of a curvilinear *V*-radiator with loads.

7

Method of a Complex
Potential for Cylindrical Problems

7.1 METHOD OF A COMPLEX POTENTIAL FOR
CALCULATING FIELDS

Problem of calculating an electrostatic field is the problem of determining a field strength at all points in accordance with the electric charges of the conducting bodies. This problem may be completely solved if we are to find the potential as a function of the coordinates. If the potential in each point is known, in the case of two conducting bodies one can find a capacity between them (as the ratio of charge of one body to difference of their potentials). In a system of few conducting bodies one can determine self- and mutual potential coefficients and after that partial capacitances.

If any problem of calculating an electrostatic field of charged conducting bodies located in homogeneous and isotropic dielectric is solved, then its solution can be used for a solution of other problems. For example, the method of electrostatic analogy allows to find constant electric fields and currents in a homogeneous weakly conducting medium, if the shape and geometrical dimensions of placed in it bodies with a high conductivity coincides with a shape and dimensions of bodies placed in a dielectric. The principle of correspondence allows building a picture of magnetic field created by constant linear currents on the base of a picture of electric field created by linear charges, if currents and charges are distributed in a space identically. The only difference between these pictures consist in a fact that lines of equal magnetic potential are located on a place of electric field lines and magnetic field lines are located on a place of lines of equal electric potential [63].

Electrostatic problem considers usually a system of charged conducting bodies surrounded with a dielectric, in which volumetric charges are absent. Either potential U_n of each body n or total charges q_n of bodies are given. A distribution of a potential in a space is unknown. A charge distribution along the surface of each body is also unknown. It is the main difficulty of a problem, especially in the case of inhomogeneous medium.

If a distribution of charges is known, then in a homogeneous medium with permittivity ε it is possible to find the potential U, which is defined by all charges located in a finite volume v of a space from the expression

$$U = \frac{1}{4\pi\varepsilon} \int\limits_{(v)} \frac{\rho_v dv}{r}, \tag{7.1}$$

where ρ_v is a volume density of a charge, r is a distance from an observation point to an integration point (point of charge location). This integral is solution of Poisson equation.

If a distribution of charges is unknown, then one must ascertain necessary and sufficient requirements, under execution of which a field is determined by only one way (uniqueness theorem). It has following requirements. First, in the case of a homogeneous medium, since charges are absent in it, a field must satisfy the Laplace's equation, i.e. in the rectangular coordinates system (x, y, z)

$$\frac{\partial^2 U}{\partial x^2} + \frac{\partial^2 U}{\partial y^2} + \frac{\partial^2 U}{\partial z^2} = 0. \tag{7.2}$$

Second, the surfaces of conducting bodies must be surfaces of equal potential, i.e. on the each surface $U = U_n = $ const. Third, if total charges q_n of bodies are given, then for each body the condition must be fulfilled

$$q_n = -\int\limits_{(S_m)} \varepsilon \frac{\partial U}{\partial n} dS, \tag{7.3}$$

where S_m is the surface of the body m; n is the normal to that surface.

The problem of calculating the electric field of charged bodies gets simplified, if all the magnitudes characterizing the field depend only on two coordinates. The field of several infinitely long and parallel to each other cylindrical wires with the charges uniformly distributed along their length (the plane-parallel field) conforms to such condition, all lines of the field and also lines of the equal potential lie in the planes

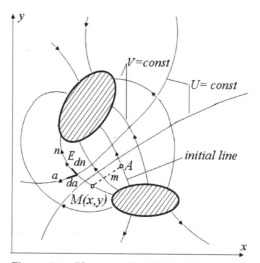

Figure 7.1 Plane-parallel field of two wires.

of cross-sections, and the field picture is the same in all such planes. If axis z of the rectangular coordinates system is parallel to the wire's axes, the potential U of the plane-parallel field is a function of only two coordinates, x and y.

The cross-sections of two wires and the picture of a field around them, i.e. lines of the equal potential and lines of the field are represented in Figure 7.1. The surfaces of the equal potential are the cylindrical surfaces with generatrix, parallel to the axis z. The lines of the equal potential in the plane xOy are defined by equations of the type

$$U(x,y)=\text{const.} \tag{7.4}$$

Let the some line of a field is regarded in the capacity of initial line. If we shall connect an arbitrary point $M(x.y)$ with any point A of the initial line, we obtain the curvilinear segment MmA. This segment, when it moves in parallel to it in direction of the axis z and covers the distance l, describes a certain surface. Let Ψ_E is the flux of the electrical field $\vec{E}$ through this surface. The function

$$V(x,y) = \Psi_E / l, \tag{7.5}$$

is the flux per unit length of wires. It is called by flux function. The magnitude $V(x, y)$ has the same value for all points located along the same line of field. That is why the equation

$$V(x,y)=\text{const}, \tag{7.6}$$

which defines the aggregate of such points, is called by the equation of the field line.

Note that equations (7.4) and (7.6) define two families of curves intersecting at right angle. Let dn be an element of the length of the field line and da an element of the length of the line with equal potential, i.e. dn and da are mutually perpendicular. We assume that coordinates n and a grow in directions shown in Figure 7.1. The potential U increases in direction opposite to the vector $\vec{E}$, i.e. in the direction of decreasing the coordinate n. It is considered customary that the function of flux increases in the same direction, in which a increases. Under these conditions the electric field strength may be expressed in terms of U and V in the form of

$$E = -\frac{\partial U}{\partial n} = \frac{\partial V}{\partial a}, \tag{7.7}$$

or, in Cartesian coordinates,

$$E_x = -\frac{\partial U}{\partial x} = \frac{\partial V}{\partial y}, \quad E_y = -\frac{\partial U}{\partial y} = -\frac{\partial V}{\partial x}, \tag{7.8}$$

from whence after repeat differentiation it is easy to be convinced that both functions U and V satisfy the Laplace's equation.

Thus, the functions U and V satisfying the equations (7.8) correspond to the first requirement of above-mentioned theorem of uniqueness.

We shall regard the cross-section plane of the plane-parallel electrostatic field like the plane of the complex variable $z = x + jy$. If $\varsigma(z) = \xi(x, y) + j\eta(x, y)$ is a regular analytical function of the complex variable z, then in accordance with conditions of Koshi-Riman

$$\frac{\partial \eta}{\partial x} = -\frac{\partial \xi}{\partial y} \frac{\partial \eta}{\partial y} = \frac{\partial \xi}{\partial x}. \tag{7.9}$$

It is easy to be convinced that the equalities (7.9) are similar to the equalities (7.8), i.e. one can accept that $\xi = V$, $\eta = U$ and use the regular analytical function

$$\varsigma(z) = V(x, y) + jU(x, y), \tag{7.10}$$

which is called by the complex potential of the field.

As it follows from the foregoing, if $\varsigma(z)$ is the regular analytical function of the complex variable z in some area, then its real and imaginary part and consequently the whole function satisfy the Laplace's equation and accordingly they satisfy the first requirement of the uniqueness theorem. This means that the problem of a field calculation is solved, if the function $\varsigma(z)$ satisfying the boundary conditions at the wires surface is found.

The detailed description of the method of complex potential and examples of its use in solving problems about parallel cylindrical wires located in a homogeneous medium are given in [63]. In particular there it is shown that for the solitary wire with a circular cross-section and with the charge τ per unit length located in the homogeneous medium with permittivity ε, its complex potential has the form

$$\varsigma(z) = -j\frac{\tau}{2\pi\varepsilon}\ln z + C, \tag{7.11}$$

where $C = C_1 + jC_2$ is the constant magnitude, coefficients C_1 and C_2 of which depend on the selecting the initial field line and the line of zero potential.

The complex potential of the line consisting of two located at the distance $2b$ from each other wires, the linear charge densities of which are accordingly τ and $-\tau$, is equal to

$$\varsigma(z) = -j\frac{\tau}{2\pi\varepsilon}\ln\frac{z+b}{z-b} + C. \tag{7.12}$$

In this and follow chapters the method of complex potential is generalized, firstly, for piecewise homogeneous media and, secondly, for three-dimensional structures.

7.2 APPLICATION OF A METHOD OF COMPLEX POTENTIAL TO PIECEWISE HOMOGENEOUS MEDIA

The problem of a field calculation becomes complicated in the case of the heterogeneous (in particular, piecewise-homogeneous) medium [64]. Such problem arises, if the solitary wire is located at the boundaries of the two media, for example, air and dielectric, and the relative permittivity ε_r of dielectric is different from 1 (Figure 7.2). The wire cross-section is considered as a circle.

It should be noted that the potential at the wire surface is constant. The potential of the infinitely far point may not be equal to zero, since in that case the potential of infinitely long wires at the all finite distances will be infinitely great. Nevertheless it is obvious that at the great distances from charged wire the potential in the all directions is the same. And therefore suppositions that the lines of equal potential are circumferences with the center in the origin of coordinates and that the surfaces of equal potential are the surfaces of the circular cylinders are natural.

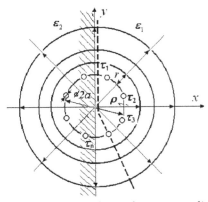

Figure 7.2 Field of a wire located at two media boundary.

The wire surface may be replaced by a system of N equidistant filaments with linear densities τ_n of charges (n is the filament number), the sum of which is equal to τ (to the linear density of wire charge). The potential of each filament at distance r from its axis is

$$U_n = -(\tau_n/2\pi\varepsilon_n)\ln\ (r/r_p). \tag{7.13}$$

Here, ε_n is the permittivity of a medium around the filament, r_p is the distance to the surface of zero potential. In order that the potentials of all filaments may be identical on the surface of circular cylinder with the radius $\rho = r + a$, where a is the wire radius, the ratio τ_n/ε_n must be the same for all the n:

$$\tau_n/\varepsilon_n=\text{const}(n), \tag{7.14}$$

i.e. the surface density of the wire charge must be proportional to the permittivity of a medium adjacent to a given part of the surface. Here the word "adjacent" signifies that the medium occupies volume from the wire surface to infinity within an angle equal to the arc length of the surface part, which is adjacent to this medium.

Formula (7.11) describing a field structure of the solitary wire with the circular cross-section, which is located in homogeneous medium, may be written in the form of a more general expression:

$$\varsigma(z) = Aj \ln z + C, \tag{7.15}$$

where A is constant magnitude. Using designation $z = \rho e^{j\varphi}$, we obtain:

$$V(z) = -A\varphi + C_1,\ U(z) = A \ln \rho + C_2, \tag{7.16}$$

from whence the equations of field lines and of equal potential lines take the form accordingly:

$$\varphi = \text{const},\ \rho = \text{const}. \tag{7.17}$$

The field structure of the wire located at the boundary of the two media has a similar character. In particular, as it is indicated earlier, here the lines of equal potential are circumferences with $\rho = \text{const}$. Therefore for the complex potential of such field we shall use the expression (7.15).

The constant A in this expression is determined in accord with the fact that in going around the cross-section of the wire along the shorted contour, the angle φ increases by 2π, and the function V increases by Ψ_E/l, where Ψ_E is the flux of the vector $\vec{E}$ through the cylindrical surface covering the wire segment of length l. In the case of the heterogeneous medium it is expedient to replace Ψ_E by Ψ_D, i.e. to replace the flux of vector $\vec{E}$ by the flux of vector $\vec{D}$, where $\vec{D}$ is the vector of electrical displacement. Besides, one must take into account that the strength of this flux in the different media may be diverse, i.e.

$$\Delta V = \frac{1}{l}\sum_i \Psi_{Di} / \varepsilon_i, \tag{7.18}$$

where i is the medium number.

If ρ_v is a volume density of a wire charge, then integration of Maxwell equation

$$div\vec{D} = \rho_V, \tag{7.19}$$

with respect to wire volume v gives for left part of equation

$$\int_{(v)} div\vec{D}\,dv = \oint_{(S)} \vec{D}\,d\vec{S} = \sum_i \oint_{S_i} \vec{D}\,dS_i,$$

for the right part

$$\int_{(v)} \rho_v\,dv = \sum_i \int_{S_i} \sigma_i\,dS.$$

Here S is the wire surface, S_i is the area of this surface adjacent to i-medium, σ_i is a density of a surface charge on this area. Equating both sides of the equation, we find

$$\sum_i \Psi_{Di} = \sum_i q_i, \tag{7.20}$$

i.e., the flux Ψ_{Di} of the displacement vector through the wire surface into i-medium is equal to the charge q_i per unit length of the surface area adjacent to this medium.

In accordance with (7.18)

$$\Delta V = \frac{1}{l}\sum_i q_i / \varepsilon_i. \tag{7.21}$$

If to introduce such a quantity ε_e that

$$\sum_i q_i / \varepsilon_i = q / \varepsilon_e, \tag{7.22}$$

where q is the total wire charge per unit of its length, then as one can see from (7.22), the quantity ε_e has meaning of the equivalent permittivity of the heterogeneous medium. Accordingly (7.14), if $\Delta\varphi_i$ is the arc length of the wire circumference, which is adjacent to the i-medium, then the equality

$$q_i / (\varepsilon_i \Delta\varphi_i) = const(i), \tag{7.23}$$

is true. It is obvious also that

$$\sum_i q_i = q. \tag{7.24}$$

It is easy to show by using (7.23) and (7.24) that

$$q_i = q\varepsilon_i \Delta\varphi_i \Big/ \Big(\sum_i \varepsilon_i \Delta\varphi_i \Big). \tag{7.25}$$

whence

$$\varepsilon_e = \frac{q}{\sum_i q_i / \varepsilon_i} = \frac{1}{2\pi} \sum_i \varepsilon_i \Delta\varphi_i. \tag{7.26}$$

If N media of the same angle width are adjacent to the wire, then

$$\varepsilon_e = \frac{1}{N} \sum_{i=1}^{N} \varepsilon_i. \tag{7.27}$$

For variant, depicted in Figure 7.2, $N = 2$, and the equivalent permittivity is equal to the arithmetic average of magnitudes ε_1 and ε_2:

$$\varepsilon_e = (\varepsilon_1 + \varepsilon_2)/2. \tag{7.28}$$

Substituting (7.22) into (7.21) and taking into account that according to (7.16) in going around the cross-section of wire along the shorted contour the function V increases by

$$\Delta V = -2\pi A, \tag{7.29}$$

we find

$$A = -\tau / 2\pi\varepsilon_e, \tag{7.30}$$

and conse.quently

$$V = \tau\varphi/(2\pi\varepsilon_e) + C_1, U = -\tau\ln\rho/(2\pi\varepsilon_e) + C_2. \tag{7.31}$$

It is necessary to note that only the magnitude ε_e is included in the equalities (7.31). That means that in this case, as in the case of the homogeneous medium, the angles between the lines of the fields are equal to each other (irrespective of the medium, in which they are located), if the increase of a flux at the transition from one line of the field to other line is identical. If increase of potential between neighbors lines of the equal potential is also the same, the radii of the circumferences of the equal potential change according to geometric progression.

Consider using the obtained results, the important practical case of a two-wire line (Figure 7.3). In the beginning we shall assume that the wires are infinitely thin. The expression (7.12) for the complex potential of the field of such a line located in a homogenous medium may be rewritten in the form

$$\varsigma(z) = Aj\ln\frac{z+b}{z-b} + C. \tag{7.32}$$

If designate $z + b = \rho_1 \exp(j\varphi_1)$, $z - b = \rho_2 \exp(j\varphi_2)$, where ρ_1 and ρ_2 are the lengths of the segments connecting the observation point M with the axes of the wires, and φ_1 and φ_2 are the angles between these segments and the axis x, we find

$$V = A(\varphi_2 - \varphi_1), \quad U = A\ln(\rho_1 / \rho_2). \tag{7.33}$$

Here it is accepted that $C = 0$. In this case the line of zero potential is the ordinate axis, and the sections of abscissa axis from the wires of the line to infinity. The lines of equal potential are the circumferences with the centers on the axis x, and the lines of the field are the circumferences, passing through the wires axes, with the centers on the axis y.

Let a circular dielectric cylinder of radius b be placed between two wires of line, and its axis is parallel to the wires and is located at the same distances from both wires. The dielectric boundary (the circumference) coincides with the line of the field. The lines of equal potential intersect it at the right angle. It means that the dielectric cylinder doesn't change the field structure.

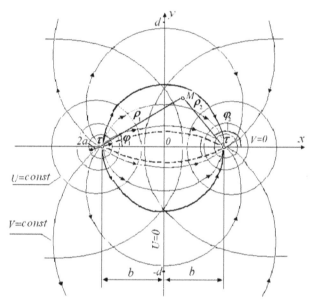

Figure 7.3 Field of two wires located along the generatrices of the dielectric cylinder.

As in the case of the solitary wire, in order to the potential near the wire was the identical in the air and in the dielectric, the surface density of the charge must be proportional to the permittivity of the medium, adjacent to a given part of the wires. So the magnitude A in the expressions (7.32)–(7.33) is defined by the equality (7.30), and the equivalent permittivity is defined by the expression (7.28).

If the wires of the real transmission line are not infinitely thin, but have circular cross-sections of the finite radius, then always one can place the axes of the wires such that the surfaces of the real wires may coincide with the surfaces of equal potential, which are the surfaces of the round cylinders.

The two-wire line may be fabricated in the shape of the round metallic cylinder located over the metal plane (Figure 7.4a), in the shape of two round cylinders with different axes, which do not encompass one another (Figure 7.4b) and in the shape of two analogous cylinders encompassing one another (Figure 7.4c). The dielectric cylinder is used in all these cases simultaneously both as an isolator and as a supporting construction. One must note that the circumference coinciding with the boundary of the dielectric cylinder goes not through axes of the round metallic cylinders, but through equivalent infinitely thin straight wires, the fields of which coincide with the fields of the round cylinders. Two equipotential surfaces of fields of thin wires coincide with the surfaces of the real wires.

The distance between the axis of the round metallic cylinder and the equivalent wire for the variant shown in Figure 7.4a is equal to

$$h - b = h - \sqrt{h^2 - R^2},$$ (7.34)

where R is the radius of the round cylinder. For the variant shown in Figure 7.4b the distances between the surface of zero potential and the axis of one round cylinder and between this surface and the equivalent wire are equal correspondingly to

$$h_1 = \frac{D^2 + R_1^2 - R_2^2}{2D}, \quad b = \sqrt{h_1^2 - R_1^2} = \frac{1}{2D}\sqrt{\left[(D - R_1)^2 - R_2^2\right]\left[(D + R_1)^2 - R_2^2\right]},$$ (7.35)

where R_1 and R_2 are the radii of the cylinders, and D is the distance between their axes. At last for the variant depicted in Figure 7.4c

$$h_1 = \frac{R_2^2 - D^2 - R_1^2}{2D}, \quad b = \sqrt{h_1^2 - R_1^2}.$$ (7.36)

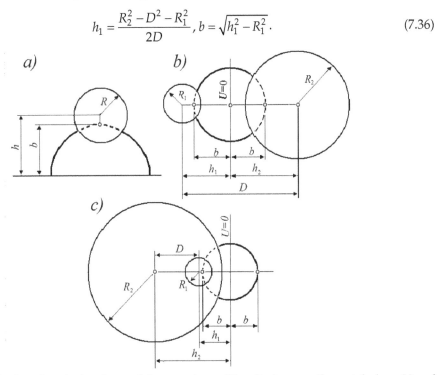

Figure 7.4 The long line in the shape of the round metallic cylinder over the metal plane (a) and in the shape of two cylinders with different axes, which do not encompass (b) and encompass (c) one another.

The capacitances of the solitary wire and of the two-wire transmission line are proportional to the permittivity ε of medium. In the case of the heterogeneous medium consisting of homogeneous layers the magnitude ε must be replaced by the equivalent permittivity—in accordance with above-presented expressions. In particular, if the solitary wire is located at the boundary of the two media (see Figure 7.2), the equivalent permittivity is equal to the arithmetic average of the magnitudes ε_1 and ε_2. This conclusion conforms completely to the known thesis, in accordance with which the capacitance between two conductors located symmetrically relative to a flat boundary of two media with the permittivity ε_1 and ε_2 is equal to half-sum of the capacitance values between the same conductors in the homogeneous media with permittivity ε_1 and ε_2 accordingly [34]. In the case of a single wire, the metal cylinder of infinite radius, axis of which coincides with the wire axis, may be accepted as the second conductor.

If the boundary of the two media goes along a broken line, whose point of sharp bend coincides with the wire center (this boundary is shown by dotted line in Figure 7.2), then in accordance with (7.26)

$$\varepsilon_e = \frac{1}{2\pi}(\varepsilon_1 \Delta\varphi_1 + \varepsilon_2 \Delta\varphi_2).$$ (7.37)

If a few media, the boundaries of which coincide with the radial surfaces, are adjacent to the solitary wire, then in the case of equal angular width of the adjacent media, the equivalent permittivity is determined by the expression (7.27), in the case of different angular width this permittivity is determined by (7.26). This conclusion coincides with the results obtained in [65, 66].

In [65] it is shown that the electrostatic field of the wires' system of the arbitrary shape located in a piecewise homogenous medium coincides with a field in a homogenous medium, if the media boundaries coincide with the surfaces of wires and the surfaces of the equal strength of the field in the homogenous medium (this condition is called the condition of invariance). Accordingly for the capacitance between the wires located in a piecewise homogenous medium [66] gives the expression

$$C = \sum_{i=1}^{N} \frac{\varepsilon_i}{\varepsilon_0} C_{i0}$$ (7.38)

Here C_{i0} is the capacitance between the medium i and the wire segment adjacent to it, if the wire is located in a homogeneous medium with permittivity ε_0.

It is obvious that the condition of invariance is performed in the case when the boundaries of the media adjacent to the solitary wire coincide with radial surfaces. If to use as the second wire the metal cylinder of the infinite radius coaxial with the solitary wire, then it is easy to obtain (7.26) and (7.27) from (7.38).

For the line from two infinitely thin wires located along the generatrices of the circular dielectric cylinder, the condition of invariance is performed also. Since the angle width of both media adjacent to each wire is equal to $\Delta\varphi_1 = \Delta\varphi_2 = \pi$, then $C_{10} = C_{20}$, i.e. the magnitude ε_e, is also determined by the equality (7.28). This proposition stays true, if the thin wires are located along the arbitrary selected generatrices, the length of arc between which is not equal to π.

If a thin dielectric insert with the angle width 2α, limited by arcs of two circumferences, the centers of which lie on the axis y in the points $y = \pm d$, is placed between the wires (the insert boundaries are shown by the dotted lines in Figure7.3), then in accordance with (7.26)

$$\varepsilon_e = \varepsilon_1 + (\varepsilon_2 - \varepsilon_1)\,\alpha/\pi,$$ (7.39)

where ε_1 is the permittivity of the air and ε_2 is the permittivity of the dielectric.

The capacitance between the metallic cylinders per unit length (see Figure 7.4) is

$$C = \frac{2\pi\varepsilon}{\ln\left\{\left[h_1/R_1 + \sqrt{(h_1/R_1)^2 - 1}\right]\left[h_2/R_2 \pm \sqrt{(h_2/R_2)^2 - 1}\right]\right\}},$$ (7.40)

where the plus sign in the second square brackets is taken when one must calculate the capacitance between the cylinders depicted in Figure 7.4b, and the minus sign corresponds to Figure 7.4c.

For two cylinders of the same radius $(R_1 = R_2 = R)$

$$C = \frac{\pi\varepsilon}{\ln\left[h/R + \sqrt{(h/R)^2 - 1}\right]} = \frac{\pi(\varepsilon_1 + \varepsilon_2)}{2\cosh^{-1} h/R}. \tag{7.41}$$

This expression coincides with the expression, which one may obtain from presented in [34] expression (8–7) for the linear capacitance of the two-wire line located in the dielectric near the cylindrical interface of the two media, if the distance between the axis of the each wire and the interface is a small magnitude of the order of the wire radius. One must note that the expression (8–7) and the expression (7.41) are true at the arbitrary location of the charged filaments (of the equivalent linear wires) along the cylindrical interface of two media, i.e. at any arc length between the filaments (not only at arc length π).

For the circular cylinder located over a metal plane (see Figure 7.4a)

$$C = \frac{2\pi\varepsilon_e}{\ln\left[h/R + \sqrt{(h/R)^2 - 1}\right]}. \tag{7.42}$$

In the particular case, if $h/R \gg 1$,

$$C = 2\pi\varepsilon_e / \ln(2h/R), \tag{7.43}$$

that coincides with the expression, which one may obtain from presented in [34] expression (8–8) for the two-wire line with isolation of the ribbon type.

7.3 SYMMETRICAL CABLE OF DELAY. THE COAXIAL CHAMBER FOR CALIBRATING INSTRUMENTS FOR MEASURING STRENGTH OF ELECTROMAGNETIC FIELD

Symmetrical delay cable is an example of a specific device for calculating electrical characteristics of which it is necessary to take into account the medium heterogeneity. An ordinary coaxial delay cable has an interior wire in the shape of a helix and an exterior wire in the shape of the circular metallic cylinder. In contrast to it the symmetrical delay cable has two interior helical wires.

In principle the helical wires may be wound on two parallel dielectric cores or on a common core. An inductance in both cases is approximately the same, but a capacitance between the wires in the case of two parallel cores is many times less. So a linear delay

$$T = \sqrt{\Lambda C} \tag{7.44}$$

is small, and a wave impedance of a cable

$$W = \sqrt{\Lambda/C} \tag{7.45}$$

is very big (in the presented expressions Λ is a cable inductance per unit length, and C is a cable capacitance per unit length). For this reason in symmetrical cables of delay, two isolated wires are wound on a common core. The wires are wound in opposite directions: otherwise they will form a bifilar helix with low inductance.

It is necessary to remind that the capacitance per unit length of the symmetrical cable is equal to

$$C = C_{12} + C_{11}/2, \tag{7.46}$$

where C_{12} is the partial capacitance between helical wires, and C_{11} is the partial capacitance between each wire and external screen.

Thus, for the calculation of the delay time and the wave impedance of the cable, one needs to calculate the capacitance between the two coaxial cylindrical helices of the equal radius wound in opposite directions and also the capacitance between each helix and the screen.

Calculation of capacitance C_{11} is simple. At dense winding, C_{11} is the capacitance per unit length between two coaxial circular cylinders. Radius of the first cylinder is equal to the screen radius, and radius of the second cylinder is equal to the winding radius. At the rare winding, it is the capacitance between the straight wire and the metal cylinder, whose radius is equal to the radius of the screen.

The calculation of the capacitance C_{12} is a result of solving two problems. The first problem is the calculation of the capacitance between the wires located in the homogeneous medium (air). This problem is considered in Section 7.4. The second problem is an account of influence of the heterogeneity of the medium, since in fact the dielectric cylinder (core of polyethylene) with the permittivity differing from the permittivity of the air is placed inside the helices. About this problem based on the results of previous section one can say the following.

As shown in section 7.2, the capacitance of the long line, the wires of which are located along the generatrices of the circular dielectric cylinder, is proportional to the equivalent permittivity, which is equal to the arithmetic average of the permittivity of the cylinder and the permittivity of surrounding medium (air). This proposition stays true if the thin wires are located along arbitrarly selected generatrices, the length of arc between which is not equal to π.

From here one can draw with a high degree of reliability a conclusion that at placing polyethylene or any different core inside the helices, the capacitance between them increases in proportion to the same equivalent permittivity.

A coaxial chamber for calibrating instruments for measuring the strength of an electromagnetic field is another example of an analogous device (of the same kind). The coaxial chamber is the fraction of coaxial line with increased dimensions, in which the electromagnetic wave is excited. The coaxial chamber secures a high degree of screening measuring devices. The electromagnetic field in it is uniform. The problem of reflected waves doesn't arise here.

The chamber for calibrating instruments for measuring the strength of electromagnetic field in the air is shown in Figure 7.5. Its central conductor 2 is made in the shape of a cylindrical helix. The conical transitions 1 at the chamber ends provide connections with the standard coaxial connectors, one of which (input 9) is joined to a generator 10, and another (output 7) is joined to a matched load 8. The coaxial line and the conical transitions are so constructed that the wave impedance is constant.

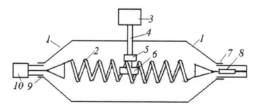

Figure 7.5 Coaxial chamber with one helical conductor for calibrating instruments for measurement of the magnetic field in the air.

A calibrated measuring instrument 6 is introduced into the chamber and is placed inside the volume confined by the helix. Antenna 6 as a rule is manufactured in the shape of a multiturn loop of finite radius. Signal of the instrument through the preliminary amplifier 5 and the coupling line 4 feeds the recording device 3.

The described circuit of the measurement gives an error caused by the fact that the strength of the magnetic field inside the volume, confined by the helix, has not only longitudinal (H_z), but also parasitic transverse component (H_p), which distorts the measurement results, since the emf created at the ends of the loop antenna, contains terms proportional both components.

In order to remove the radial component of the magnetic field, it is necessary to make the central conductor in the shape of the coaxial chamber, which is shown in Figure 7.6. The chamber is built as two helical conductors 12 and 13 with opposite direction of winding, and one conductor is connected to a generator through phase inverter 14, which changes the phase of current by π. The fields created by the currents of both conductors are summed up.

Consequently the components H_p cancel one another, and the components H_z are added together and form practically the uniform magnetic field.

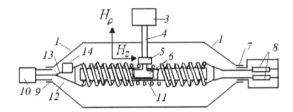

Figure 7.6 Coaxial chamber with two helical conductors for calibrating measuring devices of the magnetic field in the conductive liquid.

In the last years a development of methods and facilities for the measurements of the electromagnetic field strength in the conductive liquids (the sea water, soil, etc.) acquired important significance. The measurement results depend on the medium conductivity, which may vary within wide limits. Because of this the preliminary calibration of instruments necessary here is extremely.

The coaxial chambers for such calibration in the conductive liquid are described in [67]. In particular, for calibrating the measuring devices of the electric field, the part of the central conductor of the coaxial chamber is made in the shape of a cylindrical tank with dielectric walls. The tank is filled with the conductive liquid, and the measuring antenna is placed inside the tank along its axis.

When calibrating instruments for measurement of the magnetic field strength it is expedient to place the tank 11 with the conductive liquid inside the volume confined by the cylindrical helices (see Figure 7.6). The tank diameter must be less than the thickness of a skin layer in order to exclude the error, which is caused by attenuation of the electromagnetic field in the radial direction.

The equivalent circuit of the coaxial chamber with two helical conductors excited in anti-phase is presented in Figure 7.7. The equivalent circuit of the symmetrical delay cable has an analogous view. In the both cases it is asymmetrical line of two wires located above the ground (above the outer screen). But in the case of the coaxial chamber, the additional reactance jQ per unit length caused by the surface impedance is contained in each wire of the line:

$$jQ = Z_1 Z_2 / [2\pi a_1 (Z_1 + Z_2)]. \tag{7.47}$$

Here Z_1 is the surface impedance of the helical wire, Z_2 is the surface impedance of the conductive liquid, and a_1 is the helix radius.

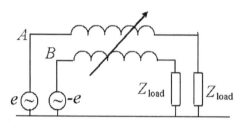

Figure 7.7   Equivalent circuit of the line with two helical conductors.

The impedance Z_1 is introduced in the following way. The outer surface of the helix is mentally replaced by a metal coating. The helix turns into the cylindrical wire of radius a_1. The slowing of wave along this wire is taken into account by means of incorporating a linear impedance in the obtained conductor. If to regard this impedance by purely reactive (without account of losses in the helical wire), then at low frequencies it has an inductive character: the surface impedance of the helical wire is equal to

$$Z_1 = j 2\pi a_1 \omega \Lambda , \tag{7.48}$$

where ω is the circular frequency,

$$\Lambda = \pi \mu_0 \omega^2 a_1^2 K_a / L^2 \tag{7.49}$$

is the inductance per unit length of the helix, μ_0 is the permeability of the free space – ω is the number of winds, L is the length of winding, K_a is the coefficient, which is equal to 1, if $a_1 \ll L$, and decreases smoothly, when a_1/L increases [68].

With increasing frequency, the self-capacitance of the helix becomes noticeable, i.e. the helix behaves as a parallel tuned circuit with the resonant circular frequency ω_0:

$$Z_1 = j \frac{2\pi a_1 \omega}{1 - (\omega / \omega_0)^2} \Lambda . \tag{7.50}$$

If the tank with the conductive liquid is placed inside the volume confined by the cylindrical helix, this is equivalent to a connection in the wire of the additional surface impedance Z_2, which is equal to

$$Z_2 = \frac{k_1 J_0 (k_1 a_1)}{\sigma_1 J_1 (k_1 a_1)}. \tag{7.51}$$

Here k_1 is the propagation constant in the conductive liquid, and σ_1 is the conductivity of the liquid. The surface impedance Z_2 is incorporating in parallel with the impedance Z_1. It is considered that the radius of the tank the coincides with the helix radius.

The structure from two wires with different surface impedances located above the ground has two different propagation constants (see Section 3.5). In the partial case, when the wires are the same and have the identical surface impedance, then regardless of the boundary conditions (loads at the ends of the line) the components of the currents and voltages with the propagation constant k_i are the same in both wires (in-phase wave), and the components with the propagation constant k_a (anti-phased wave) are equal in magnitude and opposite in sign, i.e.

$$k_1 = \sqrt{k_2^2 + \omega Q(\beta_{11} + \beta_{12})}, \, k_a = \sqrt{k_2^2 + \omega Q(\beta_{11} - \beta_{12})}, \tag{7.52}$$

where k_2 is the propagation constant of wave in the medium between the helical wires and the external screen, and β_{11} and β_{12} are accordingly the self (for the each wire) and the mutual (between the wires) coefficients of electrostatic inductions.

In the circuit depicted in Figure 7.7 two emf, equal in magnitude and opposite in sign, excite two helical wires, creating only anti-phase currents. The impedance Z_{AB} between points A and B is the input impedance of the two-wire long line with the propagation constant k_a and the wave impedance

$$W_a = 2k_a / [\omega(\beta_{11} - \beta_{12})]. \tag{7.53}$$

If we turn from the coefficients of electrostatic inductions to the partial capacitances C_{ik}, we shall obtain:

$$k_a = \sqrt{k_2^2 + \omega Q(C_{11} + 2C_{12})}, W_a = k_a / [\omega(C_{11} / 2 + C_{12})]. \tag{7.54}$$

Here, as in expression (7.46), C_{12} is the mutual capacitance per unit length between wires and C_{11} is the analogous capacitance between each wire and the ground (outer screen). The mode of traveling wave will exist in the two-wire line, if the loading impedance $2Z_{load}$ will be equal to the wave impedance W_a.

As mentioned earlier, the equivalent circuits of the coaxial cable of delay and the chamber for calibrating meters of the magnetic field strength are the same. In both cases there are two coaxial cylindrical helices of equal radius wound in opposite directions and located in the metal screen. In the equalities (7.44)–(7.45) the inductance of the helix is taken into account by means of the cable inductance Λ, in the equality (7.54) the inductance of the helix is taken into account by means of the surface impedance Z_1 included in the impedance jQ. Additionally in (7.54) the properties of the dielectric cylinder (of the tank with the conductive liquid) located inside the volume confined by the cylindrical helices (in particular the losses in this cylinder) are taken into account.

For the calculation of the wave impedance and the propagation constant of the equivalent long line one must know the partial capacitances C_{11} and C_{12}. The problem of calculating the capacitances, as already it was said during consideration of symmetrical cable of delay, breaks down into two problems. The first problem is the calculation of the capacitances between wires located in a homogeneous medium. This problem has independent importance, since its solution is necessary for calibrating instruments, measuring the magnetic field strength in the air and producing suitable basic standards. It is considered in the follow section.

The second problem is the account of heterogeneous medium influence. The capacitance between helical wires owing to a presence of a dielectric core (of a conductive liquid) inside the helices increases proportionally with an equivalent permittivity. At that the permittivity is regarded as the real magnitude, since the imaginary component of ε (losses in the conductive medium) is taken into account by means of introducing the surface impedance Z_2.

7.4 CAPACITANCE BETWEEN WIRES OF A TWO-THREAD HELIX

The given section is devoted to calculating a capacitance between two coaxial cylindrical helices of the same radius wound in opposite directions (Figure 7.8a). It is considered that the wire diameter $2a$ is small in comparison with a transverse and longitudinal dimension of the helix.

It is known [34] that, if a system consists of two identical wires and a sum of theirs charges is equals zero (i.e., the system is electro neutral), the capacitance between wires coincides with the mutual partial capacitance and is equal to

$$C_l = 0.5/(\alpha_{11} - \alpha_{12}), \qquad (7.55)$$

where α_{11} and α_{12} are the self- and mutual potential coefficients of the wires respectively.

We use the cylindrical system of coordinate (ρ, φ, z) with axis z along the helix axis and with origin O on the plane going through the initial points of the wires. The letter x designates the coordinate, which is counted along the wire. In the arbitrary point of the helical wire

$$x = \sqrt{z^2 + \rho_0^2 \varphi^2}, \qquad (7.56)$$

where ρ_0 is the radius of the cylindrical surface, along which the helix is reeled, and φ is the total rotation angle from the initial point of the wire to the point with coordinate z.

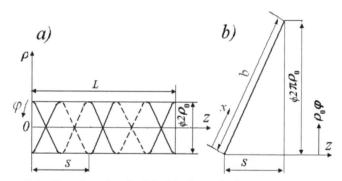

Figure 7.8　Two-thread cylindrical helix with the opposite winding (*a*) and the involutes of the first coil (*b*).

In Figure 7.8*b* the involute of the first coil of the helix is given. As one can see from the figure

$$z/x = s/b, \quad \rho_0\varphi/x = 2\pi\rho_0/b, \qquad (7.57)$$

where s is the helical pitch, and $b = \sqrt{s^2 + (2\pi\rho_0)^2}$. From here

$$x = zb/s, \quad \varphi = 2\pi x/b. \qquad (7.58)$$

If L is the helix length, then the length of the each helical wire is

$$l = Lb/s. \qquad (7.59)$$

In accordance with the method of Howe, we consider that the linear charge density τ is the same along the entire wire length:

$$\tau = q/l, \qquad (7.60)$$

where q is the all charge of the wire. Then the potential in the point M with coordinates (ρ_0, φ', z'), excited by the charge τdx of the wire element dx, is equal to

$$dU_M = \tau dx/(4\pi\varepsilon_0 R_1). \qquad (7.61)$$

Here ε_0 is the permittivity of the free space, and R_1 is the distance from the point M to the element dx with coordinates (ρ_0, φ, z). For simplicity we regard that a potential is determined by the filament charge located on the wire axis, i.e.

$$R_1 = \sqrt{(z-z')^2 + 4\rho_0^2 \sin^2 \frac{\varphi-\varphi'}{2} + a^2} = \frac{s}{b}\sqrt{(x-x')^2 + \left[\frac{2\rho_0 b}{s} \sin \frac{\pi(x-x')}{b}\right]^2 + \frac{a^2 b^2}{s^2}}. \quad (7.62)$$

The total potential of all charge q in the point M is

$$U_M(x') = \frac{q}{4\pi\varepsilon_0 l} \int_0^l \frac{dx}{R_1}. \quad (7.63)$$

Averaging this magnitude over the wire length

$$U_{11} = \frac{1}{l}\int_0^l U_M(x')dx', \quad (7.64)$$

and dividing into q, we calculate the potential coefficient α_{11}:

$$\alpha_{11} = U_{11}/q = \frac{1}{4\pi\varepsilon_0 Ll}\int_0^l dx' \int_0^l \frac{dx}{\sqrt{(x-x')^2 + \left[\frac{2\rho_0 b}{s}\sin\frac{\pi(x-x')}{b}\right]^2 + \frac{a^2 b^2}{s^2}}}. \quad (7.65)$$

When calculating α_{12} we consider that two helical wires are wound in the opposite directions. Besides, the ends of the wires in the initial cross-section are shifted along the perimeter of the cross-section by π relatively one another. Therefore, if relationships (7.58) are true for one wire, then for the second wire in the observation point M we have

$$x' = z'b/s, \quad \varphi' = \pi - 2\pi x'/b. \quad (7.66)$$

Accordingly, for the distance R_2 from the point M to the element dx we obtain instead of (7.62)

$$R_2 = \frac{s}{b}\sqrt{(x-x')^2 + \left[\frac{2\rho_0 b}{s}\cos\frac{\pi(x+x')}{b}\right]^2 + \frac{a^2 b^2}{s^2}}. \quad (7.67)$$

Calculating the total potential of all charge q of the adjacent wire in the point M and averaging this magnitude over the considering wire, we find:

$$\alpha_{12} = \frac{1}{4\pi\varepsilon_0 Ll}\int_0^l dx' \int_0^l \frac{dx}{\sqrt{(x-x')^2 + \left[\frac{2\rho_0 b}{s}\cos\frac{\pi(x+x')}{b}\right]^2 + \frac{a^2 b^2}{s^2}}}. \quad (7.68)$$

Integrals (7.65) and (7.68) cannot be expressed in terms of elementary functions, since the radicands of their integrands involve besides $(x - x')^2$ the squares of trigonometric functions, arguments of which also depend on x and x'. Therefore one may determine them only numerically. The double integral (7.65) may be reduced to the ordinary one. If we introduce the notations

$$A = \frac{1}{4\pi\varepsilon_0 Ll}, \quad f(t) = 1/\sqrt{t^2 + \left[\frac{2\rho_0 b}{s}\sin\frac{\pi t}{b}\right]^2 + \frac{a^2 b^2}{s^2}},$$

then α_{11} is equal to

$$\alpha_{11} = A\int_0^l dx' \int_0^l f(x - x')dx.$$

Since $f(t)$ is the even function relative to argument t, and the region of integration is the square l cm on side, we obtain:

$$\alpha_{11} = 2A\int_0^l dx' \int_0^{x'} f(t)dt.$$

Changing the order of integration, we find:

$$\alpha_{11} = 2A\int_0^l f(t)dt \int_t^l dx' = 2A\int_0^l (l - t)f(t)dt. \qquad (7.69)$$

If the wire radius a is small, then if t tends to zero (x tends to x') the integrand in the expression (7.65) rises sharply and complicates the numerical integration. Reducing the double integral to the ordinary one substantially facilitates the calculation.

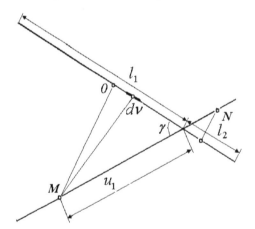

Figure 7.9   The node of crossed straight wires.

For calculating the capacitance between two coaxial cylindrical helices one can propose an approximate method as the alternative to a numerical method. The approximate method is based on the fact that this capacitance is caused by multiple crossing of one wire with the second wire. In order to calculate this capacitance, one must sum up the capacitances of the nodes of crossing. One can imagine the each node of crossing in the shape of four wire segments with lengths l_1 and l_2 and connect these segments in pairs (Figure 7.9) with the total length

$$l_1 + l_2 = l_0 = l/(2w), \qquad (7.70)$$

where w is the number of the coils. If to consider for simplicity that the wire segments are straight, we obtain

$$C_l = \sum_{n=1}^{N} C_{l_0}^n = 2wC_{l_0}^{(n)}. \tag{7.71}$$

Here n is the node number, N is the number of the nodes (it is equal to $2w$), and $C_{l_0}^n$ is the capacitance between two crossed straight wires of the node n. Similarly (2.43)

$$C_{l_0}^n = 0.5 / \left[\alpha_{11}^{(n)} - \alpha_{12}^{(n)} \right], \tag{7.72}$$

where $\alpha_{11}^{(n)}$ and $\alpha_{12}^{(n)}$ are the self and mutual potential coefficients of wires of the node n. If the radius a of the wire is small in comparison with its length l_0, then

$$\alpha_{11}^{(n)} = \frac{1}{2\pi\varepsilon_0 l_0} \left[\ln(2l_0 / a) - 1 \right]. \tag{7.73}$$

In order to find the value of $\alpha_{12}^{(n)}$, one must calculate potentials $U_M(u)$ and $U_N(u)$ in the points of left and right segments of one wire and average these potentials over the entire wire length. In the common case, if the segments lengths are different, we obtain (see Figure 7.9):

$$U_M(u_1) = q / (4\pi\varepsilon_0 l_0) \left\{ \int_0^{l_2 + u_1 \cos\gamma} dv / \sqrt{v^2 + u_1^2 \sin^2\gamma} + \int_0^{l_1 - u_1 \cos\gamma} dv / \sqrt{v^2 + u_1^2 \sin^2\gamma} \right\} =$$

$$\frac{q}{4\pi\varepsilon_0 l_0} \left[\sinh^{-1} \frac{l_2 + u_1 \cos\gamma}{u_1 \sin\gamma} + \sinh^{-1} \frac{l_1 - u_1 \cos\gamma}{u_1 \sin\gamma} \right].$$

Here γ is the angle between the wires. Similarly,

$$U_N(u_2) = \frac{q}{4\pi\varepsilon_0 l_0} \left[\sinh^{-1} \frac{l_1 + u_2 \cos\gamma}{u_2 \sin\gamma} + \sinh^{-1} \frac{l_2 - u_2 \cos\gamma}{u_2 \sin\gamma} \right],$$

i.e.

$$U_{12} = 1 / l_0 \left[\int_0^{l_1} U_M(u_1) du_1 + \int_0^{l_2} U_N(u_2) du_2 \right] =$$

$$\frac{q}{2\pi\varepsilon_0 l_0} \left[\sinh^{-1} \left(\tan\frac{\gamma}{2} \right) + \frac{l_1}{l_0} \sinh^{-1} \frac{l_2 / l_1 + \cos\gamma}{\sin\gamma} + \frac{l_2}{l_0} \sinh^{-1} \frac{l_1 / l_2 + \cos\gamma}{\sin\gamma} \right],$$

and accordingly

$$\alpha_{12}^{(n)} = \frac{U_{12}}{q} = \frac{1}{2\pi\varepsilon_0 l_0} \left[\sinh^{-1} \left(\tan\frac{\gamma}{2} \right) + \frac{l_1}{l_0} \sinh^{-1} \frac{l_2 / l_1 + \cos\gamma}{\sin\gamma} + \frac{l_2}{l_0} \sinh^{-1} \frac{l_1 / l_2 + \cos\gamma}{\sin\gamma} \right]. \tag{7.74}$$

From (7.72)–(7.74),

$$C_{10}^{(n)} = \pi\varepsilon_0 l_0 \left[\ln\frac{2l_0}{a} - 1 - \sinh^{-1} \left(\tan\frac{\gamma}{2} \right) - \frac{l_1}{l_0} \sinh^{-1} \frac{l_2 / l_1 + \cos\gamma}{\sin\gamma} - \frac{l_2}{l_0} \sinh^{-1} \frac{l_1 / l_2 + \cos\gamma}{\sin\gamma} \right]^{-1}. \tag{7.75}$$

If $l_1 = l_2$, then taking (7.70)–(7.71) into account we obtain:

$$C_l = \pi\varepsilon_0 l\left[\ln\frac{1}{aw} - 1 - \sinh^{-1}\left(\tan\frac{\gamma}{2}\right) - \sinh^{-1}\left(\cot\frac{\gamma}{2}\right)\right]^{-1}. \qquad (7.76)$$

Here, as it is easy to make sure,

$$w = l/s, \quad \tan(\gamma/2) = 2\pi\rho_0/s. \qquad (7.77)$$

The calculations and measurements results of the capacitance between the wires of a two-thread helix with dimensions (in meters): $L = 0.32$, $2a = 1.5\cdot10^{-3}$, $\rho_0 = 0.045$ and with the different number w of the coils—are presented in Figure 7.10. Curve 1 is obtained by the numerical method in accordance with the expressions (7.55), (7.68) and (7.69), the curve 2 is obtained by the approximate method, in accordance with the expression (7.76). The measured values are marked by the circles. As one can see from figure, both calculation methods give the similar results (the difference is from 2 to 4%), which coincide well with the experiment. That indicates the applicability of the approximate method for the calculation of the capacitance between the wires.

When the wire diameter rises, the accuracy of the approximate method decreases.

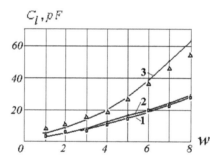

Figure 7.10 Dependence of the helix capacitance from the number of the coils.

The obtained results show the applicability of the approximate method not only for the calculating the capacitance between two coaxial cylindrical helices, but also for the calculation of the capacitance of itself nodes of crossing. The expression (7.75) permits to calculate the capacitance between two crossed wires located in the free space and in the homogeneous medium with permittivity ε or near the interface of two media, whose equivalent permittivity is equal to ε_e (with help of replacement ε_0 to ε or ε_e).

One can also generalize the approximate method to the case of wires with a sheath (Figure 7.11a). If the sheath radius is small in comparison with the transverse and longitudinal helix dimensions, i.e. if the sheath thickness is close to the wire radius, then one can consider that the wire surface and the outer surface of the sheath are equipotential surfaces. So the capacitance C_l between wires consists of three capacitances connected in series:

$$\frac{1}{C_l} = \frac{1}{C_{1l}} + \frac{1}{C_{2l}} + \frac{1}{C_{3l}}, \qquad (7.78)$$

where C_{1l} and C_{3l} are the capacitances of the capacitors created by the external and internal surfaces of each sheath, and C_{2l} is the capacitance between the helix wires, whose radius is equal to the outer radius a_1 of the sheath, i.e.

$$C_{1l} = C_{3l} = 2\pi\varepsilon_1 l / \ln(a_1 / a), \qquad (7.79)$$

(here ε_1 is the absolute permittivity of the sheath), and C_{2l} is calculated in accordance with (7.76)—with substitution a_1 for a.

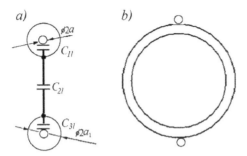

Figure 7.11 The cylindrical helix from the wires coated by a sheath (*a*) or wound around a dielectric frame (*b*).

If the two-thread helix is wound around a thin dielectric frame of a cylindrical shape (Figure 7.11*b*), then in the first approximation the capacitance between wires is equal to the capacitance between the helix wires in the sheath, the thickness of which is similar to the frame thickness.

The calculation results for the capacitance between the wires of two-thread helix with the above-indicated parameters, which is wound around a cylindrical frame of the thickness $a_1 - a = 0.005$ m with the relative permittivity $\varepsilon_{r1} = 2.6$, are presented in Figure 7.10 as curve 3. The measured values are marked by the triangles. Coincidence of the calculation with the experiments is sufficiently good.

7.5 CROSSED WIRES NEAR TO MEDIA INTERFACE

Two-thread helices wound around a solid dielectric cylinder (Figure 7.12*a*) is a matter of particular interest [69]. As it was shown in Section 7.2, if the helical wires lie on the interface of two media with permittivity ε_1 and ε_2, then it is necessary, when calculating potential coefficients and capacitances, to take as a permittivity of a medium an equivalent magnitude ε_e, which is equal to an arithmetic average. However if the wires of the helix are located over a cylinder surface (Figure 7.12*b*), such a method is inapplicable.

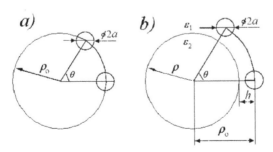

Figure 7.12 The cylindrical helix of the wires wound around a dielectric cylinder (*a*) and located over a cylinder surface (*b*).

Let us use the known expression for the capacitance per unit length of a two-wire line located in a dielectric not far from a cylindrical interface of two media [34]:

$$C_l = \cfrac{\pi \varepsilon_1}{\cosh^{-1} \cfrac{2\rho_0 \sin \theta / 2}{a} + \cfrac{m}{2} \ln \cfrac{\rho_0^4 - 2\rho_0^2 \rho^2 \cos \theta + \rho^4}{\left(\rho_0^2 - \rho^2\right)^2}}. \tag{7.80}$$

Here a constant m is equal to $m = (\varepsilon_1 - \varepsilon_2)/(\varepsilon_1 + \varepsilon_2)$. The meaning of the other designations is understandable from Figure 7.12. It is considered that in this expression the radius a of the wire is small in comparison with other characteristic dimensions. If the distance $h = \rho_0 - \rho$ between the axis of each wire and the interface is the magnitude of the order of a, then the second addend of the denominator is equal to

$$\frac{m}{2} \ln \left[1 + \left(\frac{2\rho_0 \rho \sin \theta / 2}{\rho_0^2 - \rho^2} \right)^2 \right] \approx m \ln \left(\frac{\rho_0}{h} \sin \theta / 2 \right).$$

If the medium is homogeneous ($\varepsilon_1 = \varepsilon_2$) this summand is equal to zero. For a heterogeneous medium, using the conception of equivalent permittivity, one can represent a capacitance C_l in the following form:

$$C_l = \frac{\pi \varepsilon_e}{\cosh^{-1} \left(\dfrac{2\rho_0}{a} \sin \theta / 2 \right)}. \tag{7.81}$$

Hence

$$\varepsilon_e = \frac{C_l}{\pi} \cosh^{-1} \left(\frac{2\rho_0}{a} \sin \theta / 2 \right) = \varepsilon_1 \left[1 + m \ln \left(\frac{\rho_0}{h} \sin \theta / 2 \right) / \cosh^{-1} \left(\frac{2\rho_0}{a} \sin \theta / 2 \right) \right]^{-1}. \tag{7.82}$$

If the wires of the two-thread helix are located above the media interface, then, as one can see from (7.82), the magnitude ε_e depends not only on ε_1 and ε_2, but also on the geometrical dimensions of the structures, in particular at a distance $R \approx 2\rho_0 \sin(\theta/2)$ between wires. If wires segments are shifted along a cylinder axis, it is necessary to take into account this shift during determination of a distance R. For ε_e we obtain:

$$\varepsilon_e = \varepsilon_1 \left[1 + m \frac{\ln(R / 2h)}{\cosh^{-1}(R / a)} \right]^{-1}. \tag{7.83}$$

Here R is defined from (7.62) or (7.67) depending on the fact where the point M and the element dx are located—on one wire or on the different wires.

If to substitute the magnitude ε_e in the expression (7.61) for the potential produced in the point M by the charge τdx of the element dx:

$$dU_M = \frac{\tau}{4\pi\varepsilon_1 R} \left[1 + m \frac{\ln(R / 2h)}{\cosh^{-1}(R / a)} \right] dx. \tag{7.84}$$

then for the self-potential coefficient we shall obtain instead of (7.65)

$$\alpha_{11} = \frac{1}{4\pi\varepsilon_1 l^2} \int_a^l dx' \int_a^l \frac{1}{R_1} \left[1 + m \frac{\ln(R_1 / 2h)}{\cosh^{-1}(R_1 / a)} \right] dx, \tag{7.85}$$

or after reducing the double integral to the ordinary one,

$$\alpha_{11} = \frac{1}{2\pi\varepsilon_1 l^2} \int_a^l \frac{l-t}{R_1(t)} \left\{ 1 + m \frac{\ln[R_1(t)/2h]}{\cosh^{-1}[R_1(t)/a]} \right\} dx, \tag{7.86}$$

where

$$R_1(t) = \frac{s}{b} \sqrt{t^2 + \left(\frac{2\rho_0 b}{s} \sin \pi t / b \right)^2 + \frac{a^2 b^2}{s^2}}.$$

In these expressions the magnitude a is taken as the lower integration limit, that permits to get rid of singularity, which is caused by the second addend of integrand and does not have a physical sense. Accordingly for the mutual potential coefficient we obtain instead of (7.68)

$$\alpha_{12} = \frac{1}{4\pi\varepsilon_1 l^2} \int_a^l dx' \int_a^l \frac{1}{R_2} \left[1 + m \frac{\ln(R_2/2h)}{\cosh^{-1}(R_2/a)} \right] dx. \tag{7.87}$$

We find the capacitance C_l as usual from (7.55).
  If inequalities

$$h, a \ll \rho_0, \rho, \rho_0 \sin \theta / 2, \ln(h/a) \ll \ln(p_0/a), \tag{7.88}$$

are correct, then the capacitance between the wires located above a cylinder surface doesn't differ from the capacitance between the wires, which lie along the interface.
  The calculation results for the capacitance C_l between the wires of two-thread helices with parameters (in meters): $L = 0.3$, $2a = 1.8 \cdot 10^{-3}$, $\rho = 0.03$ for the different number w of the coils are presented in Figure 7.13. Solid curves are obtained in accordance with expressions (7.55), (7.86) and (7.87). Curve 1 is plotted for the helix located in the free space, the curve 2—for the helix wound around the cylinder with relative permittivity $\varepsilon_{r2} = 2.8$ (the axes of the wires coincide with the media interface), and the curves 3 and 4 —for the wires located above the interface at the height of $h = 0.75 \cdot 10^{-3}$ and $h = 2.25 \cdot 10^{-3}$.

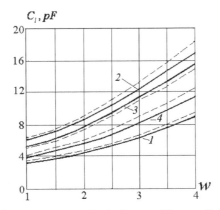

Figure 7.13 The capacitance of a helix would on a dielectric cylinder depending on the number of coils.

  In Section 7.4 the approximate calculation method for the capacitance between the coaxial cylindrical helixes located in the air is described. This method is based on the assumption that the capacitance is equal to the sum of capacitances of nodes of crossing.

For the approximate estimation of the capacitance between the helices located above the cylindrical interface it is necessary to increase the mentioned capacitance by a factor $\varepsilon_1/(1 + Bm)$, where $B = 1 - \ln(4h/a)/\ln(4\rho_0/a)$. The results of this estimation are presented in Figure 7.13 by dotted lines.

In the case of conical helices with constant winding angle (Figure 7.14), the angle γ between wires at the node of crossing doesn't change along the helices, and the radius $\rho(z)$ and pitch $s(z)$ of each helix increases in a linear fashion when z rises:

$$s(z)/s(0) = \rho(z)/\rho(0) = 1 + z/z_0, \tag{7.89}$$

where z_0 is the distance from the helix start to the cone vertex.

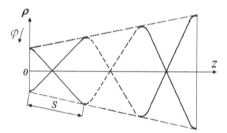

Figure 7.14   A conical helix.

The lengths $l_0^{(n)}$ of wires, which intersect in nodes, increase in the same fashion. If to increase the wire radius in the same fashion when z rises, then the potential coefficients $\alpha_{12}^{(n)}$ and $\alpha_{12}^{(n)}$ will be vary in inverse proportion to the length $l_0^{(n)}$, and it will be necessary in calculating C_l to replace the magnitude $\ln(l/aw)$ in the expression (7.76) by the magnitude $\ln[2\,l_0^{(n)}/a^{(n)}]$, which is the same in all segments. However wire radii in conical helices as a rule don't change along a wire, and that introduces an additional calculation error. This error is relatively small, since it has a logarithmic character of dependence on a wire radius.

7.6   CAPACITANCES IN A SYSTEM WITH A FEW CONDUCTORS

Generalizing the concept of "capacitance between two conductors" in the case of a system with an arbitrary but finite number of conductors, we arrive at the concepts of a self-and mutual partial capacitances. A ratio of conductor charge to its potential, if all conductors of the system, including the conductor under study, have the same potential, it is called self-partial capacitance of the conductor in a many-body system. A ratio of one conductor charge to a potential of another conductor, if all conductors of a system, except the latter one, have zero potential, is called the mutual partial capacitance between the conductors.

The relationship between the charges and potentials in the system of N conductors in accordance with these definitions is expressed by a system of equations:

$$Q_i = \sum_{n=1}^{N} C_{in}(U_i - U_n), i = 1, 2, ... N, \tag{7.90}$$

where Q_i and U_i are the charge and the potential of the conductor i, C_{ii} is its self-partial capacitance, C_{in} is the mutual–partial capacitance between conductors i and $n(i \neq n)$. Here $C_{in} = C_{ni}$.

One can add to a system of N conductors, occupying a finite volume, conductor $(N+1)$ in the shape of a sphere of infinite radius. Let this conductor have zero potential. In the resulting system the self-partial capacitance of any conductor, except conductor $(N+1)$, may be interpreted as a mutual–partial capacitance between this conductor and the sphere.

The system of equations (7.90) can be transformed by means of uniting terms with the factor U_i and obtaining expressions relating charges and potentials of conductors:

$$Q_i = \sum_{n=1}^{N} \beta_{in} U_n, i = 1, 2, ... N,$$

(7.91)

where β_{in} is called the coefficient of electrostatic induction. Another form of writing these relations is

$$U_i = \sum_{n=1}^{N} \alpha_{in} Q_n, i = 1, 2, ... N.$$

(7.92)

The values α_{in} are called the potential coefficients.

The coefficients β_{in} and α_{in} are related as follows:

$$\beta_{in} = \Delta_{in} / \Delta_N,$$

(7.93)

where $\Delta_N = |\alpha_{in}|$ is the $N \times N$ determinant, and Δ_{in} is the cofactor of the determinant Δ_N. The coefficients β_{in} and C_{in} are related also:

$$C_{in}|_{i \neq n} = -\beta_{in}, C_{in}|_{i=n} = \sum_{n=1}^{N} \beta_{in}.$$

(7.94)

In the specific case when the system consists of a single conductor, the concept of self-partial capacitance coincides with the concept of the capacitance of the isolated conductor. When such conductor consists of several (N) connected conductors, their capacitance is equal to

$$C_0 = \sum_{n=1}^{N} C_{nn}.$$

(7.95)

If $N = 2$, then

$$C_0 = C_{11} + C_{22}, C_{11} = \beta_{11} + \beta_{12}, C_{22} = \beta_{21} + \beta_{22},$$

where

$$\beta_{11} = \alpha_{22} / (\alpha_{11}\alpha_{22} - \alpha_{12}^2), \beta_{12} = \beta_{21} = -\alpha_{12} / (\alpha_{11}\alpha_{22} - \alpha_{12}^2),$$

$$\beta_{22} = \alpha_{11} / (\alpha_{11}\alpha_{22} - \alpha_{12}^2),$$

i.e.,

$$C_0 = \frac{\alpha_{11} + \alpha_{22} - 2\alpha_{12}}{\alpha_{11}\alpha_{22} - \alpha_{12}^2}.$$

(7.96)

As an example, let us consider the open at the end two-wire long line with conductors of different lengths (l_1 and $l_2 = l_1 - l$) and of the same radius a. This line was described in Section 3.3 (see Figure 3.11). As it is seen from Figure 3.11b, the considering structure consists of three conductors denoted in the figure by Roman numerals: I is the short wire, II is the parallel to it segment of the long wire with the same length, and III is the additional segment of the long wire (with the length l). The load of the line (1) is the capacitance C between the conductors I and III. It cannot be calculated directly as the capacitance between the conductors I and III, located in free space, at least because that presence of the conductor II significantly changes it. It is necessary firstly to find the capacitance C_Σ between the conductor I and the totality of conductors II and III, after that to find the capacitance C_0 between the conductors I and II and to subtract it from the first one: $C = C_\Sigma - C_0$. The calculation must take into account the self-partial capacitances of conductors in accordance with the expression (3.23).

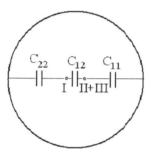

Figure 7.15   Partial capacitances in a system of two wires.

For example, calculating the capacitance between parallel conductors of equal length means that not only the mutual partial capacitance between the mentioned conductors, but the parallel to it capacitance of a series connected partial capacitances between each conductor and the sphere of infinite radius are taken into account. If under the calculation only the capacitance between the conductors are taken into account, then the calculated value of l_0 for $l = 0$ will not be different from zero. This result is caused by the approximate nature of the theory of two-wire long line, which is based on the postulate about the extremely small distance between the wires, axes. If this distance tends to zero, the partial capacitance between the conductors becomes infinitely great, and the parallel capacitance due to the self-partial capacitance of each conductor to the sphere of infinite radius can be neglected.

A widespread misconception that the field of a long line is concentrated mainly between the wires is also caused by the approximate theory. In fact, only half energy flux is concentrated inside the imaginary cylinder passing through the equivalent thin filaments (see Section 8.7).

Method of Complex Potential for Three-Dimensional Problems

8.1 COMPARISON OF CONICAL AND CYLINDRICAL PROBLEMS. THE FIELD OF A LONG LINE FROM THE CONVERGING STRAIGHT WIRES

As was pointed out in the Introduction, the problem of calculating electric fields of charged bodies is substantially simplified, if the all geometrical dimensions depend only on two coordinates (such a field is called a plane-parallel field). A three-dimensional problem is solved only in a few particular cases, whereas a two-dimensional problem was considered more widely—for the different number of wires and for various shape of their cross-section. In this connection an attempt to use the solution results of two-dimensional problems for electrostatic fields' calculations in three-dimensional problems, when a mutual location of metal bodies reminds the two-dimensional variant, is of interest.

The field calculation of two infinitely long charged filaments converging to one point (Figure 8.1a) is an example of such a problem. The linear charge densities of both filaments are the same in magnitudes and opposite in sign

$$\tau_1 = -\tau_2 = \tau. \tag{8.1}$$

Analogue of this three-dimensional problem is the two-dimensional problem for two parallel filaments (Figure 8.1b). The scalar potential of two such filaments in accordance with (1.12) is equal to

$$U(x,y) = \frac{\tau}{2\pi\varepsilon} \ln \frac{\rho_2}{\rho_1}, \tag{8.2}$$

where ε is the dielectric permittivity of the medium, ρ_1 and ρ_2 are the distances from the observation point M to the filament axes. At that

$$\rho_2^2 = (b - x)^2 + y^2, \; \rho_1^2 = (b + x)^2 + y^2.$$

Here b is the half distance between the axes of the filaments.

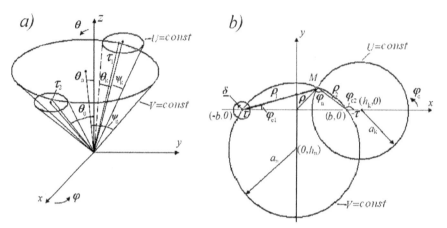

Figure 8.1 Three-dimensional (a) and two-dimensional (b) problems of two infinitely long charged filaments.

As has been mentioned earlier, lines of equal potential U = const in the plane field of two charged filaments are the circumferences with the centers on the axis of the abscissas. From here in particular it follows that the field of two parallel no coaxial metallic cylinders has the same character, since one can always locate the axes of the equivalent filament so that in their field two surfaces of equal potential coincide with the surfaces of the metallic cylinders (Figure 8.2a). Lines of force V = const are the circumferences with the centers on the axis of ordinates.

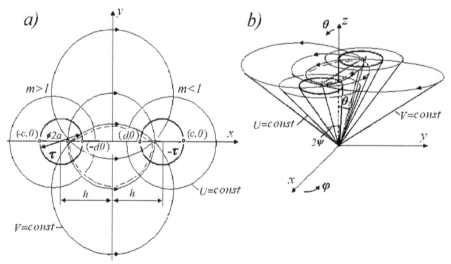

Figure 8.2 The two-dimensional problems for two metal cylinders (a), and the three-dimensional problem for two metal cones (b).

One must remember that in accordance with the uniqueness theorem, the solution of an electrostatic problem must satisfy Laplace's equation, and that surfaces of conductive bodies must coincide with the surfaces of equal potential. The three-dimensional problem of two convergent charge lines (see Figure 8.1a) is a particular case of a conical problem, in which conductive bodies have a shape of a cone with the top in the origin (Figure 8.2b). The conical and cylindrical problems are compared with each other in [70], where it is shown that Laplace's equation remains correct in going from one problem to another problem, if the replaceable variables are related by equalities:

$$\rho = \tan \frac{\theta}{2}, \quad \varphi_c = \varphi \cdot \tag{8.3}$$

Here ρ and φ_c are cylindrical coordinates, and θ and φ are spherical coordinates.

The result of such transformation of variables is mapping of spherical surface of an arbitrary radius R on the plane (ρ, φ_c). Here the line of that surface intersection with any circular cone (with a vertex in the origin) transforms into the circumference. Therefore the three-dimensional conical problem may be reduced to a two-dimensional one, in which the coordinates of the conductive bodies are related with the coordinates of the bodies by equalities (8.3).

The case of the two convergent charged filaments located at an angle $2\theta_0$ to each other in the plane xOz (see Figure 8.1a) corresponds if to base on the analogy between the conical and the cylindrical problems to two parallel filaments spaced at the distance spaced at the distance $2b = 2\tan(\theta_0/2)$ from each other (see Figure 8.1b). The case of the two metal cones with the angle 2ψ at the vertex of each cone and with the angle $2\theta_1$ between the cones axes (see Figure 8.2b), corresponds to two metallic cylinders of the radius $a = (c - d)/2$, the distance between axes of which (see Figure 8.2a) is equal to $2h = c + d$. Since according (8.3)

$$c = \tan\left[\frac{\theta_1 + \psi}{2}\right], \quad d = \tan\left[\frac{\theta_1 - \psi}{2}\right], \tag{8.4}$$

it is not difficult to make sure that

$$a = \frac{\sin \psi}{\cos \theta_1 + \cos \psi}, \quad h = \frac{\sin \theta_1}{\cos \theta_1 + \cos \psi}. \tag{8.5}$$

In particularly, it follows from (8.4) that

$$\theta_1 = \tan^{-1} c + \tan^{-1} d = \tan^{-1} \frac{2h}{1 - (h^2 - a^2)},$$

$$\psi = \tan^{-1} c - \tan^{-1} d = \tan^{-1} \frac{2a}{1 - (h^2 - a^2)}. \tag{8.6}$$

It is necessary to emphasize that upon the transition from the cone to the cylinder the cone axis doesn't coincide with the cylinder axis.

The scalar potential of the electric field created by the two convergent charged filaments by analogy with (8.2) is equal to

$$U(\theta, \psi) = \frac{\tau}{2\pi\varepsilon} \ln\left(\frac{\rho_2}{\rho_1}\right), \tag{8.7}$$

where

$$\rho_1^2 = \left[\tan\left(\frac{\theta_0}{2}\right) + \cos\varphi \, \tan\left(\frac{\theta}{2}\right) \right]^2 + \sin^2\varphi \, \tan^2\left(\frac{\theta}{2}\right),$$

$$\rho_2^2 = \left[\tan\left(\frac{\theta_0}{2}\right) - \cos\varphi \, \tan\left(\frac{\theta}{2}\right) \right]^2 + \sin^2\varphi \, \tan^2\left(\frac{\theta}{2}\right).$$

In the given case, the surfaces of equal potential U = const are the circular cones, the axial lines of which lie in the plane xOz. Each surface satisfies an equation:

$$\frac{\rho_2}{\rho_1} = m = \text{const.} \tag{8.8}$$

For the plane problem the line of equal potential is the circumference with the center in the point $h_m = (1 + m^2)b/(1 - m^2)$ and with the radius $a_m = 2mb/|1 - m^2|$. Using these magnitudes and the equality (8.6), we find the angle θ_m between the axis z and the axis of an equipotential circular cone in the conical problem

$$\theta_m = \tan^{-1}\frac{2h_m}{1-(h_m^2 - a_m^2)} = \tan^{-1}\left(\frac{1+m^2}{1-m^2}\tan\theta_0\right), \tag{8.9}$$

and also the angle ψ_m between the cone generatrix and its axis

$$\psi_m = \tan^{-1}\frac{2a_m}{1-(h_m^2 - a_m^2)} = \tan^{-1}\frac{2m\tan\theta_0}{|1-m^2|}. \tag{8.10}$$

The surfaces of field strength V = const, in which field lines are located, are also circular cones. Actually, in the plane problem the flux function appears as

$$V = -\frac{\tau}{2\pi\varepsilon}(\varphi_{c2} - \varphi_{c1}). \tag{8.11}$$

The sense of the angles φ_{c2} and φ_{c1} is clear from Figure 8.1b. The equation of the field line

$$\varphi_{c2} - \varphi_{c1} = \varphi_n = \text{const} \tag{8.12}$$

is the equation of a circumference, the radius of which is equal to $a_n = b/\sin\varphi_n$, and the axis is at a distance $h_n = b \cot \varphi_n$ from axis z. In the conical problem the axial lines of the circular cones V = const also lie in the plane yOz forming with axis z the angle

$$\theta_n = \tan^{-1}(\cot \varphi_n \tan \theta_0), \tag{8.13}$$

and the angle ψ_n between the cone axis and generatrix is equal to

$$\psi_n = \tan^{-1}\left(\frac{\tan\theta_0}{\sin\varphi_n}\right). \tag{8.14}$$

If to reduce a conical problem to a cylindrical, it permits to calculate the capacitance per unit length and the wave impedance of the long line consisting of two convergent filaments or cones. It is known, for example, that the capacitance per unit length of the line consisting of two conductors with radius a located at the distance $2h$ from each other is equal to

$$C_l = \frac{\pi\varepsilon}{\ln\left[\dfrac{h}{a} + \sqrt{(h/a)^2 - 1}\right]} = \frac{\pi\varepsilon}{ch^{-1}(h/a)}, \tag{8.15}$$

and the wave impedance is

$$W = \frac{1}{(cC_l)} = 120\,ch^{-1}(h/a). \tag{8.16}$$

Here, c is the light velocity.

For two convergent cones we obtain in accordance with (8.5)

$$C_1 = \frac{\pi\varepsilon}{ch^{-1}(\sin\theta_1/\sin\psi)}, \qquad W_1 = 120ch^{-1}\frac{\sin\theta_1}{\sin\psi}. \tag{8.17}$$

In particular for a dipole with an angle 2α between axes of conic arms (see Figure 8.3), $\sin\psi = a/L$, $\theta_1 = \alpha$, i.e.

$$W_2 = 120ch^{-1}\left(\frac{L\sin\alpha}{a}\right). \tag{8.18}$$

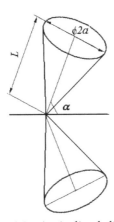

Figure 8.3   An inclined dipole.

8.2   NON-CLOSED COAXIAL SHELLS. A SLOTTED ANTENNA ON A CONE AND ON A PYRAMID

The case of two convergent charged shells, located along a surface of a circular cone with the angle $2\theta_0$ at vertex (Figure 8.4a) is of specific interest. Let the arc length in a cross-section of a charged shell be equal to 2α. The line of two coaxial cylindrical shells (Figure 8.4b) of radius $a = \tan(\theta_0/2)$ with the same arc length in a cross-section, corresponds to this case.

One can obtain a concept about the character of an electrostatic field of two cylindrical shells, if to sum fields of pairs consisting of symmetrically located parallel filaments 1-1', 2-2', 3-3' and so on (see Figure 8.4b) are equal in magnitude and opposite in sign charges. The lines of equal potential for each pair are the circumferences with the centers on the curve line passing through the filaments. The envelope of the circumferences with the same value of constant m is the line of equal potential for the field of the shells. It is a curve line of a complicated shape, extended along both sides of each shell and smoothly bent around its ends. The axis of structure symmetry, i.e. y-axis, is also one of the equipotential lines.

The lines of field strength for each pair of the filaments are circumferences with centers on the symmetry axis y, which pass through the filaments. Two lines coincide with the circumference, on which the shells are situated, i.e. these lines close two gaps between the shells. Field lines inside this circumference connect the symmetrically placed filaments with each other and cross the lines of equal potential at right angles. The field structure in the case of two convergent shells (see Figure 8.4a) is of a similar nature, the only difference being that the surfaces of equal potentials U = const and the surfaces of field strength V = const coincides with the conic surfaces rather than with cylindrical ones.

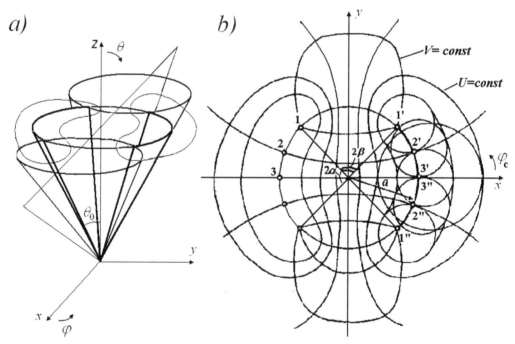

Figure 8.4   Three-dimensional (a) and two-dimensional (b) problem of non-closed coaxial shells.

The capacitance C_l per unit length and the wave impedance W of the long line formed by two cylindrical shells [34, 71] are equal to

$$C_l = \varepsilon K(\sqrt{1-k^2})/K(k), \quad W = 120\pi K(k)/K(\sqrt{1-k^2}),\qquad (8.19)$$

where $K(k)$ is the complete elliptic integral of the first kind of the argument

$$k = \tan^2(\beta/2).\qquad (8.20)$$

Here 2β is the angular width of the slot, $2\alpha = \pi - 2\beta$ is the angular width of the metal shell, i.e. C_l and W depend only on angular slot width 2β and hence on angular width

2α of the cross-section of the metal shell. Magnitudes C_l and W, as it follows from (8.19), are independent of cylinder radius a. This means that both expressions are correct for a conical structure with the identical cross-section.

Note that this section considers the structure, which consists of two metal plates and two slots and is located on a circular cone. This is a particular case of a structure. The general problems of calculating wave impedances of these and similar structures, located on planes, cones, and pyramids' sides, with different numbers of metal plates and slots are considered in Chapter 9.

Magnitudes C_l and W are constant along the conic line, i.e. two-wire line of the convergent shells is a uniform line. As is known, an input impedance of a uniform two-wire line, when its length infinitely increases, tends to its wave impedance. Therefore, the input impedance of an infinitely long line excited by a generator situated near the cone vertex is

$$Z_l(k) = 120\pi\, K(k)/K(\sqrt{1-k^2})\,. \tag{8.21}$$

One can consider the structure under study on the one hand as a two-wire line and on the other hand as an antenna. An antenna is a symmetrical V-radiator, with the arms shaped as two convergent metal shells located along the surface of a circular cone. Finally, we can consider this structure as a slot antenna in a conic screen. If $Z_E(2\alpha)$ is the input impedance of a metallic (electric) radiator with angular arm width 2α, and $Z_S(2\beta)$ is the impedance of a slot antenna with width 2β, then, if the structure length is great, we find

$$Z_E(\alpha) = Z_S(\beta) = Z_l(k) = 120\pi\, K(k)\big/K(\sqrt{1-k^2})\,. \tag{8.22}$$

The same slots in a metal cone are shown in Figure 8.5. In the first variant (see Figure 8.5a) the slot edge coincides with the cone generatrix. This variant was considered earlier. In the second variant (see Figure 8.5b), the slot edge is a helix, and the metallic cone with the slot in it forms a two-thread helix, excited near the vertex. The angular slot width is considered to be constant. Strictly speaking, expressions (8.19)–(8.22) are true only for the variant of the slot antenna shown in Figure 8.5a. But if the angular width of the metal shell in both variants is same, one can say with a high degree of probability that the capacitances per unit length, the wave impedances, and consequently the input impedances of radiators are the same in both cases.

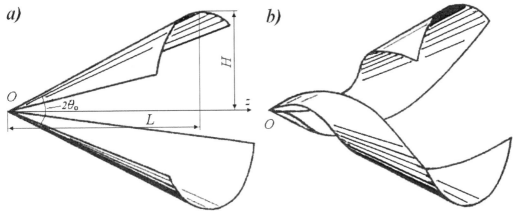

Figure 8.5 The slot antennas at the cone with straight-line (a) and helical (b) edges.

It is useful to compare the input impedances of metallic and slot radiators with the same width. If, for example, the metal shell width is $2\alpha = 2\pi/3$, then $\beta/2 = \pi/12$, $k^2 = 0.00515$ and $K(\sqrt{1-k^2})/K(k) = 2.56$, i.e. $Z_E(2\pi/3) = 120\pi/2.56$. In fact, the slot impedance is the input impedance of the metallic radiator situated next to it. If the slot width is $2\beta = 2\pi/3$, then $\beta/2 = \pi/6$, $k^2 = 0.111$, $K(\sqrt{1-k^2})/K(k) = 1.56$. Accordingly, $Z_S(2\pi/3) = 120\pi/1.56$. Therefore, the impedances of metallic and slot radiators of the same width $2\pi/3$ are related to each other by the expression $Z_E(2\pi/3) \cdot Z_S(2\pi/3) = (120\pi)^2/(2.56 \cdot 1.56)$, from whence it follows

$$Z_S = (60\pi)^2 / Z_E. \tag{8.23}$$

Here Z_E is the input impedance of the metallic radiator, which is identical to the slot in shape and dimensions. It is easily verified that (8.23) holds for radiators of any width.

Radiators with the same width of the metal shell and the slot are of particular interest. Setting $2\alpha = 2\beta = \pi/2$, we obtain: $k^2 = 0.0294$ and $K(\sqrt{1-k^2})/K(k) = 2.0$, i.e.

$$Z_E(\pi/4) = Z_S(\pi/4) = 60\pi. \tag{8.24}$$

As can be seen from (8.24), the infinite long radiator mounted on a cone has a constant and purely resistive input impedance, and hence the high level of matching with the cable in an unlimited frequency range. If the radiator is of finite size, the frequency range is limited, but remains sufficiently wide.

These results do not answer the question why, when the width of metallic and slot radiators changes, the product of their wave impedances does not change, i.e. expression (8.23) remains true. In accordance with (8.24), if $\alpha = \beta = \pi/4$, then $k = \tan^2(\pi/8)$, $K(\sqrt{1-k^2})/K(k) = 2.0$, and $Z_E(\pi/4) = Z_S(\pi/4) = 60\pi$. The function $K(k)$ is the complete elliptic integral of the first kind of the argument k. It is calculated by the formula

$$K(k) = \int_0^1 \frac{dt}{\sqrt{(1-t^2)(1-k^2 t^2)}}. \tag{8.25}$$

We shall change the angular width of the radiators. Let $\alpha/2 = \pi/8 + \delta$, and $\beta/2 = \pi/8 - \delta$, where $\delta \ll \pi/8$. Then the magnitude k_m for the metallic radiator is equal to

$$k_m = \tan^2(\pi/8 + \delta).$$

Introducing the notation $T = \tan(\pi/8)$, it is easy to show that

$$\tan(\pi/8 + \delta) \approx T + \delta(1 + T^2),$$

and

$$k_m^2 = \tan^4(\pi/8 + \delta) \approx T^4 + 4\delta T^3(1 + T^2),$$

i.e.

$$\sqrt{1 - k_m^2 t^2} = \sqrt{1 - T^4 t^2 - 4\delta T^3(1+T^2)t^2} = \sqrt{1 - T^4 t^2}\left[1 - \frac{2\delta T^3(1+T^2)t^2}{1 - T^4 t^2}\right],$$

and

$$\sqrt{1 - (1 - k_m^2)t^2} = \sqrt{1 - t^2 + T^4 t^2 + 4\delta T^3(1+T^2)t^2} = \sqrt{1 - t^2 + T^4 t^2}\left[1 - \frac{2\delta T^3\left(1+T^2\right)t^2}{1 - t^2 + T^4 t^2}\right].$$

Then

$$K(k_m) = \int_0^1 \frac{dt}{\sqrt{(1-t^2)(1-k_m^2 t^2)}} = K(T) + 2\delta T^3 (1+T^2) \int_0^1 \frac{t^2 dt}{\left(1-T^4 t^2\right)\sqrt{1-t^2}\sqrt{1-T^4 t^2}},$$

and

$$K\left(\sqrt{1-k_m^2}\right) = \int_0^1 \frac{dt}{\sqrt{(1-t^2)[1-(1-k_m^2)t^2]}} =$$

$$K\left(\sqrt{1-T^2}\right) + 2\delta T^3 (1+T^2) \int_0^1 \frac{t^2 dt}{(1-t^2+T^4 t^2)\sqrt{1-t^2}\sqrt{1-t^2+T^4 t^2}}.$$

Similar expressions for slot radiators, more precisely for the metallic radiators located next to it with an angular width β, are obtained replacing δ with $-\delta$. This means that

$$K(k_m)K(k_S) = K^2(T) - 4\delta^2 T^6 (1+T^2)^2 \left[\int_0^1 \frac{t^2 dt}{(1-T^4 t^2)\sqrt{1-t^2}\sqrt{1-T^4 t^2}}\right]^2,$$

$$K\left(\sqrt{1-k_m^2}\right)K\left(\sqrt{1-k_S^2}\right) = K^2\left(\sqrt{1-T^2}\right) - 4\delta^2 T^6 (1+T^2)^2 \left[\int_0^1 \frac{t^2 dt}{\left(1-t^2+T^4 t^2\right)\sqrt{1-t^2}\sqrt{1-t^2+T^4 t^2}}\right]^2.$$

Therefore in the first approximation

$$\frac{K(k_m)}{K\left(\sqrt{1-k_m^2}\right)} \frac{K(k_S)}{K\left(\sqrt{1-k_S^2}\right)} = \frac{K^2(T)}{K^2\left(\sqrt{1-T^2}\right)}, \tag{8.26}$$

i.e. in accordance with (8.23), when the angular width of metallic and slot radiators is different,

$$Z_E Z_S = Z_A^2,$$

where Z_A is the input impedance of the self-complementary radiator.

Thus, if $\alpha/2$ is distinguished from the required magnitude $\pi/8$ on small magnitude δ, the product $Z_E Z_S$ distinguishes from $(60\pi^2)$ on δ^2, that confirms the stationary property of the ratio (8.24).

It is necessary to emphasize that the last expression and the expression (8.26) have a higher degree of accuracy than a first approximation, since the neglected terms are not only proportional to δ^2, but the factor which is multiplied to δ^2, is equal to the product of the same functions $f(T) = -4T^6(1 + T^2)^2$ to the integrals with the same dependence on the value $k(T)$ in the numerator of the expression (8.26) and on the value $\sqrt{1-k^2}$ in the denominator of that expression. It should expect that the substitution of functions with precision up to δ^2 will show that difference between the left and right sides of (8.26) is a magnitude of the third order infinitesimal.

The method of calculating the wave and input impedances of metallic and slot radiators located along a surface of a circular cone through the wave impedance of an infinitely long uniform line may be applied also in other cases, for example to radiators located along pyramid sides. Let the pyramid have two metal sides (in the shape of the flat triangles) and two air sides, and also a rectangular cross-section (Figure 8.6). The wave impedance of the line formed by two triangular plates depends on the relation of the magnitudes b and d, which is constant along the line having the shape of a pyramid (b is the plate width in the given cross section, d is the distance between the plates). But the wave impedance doesn't depend on absolute values of these magnitudes. That means that this two-wire line is also uniform, and that its input impedance with increasing line length tends to the wave impedance W.

Angles α and β are related by the expression $\tan\alpha/\sin\beta = b/d$. As one can see from the given formulas, W is the function of two arguments: the ratio b/d and the angle α. If, for example, $b/d = 1$, the change of α from $15°$ to $30°$ (and accordingly β from $15.5°$ to $35.3°$) results in the change of the wave impedance from 42.3π to 45π, i.e.

$$Z_E = Z_S = (42.3 - 45)\pi.$$

Z_S is equal to the input impedance of the metal radiator located next to the slot. If $b/d = 1$, $\beta = 30°$, then $Z_S = 28.2\pi$. Z_E is the input impedance of the metal radiator, which is identical to the slot in shape and dimensions. One can show that in this case, $\sin\delta = \sin\beta\cos\alpha = 1/(2\sqrt{2})$, $\gamma = \tan^{-1}(b\tan\beta/d) = 63.4°$, i.e. $Z_E = 54.2\pi$, and $Z_E Z_S = (39.1\pi)^2$.

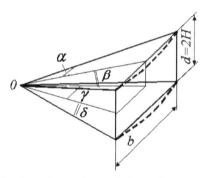

Figure 8.6   An antenna located along the pyramid sides.

If $b/d = 5$, $\beta = 15°$, then $Z_E = 127\pi$. As to Z_S, it is equal to the input impedance located next to the slot metal radiator, whose ratio b/d is equal to $1/5$, and the angle β is calculated from an equation

$$\tan\beta/\tan\gamma = d/b,$$

where $\gamma = 15°$ (with mutual replacing b by d the angles β and γ are interchanged). It is easy to be convinced that $\beta = 53.3°$, i.e. $Z_S = 105\pi$, and $Z_E Z_S = (36.6\pi)^2$.

As the calculations show, the product $Z_E Z_S$ depends on the ratio b/d, but is always smaller than $(60\pi)^2$. The smaller angle α, the closer wave impedance of the radiator located along the pyramid sides to the wave impedance of the conical radiator with the same angular width.

On the example of a conical radiator we shall consider in detail results of using solutions of two-dimensional problems for an analogous three-dimensional problem. It is known that on the cone infinitely long radiators with the same angular width of the metal shell and the slots have a constant and purely resistive input impedance. It permits to secure a high level of matching in unlimited frequency ranges. If the radiator has finite dimensions, then the frequency range is limited, but remains rather wide.

Section 7.1 describes the requirements for the task, under which the field is uniquely determined. In particular, the shape and size of the conductive bodies in both problems should be the same.

In going to the two-dimensional problem in accordance with the conditions (8.3), the radiator length remains finite, if the radiator before transition has a finite length and therefore it is incompatible with the system of infinitely long cylindrical wires. Limitation on the length of wires changes the nature of the problem: it ceases to be flat. The line of infinite length can be considered uniform since the capacitance per unit length and the line impedance are not dependent on the angle magnitude at the vertex of the cone. If the line length is finite, then in calculating the mentioned capacity it is necessary to take into account the partial capacitance of each line wire to the surface of zero potential (see Figure 7.15). Since the transverse dimensions of the line are small in comparison with its length, it is expedient to consider that the distance to the surface of zero potential is equal to double generatrix length ($2l = 2H/\sin\theta_0$). Assuming that the maximal transverse dimension of the metal shell is $2\alpha a$, we estimate the additional capacitance per unit length of the cylinder as

$$C_1 = \frac{2\pi\varepsilon}{\ln(4l/\alpha a)},$$ (8.27)

where a is the cross-section radius.

The partial mutual capacitance between the metal shells depends on the arc length 2α. If $2\alpha = \pi/2$, this capacitance in accordance with (8.19) is equal to $C_0(\pi/2) = 2\varepsilon$. When the arc length is changed, the capacitances C_0 and C_1 are changed too. The wave impedance of a transmission line formed by the metal shell and its reflection in the ground is equal to

$$W(2\alpha) = \frac{k}{\omega\left[C_0(2\alpha) + C_1(2\alpha)/2\right]} = \frac{120\pi\varepsilon}{C_0(2\alpha) + C_1(2\alpha)/2}.$$ (8.28)

Here $C_0(2\alpha)$ is the partial mutual capacitance per unit length, when the value 2α is arbitrary.

The input impedance of the slot antenna, whose shape and dimensions coincide with the shape and dimensions of the metal radiator, is equal to the input impedance of the metal radiator located near the slot, whose width is $2\beta = \pi - 2\alpha$. Its wave impedance is equal to

$$W(2\beta) = \frac{120\pi\varepsilon}{C_0(2\beta) + C_1(2\beta)}.$$ (8.29)

Here $C_0(2\beta)$ is the partial mutual capacitance per unit length, when the length of the metal shell cross section is 2β. The difference between α and β leads to different wave impedances in metal and slot radiators. In the general case we are talking about the difference in wave impedances between structures answering to expressions (8.28) and (8.29). In particular case, such difference exists between the one located on the cone structures of a metal and slot antennas with different width. It is also the difference between the antennas located along the sides of a regular pyramid (with square cross section and a value b/d, equal to 1) and along the sides of an irregular pyramid (with rectangular cross section and a value b/d, not equal to 1).

In the common case, the input impedance of the metal radiator with finite length L and angular width 2α is approximately equal to $Z_{A1} = R_{A1} - jW(2\alpha)f_1(kl)$. Here $l = L/\cos\theta_0$ is the generatrix length. The input impedance of the metal radiator situated next to the slot is approximately equal to $Z_{A2} = R_{A2} - jW(2\beta)f_2(kl)$. The different widths of these radiators in spite of having the same length causes a small difference in their impedances, which is shown schematically as a distinction between R_{A1} and R_{A2}, $f_1(kl)$ and $f_2(kl)$. If $\alpha = \beta$, the product $Z_{A1}Z_{A2}$ is equal to

$$Z_{A1}Z_{A2} = [R_{A1} - jW(2\alpha)f_1(kl)]^2 . \tag{8.30}$$

If $\alpha \neq \beta$, the product is

$$Z_{A1}Z_{A2} = R_{A1}R_{A2} - W(2\alpha)W(2\beta)f_1(kl)f_2(kl) - jR_{A1}W(2\beta)f_2(kl) - jR_{A2}W(2\alpha)f_1(kl). \tag{8.31}$$

In this case, if $\alpha > \beta$, an equivalent radius of the first radiator is greater than that of the second radiator, i.e. its partial mutual capacitance per unit length is greater, and its wave impedance is smaller. The partial capacitance of each wire to the surface of zero potential (see Figure 7.15) also decreases the wave impedance of the radiator. The function $f_1(kl)$ is changed faster than $f_2(kl)$.

When the length of the cone increases, the reactive components of the input impedances tend to be zero. If $\alpha = \beta$, then these components of both radiators change fast and in step, and after that asymptotically approaches to zero. As a result the input impedance of each infinite radiator is purely active and does not depend on the frequency. This result is consistent with (8.24). If $\alpha \neq \beta$, then in accordance with (8.25) one can suppose that the reactive components change slowly in the beginning and after that begin to oscillate synchronously in opposite phase. In this case the second item in the right part of expression (3.31) is positive and compensates the decrease in the first item caused by R_{A2} decrease.

Further one must emphasize that the performance of self-complementary radiators is not always preferable in comparison with that of radiators close thereto in shape. They would be better, if the wave impedance of the cable was equal to 60π Ohm. When using a standard cable with a wave impedance 100 Ohm, the matching level in a given frequency range may be higher, if, for example, a pyramid has not the square, but a rectangular cross section, since the change of the cross section may decrease the impedance of the antenna, bringing its value closer to the wave impedance of the standard cable (see Section 8.6).

8.3   PRINCIPLE OF COMPLEMENTARITIES OF ANTENNAS ON A CONE AND ANTENNAS WITH LOADS

The equality (8.23) is the expression for the input impedance Z_S of the slot antenna which has an arbitrary shape and is cut in a flat plate of unlimited dimensions with an infinitely small thickness (Figure 8.7). Let us remember that Z_E in this formula is the input impedance of the metal radiator, the shape and dimensions of which are the same as the shape and dimensions of the slot.

The relation (8.23) for a slot antenna in a flat metal shield is obtained by two methods: by means of the duality principle [72–75] and by means of complementary principle [76]. The complementarities principle comes from an interconnection between scattering properties of the metallic and slot radiators of the same shape and dimensions.

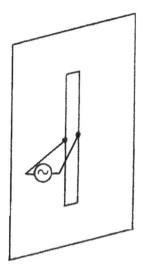

Figure 8.7 Flat slot antenna.

Two infinite plates (each plate is aggregate of the metallic and slotted antennas), in which metal antennas are replaced by slotted antennas and vice versa) are called complementary structures. As it was said formerly, the input impedances of the slotted and metal antennas of the identical shape and dimensions are related to each other by the expression (8.23).

In the case of this interconnection, the identical in shape and size metallic and slotted radiators fill the entire plane. In this case, the structure is called self-complementary. Variants of flat self-complementary radiators are shown in Figure 8.8. Characteristics of these antennas are independent from frequency [77]. The main distinctive property of such antennas is the constancy of angles between the limitative lines of the metallic elements of the antenna. The radiator thereby has the same shape at all frequencies (the radiator dimensions are proportional to the wave length).

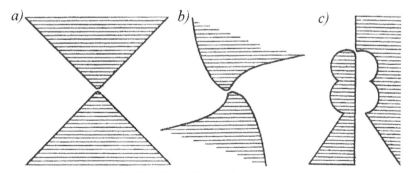

Figure 8.8 The flat self-complementary structures.

As was stated early, an input impedance Z_S of a slotted antenna is in fact an input impedance of a metallic radiator located alongside it. Impedance Z_e is an input impedance of another metallic radiator identical to a slotted antenna in the shape and size. In the particular case of a self-complementary structure they are identical, i.e. the equality (8.24) is performed irrespective of the structure shape and operation frequency.

Therefore the self-complementary structure has a constant and purely resistive input impedance in a wide frequency range.

An antenna consisting of a metallic (electrical) and a slot (magnetic) radiator of identical shape and dimensions needn't be necessarily flat. For example, the radiators depicted in Figure 8.5, which are placed at the surface of a circular cone, satisfy the condition of identity in the shape and size. Such antennas should be called volumetric self-complementary antennas. But they appeared much later, than the flat antennas, and the term itself in the beginning was interpreted otherwise. For example, in [75], the structure of two flat antennas located in mutually perpendicular planes was called a volume antenna. The author of [75] calls the volumetric antennas located on the cone as conically deformed antennas, i.e. he assumes that the electric and magnetic radiators which are elements of such antennas, do not form self-complementary antennas. At the same time he correctly considers that this structure may be used to create signals with circular polarization.

Another point of view is presented in [77], page 42: 'We can speak of a self-complementary structure in the conical case, meaning that the region of the cone covered by the metal arms is the same as that not covered except for a rotation of 90°.' In [78] it was shown firstly that structures consisting of electric and magnetic radiators of the same shape and dimensions may not be flat and secondly that the choice of surface for placement of such structures is not random. This is, in particular the case with circular cones, which in the limiting case turns into a plane.

The volumetric structure has properties similar to those of a flat structure. Really, in the general case the relation analogous to the expression (8.24) is correct for an input impedance of a symmetrical double-sided slot antenna of arbitrary shape and dimensions, which is located at a circular metal cone of an infinite length and is excited on its vertex (Figure 8.9). For derivation of such relation the duality principle is used. We consider a symmetrical magnetic V-shaped radiator (Figure 8.10a) located in a free space. Its radiation resistance R_M and the radiation resistance R_E of the electrical radiator, identical in shape and size, are related by the expression

$$R_M = (120\pi)^2 / R_E. \tag{8.32}$$

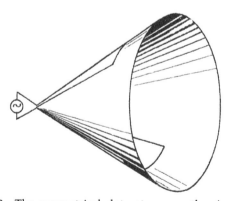

Figure 8.9   The symmetrical slot antenna on the circular cone.

This expression is well known. It follows from a comparison of the powers radiated by both antennas. If to compare the oscillating powers of both radiators, then we shall have obtain similarly to (8.24)

$$Z_m = (120\pi)^2 / Z_E, \tag{8.33}$$

where Z_M and Z_E are the input impedances of the magnetic and electrical radiators correspondingly.

In order to go from magnetic V-shaped radiators to slot antennas, we divide each arm of the magnetic radiator by a conical metallic surface (of a circular cross-section) passing through their axes. Since the shape of magnetic field lines coincides with this surface shape, the radiator field in consequence of the metallic surface insertion doesn't change. Actually, as it is shown in Section 8.1 for the field of a long line of converging straight wires, the surfaces of field strength, along which the field lines are located, are the circular cones. The magnetic field of magnetic wires has similar structure. The axis of a cone coincides with the bisector of the angle between the arms of a V-shaped radiator.

The inserted metallic surface divides the magnetic radiator into two radiators located with different its sides (inside and on the outside of the metal cone). Since as a result the magnetomotive force e_M, which excites the radiator, and its oscillating power

$$P = e_M J_M \tag{8.34}$$

do not change, the fraction of magnetic current J_M in each newly formed radiator is equal to the fraction of power radiated into each part of space.

Assume for definiteness that m is the power fraction in the smaller part of space (inside the cone) and $1-m$ is the power fraction in the greater part of space. The input admittance of the magnetic radiator located inside the cone is

$$Y_1 = e_M /(m J_M) = Y/m, \tag{8.35}$$

where $Y = 1/Z_m$ is the total admittance of the initial radiator.

Let the cross section of the magnetic radiator with the current $m J_M$ have the shape of a curved rectangle with the sides b and a, at that $b \gg a$ (Figure 8.10b). The greater side b of this cross section is parallel to the metallic surface and has the shape of an arc. The interior radiator is equivalent to a one-sided slot of width b. (It is necessary to note that the value b is changed along the axis of the slot antenna.)

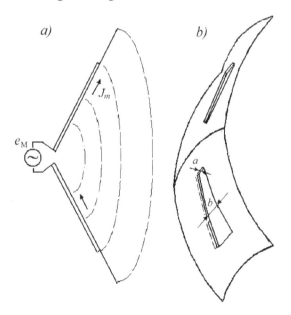

Figure 8.10 Magnetic V-shaped radiator (a) and double-sided slotted antenna (b).

The outer radiator is equivalent to the similar one-sided slot with the current $(1-m)J_M$. Its input admittance is

$$Y_2 = e_M / [(1-m)J_m] = Y/(1-m). \tag{8.36}$$

If to unite both slotted antennas into a double-sided slotted antenna and to consider that its admittance is equal to the sum of the admittances of both antennas, then

$$Y_S = Y_1 + Y_2 = Y/[m(1-m)]. \tag{8.37}$$

The input impedance of the double-sided slot is

$$Z_s = 1/Y_s$$

or taking into account (8.33) and (8.37),

$$Z_s = m(1-m)Z_M = (120\pi)^2 m(1-m)/Z_E. \tag{8.38}$$

The conical surface in the particular case of the straight magnetic radiator becomes the plane surface, which divides the space onto two equal parts. Accordingly, the fraction of current in each of the newly formed radiators is equal to $m = 1/2$, and equality (8.38) is transformed to the expression (8.24).

If the structure is self-complementary, then the metallic and slotted radiators are identical in shape and dimensions. And so

$$Z_S = Z_E = 120\pi \sqrt{m(1-m)}, \tag{8.39}$$

i.e. the volumetric self-complementary structure has properties similar to the plane structure properties. It has a constant and purely resistive input impedance, which ensures a high level of matching with a cable in the wide frequency range.

The presented consideration shows that a choice of surface for placement of the self-complementary radiation structure is not accidental. This surface is the circular cone, which in the limiting case turns into a plane. The shape of a metallic and slotted radiator, located at the circular cone, may be different – similarly to the shape of flat radiators (see Figure 8.8). The simplest shape is obtained, if the edge of a slotted antenna coincides with the cone generatrix. From (8.38) and (8.24) it follows that the magnitude m at least for infinitely long symmetrical slots of this kind located along the circular metal cone is equal to $1/2$, i.e. the identical power is radiated inside and outside irrespective of the vertex angle of the cone and width of the slotted antenna. In the general case, the question of the magnitude m is open.

The real antennas constructed in accordance with the considered principle differ from the self-complementary structures represented in Figure 8.8. First of all they have finite dimensions. Secondly the continuous metal sheet as a rule is replaced by the system of wires diverging from the input (from the feed point).

The flat metallic antenna in the shape of a triangle with the vertex angle $\pi/2$, suspended on two grounded metal supports (Figure 8.11a), is an example of that radiator. The metallic radiator is performed in the form of the system of wires divergent at same angles from the lower vertex of a triangle. This antenna with allowance for mirror image is the flat radiator consisting of metallic and slot radiators of the same shape and dimensions. The metal radiator is limited from the top by a horizontal wire, and the slotted antenna is limited from both sides by vertical shunts realized in the shape of grounded metal supports. Symmetrical variants of such radiators (Figure 8.11b) is used as stand-alone horizontal antenna and as an element of an antenna array.

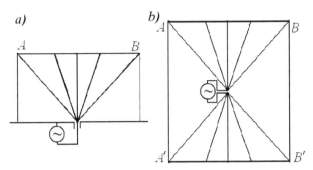

Figure 8.11 Flat vertical antenna of finite dimensions: asymmetrical (*a*) and symmetrical (*b*).

Three-dimensional antennas constructed on this principle are depicted in Figure 8.12. In them the metallic radiator with horizontal axis is located at a sharp angle to ground. This permits to increase radiator dimensions (the arm length) at the same height of supports, and also to increase its directivity. In order that dimensions of metallic and slotted radiators be equal, the distance between the supports must be twice as much as their height. Then the vertices A and B and their mirror images coincide with the vertices of a regular quadrangle (of a square).

The asymmetrical self-complementary antenna, which is depicted in Figure 8.12*a*, is located on the surface of a round cone. For simplification of construction it is expedient to replace a conical shell in each arm of a metal radiator by the flat triangle shell (see Figure 8.12*b*). This is an asymmetrical variant of an antenna located along the sides of a pyramid, which was submitted in Figure 8.6. In this case it is impossible to talk about complementary or self-complementary structure, since there is no smooth conical surface on which the metallic and slotted radiators are located. And so in a given case, the expressions (8.38) and (8.39) are inapplicable. As is shown in Section 8.2, the input impedance of such radiators is smaller than 60π, but at $b/d = 1$ it remains constant.

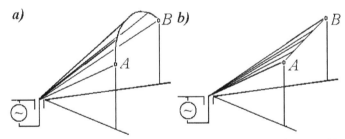

Figure 8.12 Asymmetrical antennas of finite dimensions, located on the cone (*a*), and pyramid (*b*).

The flat vertical antennas (see Figure 8.11) create the same bidirectional radiation (radiation into both sides from an antenna plane). The inclined antennas (see Figure 8.12) create unidirectional radiation with increased directivity to one side.

An interested variant of radiators placement along the faces of the pyramid was proposed in [79]. The article describes the octagonal pyramid with log-periodic radiators. Radiators are excited on the pyramid vertex and create signals of two mutually perpendicular polarizations. The author called this structure as quasi-self-complementary. In this case, the metal and slot radiators have not only different angular widths, but also different shapes.

The structure of two inclined antenna, the wires of which diverge in opposite directions from the common feed point (Figure 8.13) gives interesting opportunities. Upon in-phase excitation (see Figure 8.13*a*), the directional pattern of a structure in a horizontal plane is close to a circular one, if the angle at the vertex of the triangular plate is not small. If the structure is excited in anti-phase (see Figure 8.13*b*), its directional pattern is similar to the pattern of a horizontal dipole. In this case, the structure resembles an antenna known as a "bow-tie antenna" (see, e.g., [80]). Using this similarity, it is possible to optimize the electrical characteristics of the well-known antenna. The theory of antennas placed on the pyramid faces may become the basis for analysis of characteristics of this antenna.

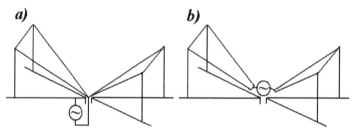

Figure 8.13   The structure of two inclined antennas with in-phase (*a*) and anti-phase (*b*) excitation.

In spite of the fact that the antenna depicted in Figure 8.11*a*, has finite dimensions, it ensures a high level of matching with a cable in the wide frequency range. But the main lobe of its vertical directional pattern deviates from the perpendicular to the antenna axes (deviates from ground) on the high frequencies, if the antenna height exceeds $0.7\lambda(\lambda/H = 1.43)$. This effect, corroborated by calculations and measurements (see Section 8.4), sets the upper limit of an antenna frequency range.

In the case of a thin radiator one can expand the antenna frequency range, if to connect in it the concentrated capacitive loads permitting to create in-phase current distribution along an antenna wire (see Section 5.2). The capacitive loads of the flat triangular metal radiator may be performed in the shape of horizontal slots. In order that the current distribution in the metallic and slotted radiator coincide with each other, the slot radiator must be divided by vertical metal plates, and the width of each plate must coincide with a width of according horizontal slot (Figure 8.14). For such flat structure, the expression (8.24) remains correct.

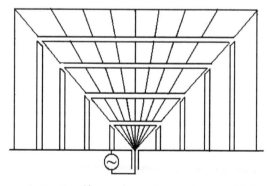

Figure 8.14   A self-complementary antenna with loads.

In derivation of equality (8.39) the metal surface was used. This surface, coincident with the shape of magnetic field lines, divided the each arm of the magnetic V-radiator. In the general case this conical surface may have a cross section in the shape of two arcs (in Figure 8.2 these arcs are shown by a dotted line). These arcs may have different radii and different centers located in the plane yOz (but not in the shape of the one circumference). It is impossible to place the self-complementary radiators on such a surface. But the equality (8.38) for the input impedance of the symmetrical double-sided slotted antenna remains correct.

8.4 COMPARISON OF PARABOLIC AND CYLINDRICAL PROBLEMS. THE FIELD OF A LONG LINE FROM THE CONVERGING PARABOLIC WIRES

As it is shown in Section 8.1, using equalities (8.3), relating the replaceable variables of the spherical and cylindrical systems of coordinates, one can reduce conical three-dimensional problem to two-dimensional problem. This result had played a big role in the development of electromagnetic theory and in its use in antenna engineering. A similar result can be obtained, if to reduce the parabolic problem to the plane problem. The generatrix of a circular cone is a straight wire. A curve line also may be used as a generatrix. If, for example, a generatrix has a parabolic shape, the metallic surface assumes the shape of a paraboloid. Calculating the field of two infinitely long charged filaments of curvilinear shape located on the paraboloid surface and converging to its vertex (Figure 8.15) is a matter of unconditional interest.

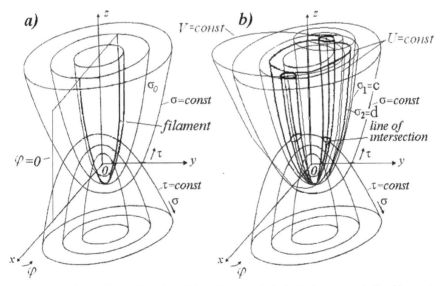

Figure 8.15 A three-dimensional problem for two infinitely long parabolic filaments (a), and solid geometrical figures (b).

Use of parabolic coordinates (σ, τ, φ) facilitates an analysis of such structures. This is a system of orthogonal curvilinear coordinates [81]. Its coordinate surfaces are firstly confocal paraboloids of rotation ($\sigma = $ const, $\tau = $ const), whose focal point coincides with the origin of the coordinates system, and secondly half-planes (φ = const) passing through the axis of rotation (see Figure 8.15). Rectangular coordinates are related to parabolic coordinates by the equalities:

$$x = \sigma\tau \cos \varphi, \, y = \sigma\tau \sin \varphi, \, z = (\tau^2 - \sigma^2)/2. \tag{8.40}$$

Their parabolic wires are located along the surface of a paraboloid, or more exactly along the curves of intersection of this surface with the half-planes, passing through the axis of rotation.

As in the case of convergent straight wires, it is expedient to reduce the calculation problem for the electrostatic field of two charged parabolic filaments to the problem of two parallel filaments (see Figure 8.1b). To this end the Laplace's equation in accordance with the uniqueness theorem must remain correct at the transition from one problem to another, and the surfaces of conductive bodies (wires) must coincide with the surfaces of equal potential.

In the cylindrical coordinates system (ρ, φ_c, z), the Laplace's equation for a potential U has the form

$$\frac{\partial}{\partial\rho}\left(\rho\frac{\partial U}{\partial\rho}\right) + \frac{1}{\rho}\frac{\partial^2 U}{\partial\varphi_c^2} = 0. \tag{8.41}$$

Here, it is taken into account that $\partial U/\partial z = 0$, i.e. the lines parallel to z-axis have the constant potential (the field is plane-parallel). In the system of parabolic coordinates the Laplace's equation has the form

$$\frac{1}{\sigma^2 + \tau^2}\left[\frac{1}{\sigma}\frac{\partial}{\partial\sigma}\left(\sigma\frac{\partial U}{\partial\sigma}\right) + \frac{1}{\tau}\frac{\partial}{\partial\tau}\left(\tau\frac{\partial U}{\partial\tau}\right) + \left(\frac{1}{\sigma^2} + \frac{1}{\tau^2}\right)\frac{\partial^2 U}{\partial\varphi^2}\right] = 0. \tag{8.42}$$

As seen from (8.42), this equation is symmetrical with respect to σ and τ, that is, the equation for each unknown quantity is true, irrespective of other equationas. In particular, for σ we obtain

$$\frac{1}{\sigma^2}\left[\frac{1}{\sigma}\frac{\partial}{\partial\sigma}\left(\sigma\frac{\partial U}{\partial\sigma}\right) + \frac{1}{\sigma^2}\frac{\partial^2 U}{\partial\varphi^2}\right] = 0. \tag{8.43}$$

Here $U(\sigma) = U(\tau)$, if other coordinates are the same. If, for example, $\partial U/\partial z = 0$, then in accordance with (8.40) $\sigma = \sqrt{\tau^2 - 2z}$, $\tau = \sqrt{\sigma^2 + 2z}$, i.e.

$$\partial\sigma/\partial z = -1/\sigma, \partial\tau/\partial z = 1/\tau, \tag{8.44}$$

and

$$\frac{\partial U}{\partial z} = \frac{\partial U}{\partial\sigma}\frac{\partial\sigma}{\partial z} + \frac{\partial U}{\partial\tau}\frac{\partial\tau}{\partial z} = 0,$$

whence

$$(1/\tau)\partial U/\partial\tau = (1/\sigma)\partial U/\partial\sigma. \tag{8.45}$$

Comparison of (8.41) and (8.43) shows that these equations coincide, if the substituted variables are related by equations:

$$\rho = \sigma, \, \varphi_c = \varphi. \tag{8.46}$$

Here ρ and φ_c are the cylindrical coordinates, and σ and φ are the parabolic coordinates. Hence, the Laplace's equation holds true in the transition from the parabolic problem to the cylindrical, if expressions (8.46) are true. The previous conclusion of the author, which requires the execution of inequality $\tau \gg \sigma$, is incorrect.

The substitution of variables in accordance with (8.46) results in the mapping of parabolic surface τ = const (for arbitrary τ) onto the plane (ρ, φ_c). The line of this surface intersection with any paraboloid σ = const is transformed into a circumference. Two parabolic filaments situated along surface $\sigma = \sigma_0$ in the plane xOz (see Figure 8.15a) are transformed into two parallel wires spaced at distance $2b = 2\sigma_0$ in the cylindrical coordinates system (see Figure 8.1b). Two continuous geometrical figures located between parabolic surfaces $\sigma_1 = c$ and $\sigma_2 = d$ (see Figure 8.15b) are transformed into two metal cylinders of radius $a = (c - d)/2$ (see Figure 8.2a), whose axes spaced at the distant $2h = c + d$.

The scalar potential of the electric field for two parabolic charged filaments situated along surface $\sigma = \sigma_0$ (with linear charge density equal to $\pm q_0$) is similar to (8.2)

$$U(\sigma,\varphi) = \frac{q_0}{2\pi\varepsilon} \ln(\rho_2/\rho_1),\qquad(8.47)$$

where

$$\rho_2^2 = (\sigma_0 - \sigma\cos\varphi)^2 + \sigma^2\sin^2\varphi, \quad \rho_1^2 = (\sigma_0 + \sigma\cos\varphi)^2 + \sigma^2\sin^2\varphi.$$

The designation q_0 is used here instead of τ in order to avoid confusion.

The surfaces of equal potential U = const in the given case are the volumetric geometrical figures, and their planes of symmetry coincide with the plane xOz. Each surface satisfies an equation

$$\rho_2/\rho_1 = m = \text{const}.$$

For the plane problem the line of equal potential is the circumference with the center in the point

$$h_m = b(1 + m^2)/(1 - m^2)$$

and with the radius

$$a_m = 2bm / |1 - m^2|.$$

Using these magnitudes, we find the coordinate σ of the parabolic surface, on which the axis of volumetric figure is located:

$$\sigma_m = h_m = \sigma_0(1 + m^2)/|1 - m^2|,\qquad(8.48)$$

and also the semi-axis length of the shorted curve formed by the intersection of this volumetric figures with the surface τ = const:

$$a_m = (\sigma_{1m} - \sigma_{2m})/2 = 2m\sigma_0/|1 - m^2|.\qquad(8.49)$$

The angle ψ_m between the paraboloid generatrix and axis is

$$\psi_m = \tan^{-1}\left\{2a_m\Big/\left[1 - \left(h_m^2 - a_m^2\right)\right]\right\} = \tan^{-1}(2m/|1 - m^2|)\tan\theta_0.\qquad(8.50)$$

The surfaces of field strength V = const, in which force lines are located, are circular paraboloids. Their axial lines lie in the plane yOz on the parabolic surfaces

$$\sigma_n = h_n = \sigma_0\cot\varphi_n,\qquad(8.51)$$

and the semi-axis lengths of the shorted curves on the surface τ = const are equal to

$$a_n = (\sigma_{1n} - \sigma_{2n})/2 = \sigma_0/\sin\varphi_{cn}.\qquad(8.52)$$

The sense of the angle φ_{cn} is clear from Figure 8.1b, where it is designated by φ_n.

Indeed in the plane problem the flux function appears as

$$V = -q_0(\varphi_{c2} - \varphi_{c1})/(2\pi\varepsilon).$$ (8.53)

The equation of the field line

$$\varphi_{c2} - \varphi_{c1} = \varphi_{cn} = \text{const}$$ (8.54)

is the equation of a circumference, the radius of which is equal to $a_n = b/\sin\varphi_{cn}$, and the axis is located from axis z to the distance $h_n = b\cot\varphi_{cn}$. In the parabolical problem the axial lines of the circular paraboloids $V = \text{const}$ also lie in the plane yOz and form with axis z the angle

$$\theta_n = \tan^{-1}(\cot\varphi_{cn}\tan\theta_0),$$ (8.55)

and the angle φ_n between the paraboloid axis and the generatrix is equal to

$$\varphi_n = \tan^{-1}(\tan\theta_0/\sin\varphi_{cn}).$$ (8.56)

From this it is clear that one of the surfaces of the equal strength created by the parabolic V-radiator coincides with the parabolic metallic surface passing through the axes of this radiator. Lines of magnetic field lie in this surface, and therefore if to insert the parabolic metallic surface passing through the axes of this radiator, the radiator field does not change.

Use of the equalities (8.46) allows to reduce the parabolic problem to the cylindrical problem and to calculate capacitance C_l per unit length and wave impedance W of a long line consisting of two volumetric figures with the parabolic axes. By analogy with (8.15), we find:

$$C_l = \frac{\pi\varepsilon}{ch^{-1}\left[(\sigma_1+\sigma_2)/(\sigma_1-\sigma_2)\right]}, \quad W = 120ch^{-1}\left[(\sigma_1+\sigma_2)/(\sigma_1-\sigma_2)\right].$$ (8.57)

The input impedance of a uniform two-wire line tends to its wave impedance. When the line length increases, the input impedance of the line tends to

$$Z_{AB} = 120ch^{-1}\frac{\sigma_1+\sigma_2}{\sigma_1-\sigma_2}.$$ (8.58)

The case of two charged converging shells located along the paraboloid surface (Figure 8.16) is of a specific interest. This structure is a variant of placing metallic and slot antennas of finite length on surface of rotation. If the arc lengths of the converging shells are equal to 2α, then this parabolic problem corresponds to the plane problem for a line from two coaxial cylindrical shells. The electrostatic field of such line is shown

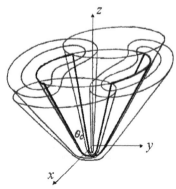

Figure 8.16 Three-dimensional problem of non-closed coaxial shells located on the paraboloid.

in Figure 8.4b. The field structure in the case of two shells located on the paraboloid surface is of a similar nature, but surfaces of equal potential U = const and surfaces of field strength V = const coincide with the parabolic surfaces rather than with the cylindrical surface.

As seen from (8.19), capacitance C_l per unit length and wave impedance W of an equivalent line are dependent only on the arc length 2β of the slot antenna in the cross section of the paraboloid and, accordingly, on arc length $2\alpha = \pi - 2\beta$ of the metallic shell. For this reason, expressions (8.19) are true also for the parabolic envelopes as for the conical. Magnitudes C_l and W are constant along the line, i.e. the line of two wires is the uniform. The wave impedance and the input impedance of an infinitely long line excited by a generator situated near the paraboloid vertex are equal to

$$W = Z_{AB}(\alpha) = 120\pi K(k)/K(\sqrt{1-k^2}),$$ (8.59)

where in the case of two metal plates (one dipole) $k = \tan^2(\beta/2)$. If $\alpha = \beta$, the radiator is self-complementary, and $Z_{AB}(\pi/4) = 60\pi$.

8.5 CHARACTERISTICS OF FLAT ANTENNAS AND THE ANTENNAS ON A CONIC SURFACE

Let us go back to the characteristics of the symmetrical self-complementary vertical flat antenna (see Figure 8.11b). The finite dimensions is main its difference from the radiator shown in Figure 8.8a. Let each arm of the symmetrical flat antenna be built as the metal sheet of the height H and the thickness $0.001H$, and the gap between the arms in the feed point is equal to $0.001H$. The electrical characteristics of this antenna depending on λ/H, are presented in Figure 8.17 (active R_A and reactive X_A components of input impedances), 8.18 (reflectivity and standing wave ratio in a cable with a wave impedance of 60π and 100 Ohm), and 8.19 (horizontal and vertical directional pattern). These characteristics were calculated using the program CST. Besides, in Figure 8.20 the pattern factor is shown—see expression (5.65). And in Table 8.1 the maximum gain and the radiation efficiency of the flat antenna are given. The gain is calculated with due account of matching losses (the wave impedance of the cable is adopted 60π Ohm).

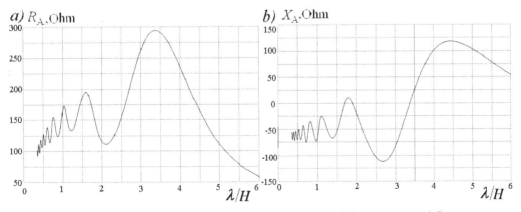

Figure 8.17   Active R_A (a) and reactive X_A (b) impedance of the symmetrical flat antenna.

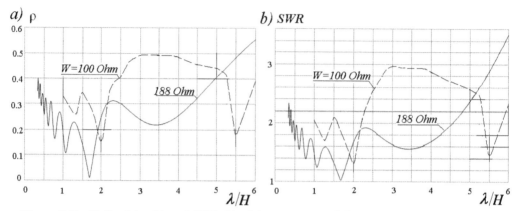

Figure 8.18 Reflectivity (*a*) and *SWR* (*b*) of the symmetrical flat antenna in the cable with wave impedance 60π and 100 Ohm.

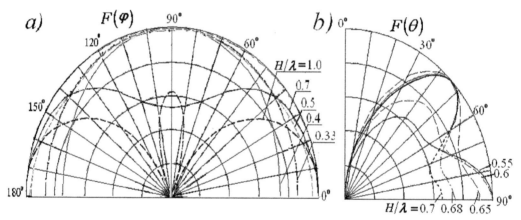

Figure 8.19 Horizontal (*a*) and vertical (*b*) pattern for $\varphi = 0°$ of the symmetrical flat antenna.

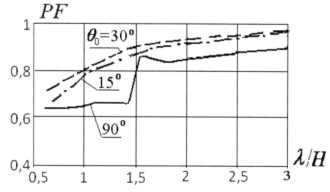

Figure 8.20 Pattern factor of the symmetrical flat and conical antennas.

Table 8.1 Maximal Gain and Radiation Efficiency of Symmetrical Flat and Conic Antennas

λ/H	Gain at θ_0			Efficiency at θ_0		
	90°	30°	15°	90°	30°	15°
1	2.8	16.1	17.8	0.97	0.96	0.98
2	1.9	4.7	4.9	0.91	0.96	0.98
2.5	2.3	3.4	3.4	0.92	0.98	0.95
3	1.9	2.5	2.5	0.93	0.90	0.91

Large radiators with equal angle width of the metallic and slotted radiator have the constant and purely resistive input impedance, i.e. permit to secure a high level of the matching in wide range of frequencies. The radiators located along the cone and operating in a predetermined frequency range allow either to obtain the given electrical characteristics when the antenna has smaller height than the height of a vertical antenna or to improve the electrical performance at the same height.

Therefore, the interest to the conic self-complementary antennas is explained by two factors: by a high level of the matching and by the opportunity to reduce the antenna height.

We compare from this point of view the electrical characteristics of a symmetrical antenna located along the surface of the cone with horizontal axis and the angle $2\theta_0 = 60°$ at the vertex (see Figure 8.5a), which are given by solid curves in Figures 8.21–8.23 depending on λ/H. Analogous characteristics are given here by dotted lines for the cone with the angle 30° at the vertex. Each arm of the antenna is built as a thin metal shell of height H and a thickness $0.01H$ located along the conic surface. The gap between the arms is equal to $0.01H$. Characteristics are calculated by the program CST.

Active R_A and reactive X_A components of input impedances of these antennas are presented in Figure 8.21. A comparison of them shows that their magnitudes depend on the angle θ_0 though for an infinite length of the wires the input impedances are the same, i.e. do not depend on the angle at the cone vertex. As it is seen from Figure 8.17 and 8.21, decreasing the arm length (when the angle $2\theta_0$ is changed from 30° to 60°) leads to displacement of maximums and minimums to the left, in the direction of small λ, and to decreasing the maximal magnitudes.

Reflectivity ρ in the cable with wave impedance 60π is given in Figure 8.22 for the angle 30° and 60° at the cone vertex. These magnitudes are compared with each other and with corresponding reflectivity of the vertical antenna with the same height, i.e. the antenna located at the angle 90° to the earth (this curve is shown by circles). The standing wave ratio of the same antennas is presented in Figure 8.23, including SWR in the cable with wave impedance 60π (Figure 8.23a) and SWR in the cable with wave impedance 100 Ohm (Figure 8.23b). As can be seen from these figures, in the operating range, reflectivity ρ of conic antennas is smaller than the corresponding magnitude of the vertical antenna. SWR of conic antennas is substantially closer to unity than SWR of the vertical antenna. In the cable with wave impedance 60π (see Figure 8.23a) SWR of the antenna with $2\theta_0 = 60°$ is smaller than SWR of the vertical antenna in the band $1.9 \leq \lambda/H \leq 2.9$, SWR of the antenna with $2\theta_0 = 30°$ is smaller also in the band $1.75 \leq \lambda/H \leq 2.9$. In the cable with wave impedance 100 Ohm, SWR of the symmetrical conic antenna with $2\theta_0 = 60°$ is smaller than the SWR of the vertical antenna in the band $2.5 \leq \lambda/H \leq 3.75$, SWR of the antenna with $2\theta_0 = 30°$ is smaller in the band $2.2 \leq \lambda/H \leq 3.8$ (see Figure 8.23b).

In Figure 8.24 are given the horizontal and vertical patterns of conic antennas. The horizontal pattern of the conic antenna is very different on the aperture side and on the opposite side, i.e. radiation of the conic antennas is increased in the direction of the cone aperture. The horizontal pattern of the flat antenna is symmetrical about the plane in which it is located.

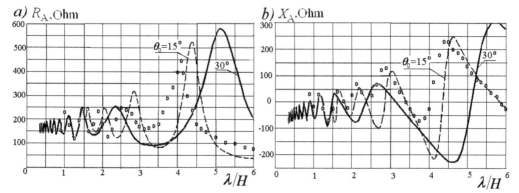

Figure 8.21   Active R_A (a) and reactive X_A (b) impedance of the symmetrical conic antennas with the angle 60° and 30° at the cone vertex.

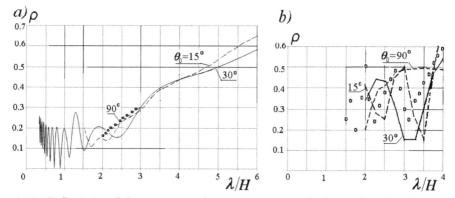

Figure 8.22   Reflectivity of the symmetrical conic antennas with the angle 60° and 30° at vertex in the cable with wave impedance 60π (a) and 100 Ohm (b).

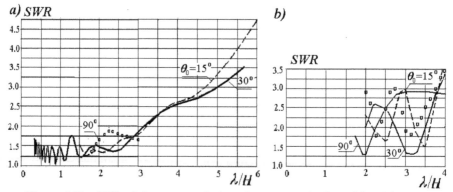

Figure 8.23   SWR of the symmetrical conic antennas in the cable with wave impedance 60π (a) and 100 Ohm (b).

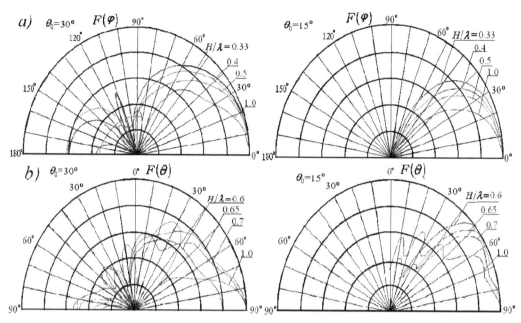

Figure 8.24 Horizontal (*a*) and vertical (*b*) pattern of the symmetrical conic antennas with the angle 60° and 30° at the cone vertex.

The vertical directional pattern of the flat antenna (see Figure 8.19) shows that main lobe of its vertical pattern deviates from the perpendicular to the antenna at λ/H about 1.5, and at higher frequencies the value of PF falls down to 0.65. This value for the conic antennas is close to 0.8, when $\lambda/H \approx 1$. Maximum gain of the conic antennas (see Table 8.1) is substantially higher than the maximum gain of the flat antenna. Really, maximum gain of the flat vertical antenna when the wave impedance of the cable is equal to 188 Ohm, in the range $1.0 \leq \lambda/H \leq 3.0$ is close to 2, while the maximal gain of the conic antenna with $2\theta_0 = 60°$ varies from 2.5 to 16, and the maximal gain of conic antenna with $2\theta_0 = 30°$ varies from 2.5 to 18 (see Table 8.1). From the same table it can be seen that the efficiency of antennas is approximately the same—from 0.9 to 0.99. The curves given in the figures, and the numerical values presented in the table are obtained using the CST program.

The calculation results were verified on models placed in an anechoic chamber. A general view of the set with the model of conic antenna is shown in Figure 8.25. The angle at the cone vertex is equal $2\theta_0 = 30°$. Experimental values of active and reactive components of the antenna impedance are given by circles in Figure 8.21. Since the input impedance of the asymmetric version of the antenna is half as many as impedance of the symmetric version, then the experimental values are doubled for comparison with the calculation. As can be seen from the figures, the coincidence between theory and experiment is not only qualitative, but also quantitative.

The measurement results for reflectivity and SWR in the cable with the wave impedance 100 Ohm are presented in Figures 8.22*b* and 8.23*b* (asymmetric antenna was excited by the cable with the wave impedance 50 Ohm). Here the agreement of the calculation and the experiment also corresponds to a measurement precision.

As mentioned in Section 8.2, the wave and input impedances of the antenna with infinite sizes do not vary with frequency. If the antenna has finite sizes, then, as the calculation and experiment show, the electrical characteristics are frequency dependent.

Figure 8.25   The general view of the set with the mock-up of conic antenna in an anechoic chamber.

8.6   CHARACTERISTICS OF ANTENNAS ON FACES OF A PYRAMID AND ON A PARABOLIC SURFACE

In Section 8.3 it was said that for simplification of construction one can perform the arms of a metallic radiator not in the shape of conical shells, but in the form of flat triangles located along the faces of a pyramid (see Figure 8.12b). This is especially important for antenna of great dimensions, i.e. at low frequencies. A set with a model of such an antenna is presented in Figure 8.26. The angle at the pyramid vertex is equal to $2\theta_0 = 30°$.

The electrical characteristics of such symmetrical antennas are given by solid curves in Figures 8.27–8.29 depending on λ/H, and also in Table 8.2. Active R_A and reactive X_A components of the input impedance of the antenna, placed along the faces of the pyramid with the square cross section and the angle 30° at vertex are presented in Figure 8.27. The same components of the input impedance of a conical antenna with the same angle at vertex are given for comparison in the same picture. The characteristics of different antennas are similar, but not identical. Besides, in Figure 8.30 the pattern factor of antennas placed along the pyramid faces is shown.

Figure 8.26   A set with the model of an antenna placed on the faces of a regular pyramid.

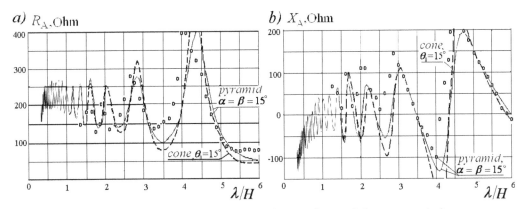

Figure 8.27 Active R_A (a) and reactive X_A (b) impedance of the symmetrical antennas located on a regular pyramid and cone.

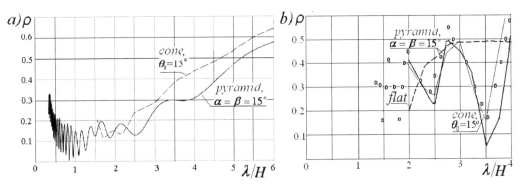

Figure 8.28 Reflectivity of symmetrical antennas located on the regular pyramid and cone in the cable with wave impedance 60π (a) and 100 Ohm (b).

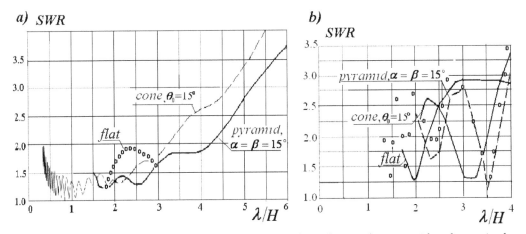

Figure 8.29 SWR of the symmetrical antennas located on the regular pyramid and cone in the cable with wave impedance 60π (a) and 100 Ohm (b).

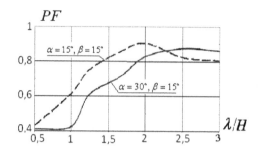

Figure 8.30 Pattern factor of the symmetrical antennas on the pyramid.

Table 8.2 Maximal Gain and Efficiency of a Symmetrical Antenna on the Pyramid

λ/H	Gain at		Efficiency at	
	$\alpha = 15°, \beta = 15°$	$\alpha = 30°, \beta = 15°$	$\alpha = 15°, \beta = 15°$	$\alpha = 30°, \beta = 15°$
1	23.3	17.6	0.97	0.98
2	6.56	10.2	0.98	0.95
2.5	5.21	6.87	0.98	0.95
3	4.08	5.5	0.93	0.94

The directional pattern of an antenna placed along the faces of a regular pyramid has in the horizontal plane the shape of an oval, elongated in a direction perpendicular to its aperture, i.e. this pattern is similar to the directional pattern of any inclined antenna. Already it was pointed out that the use of the volumetric antenna with inclined triangular sheet allows for the same height of supports to provide unidirectional radiation and to increase the radiator length, i.e. to expand its operation range to the side of lower frequencies.

The electrical characteristics of an antenna placed along the faces of the irregular pyramid are given in Figures 8.31–8.33 depending on λ/H, and also in Table 8.2. The pyramid has a rectangular cross-section. The angle at the vertex of the pyramid between the arms of the metallic antenna is equal to $60°$, and the angle at the vertex between the arms of slot antenna is equal to $30°$ (see Figure 8.6). The same characteristics of the antenna placed along the faces of a regular pyramid are given for comparison in the same figures. They show that in the given case, SWR of the antenna placed along the faces of the irregular pyramid is greater than SWR of the antenna placed along the faces of the regular pyramid in the cable with the wave impedance 60π Ohm and smaller in the cable with the wave impedance 100 Ohm.

Results of calculating antennas located along the faces of a regular and an irregular pyramid show that the characteristics of self-complementary radiators is not always preferable in comparison with characteristics of the radiators close thereto in shape. As can be seen from Figure 8.31, increasing the angular width of the metal (electric) radiator arm increases a current along it and reduces the active and reactive components of the input impedance, bringing its value closer to the wave impedance of the standard cable.

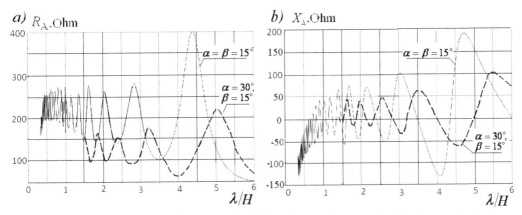

Figure 8.31 Active R_A (a) and reactive X_A (b) impedance of symmetrical antennas located on the faces of the regular and irregular pyramid.

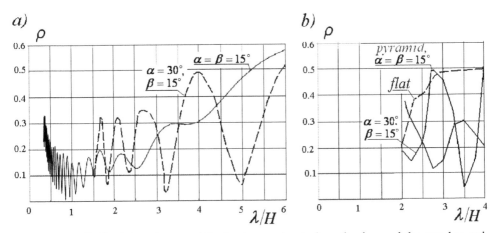

Figure 8.32 Reflectivity of symmetrical antennas located on the faces of the regular and irregular pyramid in the cable with wave impedance 60π (a) and 100 Ohm (b).

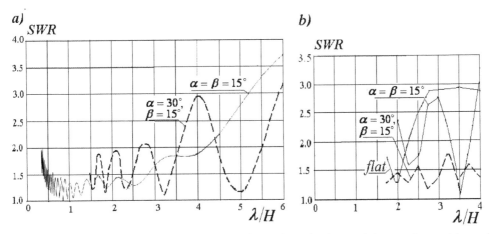

Figure 8.33 *SWR* of the symmetrical antennas located on the faces of the regular and irregular pyramid in the cable with wave impedance 60π (a) and 100 Ohm (b).

From these results it follows that the optimum radiator choice depends on the used cable. If the special cable with an wave impedance $30\pi = 94$ Ohm is employed with an asymmetric variant of the antenna, the characteristics of the antenna located along the faces of a regular pyramid are better. If the standard cable with the wave impedance 50 Ohm is used with an asymmetric variant of the same antenna, the characteristics of the other antenna are better. In order to match the conic antenna with the standard cable, whose wave impedance is smaller than the wave impedance of the antenna, one must increase the angular width of the metal radiator.

When selecting the cable wave impedance, the cable length is also essential. Let, for example, the cable attenuation in the given frequency range in conditions of traveling wave, when $SWR = 1$, be 0.6 dB/m. As can be seen from Figure 8.33, the average SWR in the range 2–4 GHz in the cable with $W = 50$ Ohm during exploitation of the asymmetric flat antenna, and the antennas located along the faces of regular and irregular pyramids is equal to 1.5, 2.0 and 2.9 respectively. Calculation shows that for mentioned SWR the equivalent attenuation in a cable of length 2 m is equal to 1.3, 1.4 and 1.6 dB respectively, and in the cable of length 10 m the equivalent attenuation is equal to 6.3, 6.5 and 7.1 dB. Efficiency of the first cable is 0.75, 0.73 and 0.7 and the efficiency of the second cable is 0.23, 0.22 and 0.19.

Together with the calculating characteristics of the radiators located along the faces of a pyramid, the active and reactive components of the input impedance, as well as the reflectivity and SWR of these antennas in the cable with wave impedance 100 Ohm were measured. The obtained results were found in fairly good conformity with the calculation.

The electrical characteristics of antennas placed on a parabolic surface are similar to the characteristics of the antennas on a cone. The close correspondence of W and k for radiators located along parabolic and conic surfaces means that for matching the parabolic antenna with the standard cable, whose wave impedance is smaller than the antenna wave impedance, one must increase the angular width of the metal radiator. That increases a current along it and reduces the active and reactive components of the input impedance, bringing their magnitudes closer to the wave impedance of the standard cable.

Let the axis of the paraboloid of rotation be located horizontally (Figure 8.34), as in the case of conic radiators considered earlier. Its radius in the aperture plane is equal to H, the distance from the paraboloid focus to the aperture plane is L, and the inclination angle of the straight line from the focal point to the aperture boundary is equal to θ_0.

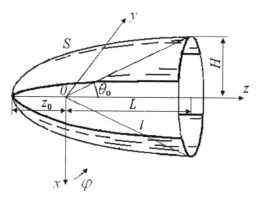

Figure 8.34 A self-complementary antenna on a circular paraboloid.

From (8.40) for the paraboloid it follows that the radius of its cross section is equal to $R = \sigma\tau$, i.e. it varies along its axis z. If to introduce the notation $\xi = \tau^2$ and to take into account that $\sigma^2 = R^2/\tau^2$, then the third expression of (8.40) permits to obtain a quadratic equation for ξ: $\xi^2 - 2z\xi - R^2 = 0$ in the arbitrary point z. Its solution gives

$$\tau = \sqrt{x} = \sqrt{z + \sqrt{z^2 + R^2}}.$$

In the point $z = L$, the radius R is equal to H, and $\tau = \tau_0 = \sqrt{L + l}$, where $l = \sqrt{L^2 + H^2}$, i.e. on the surface of a paraboloid $\sigma_0 = H/\tau_0 = H/\sqrt{L + l}$. From (8.40) $\tau = \sqrt{\sigma_0^2 + 2z}$, where from

$$R = \sigma_0 \tau = \frac{H}{L + l}\sqrt{H^2 + 2z(L + l)}. \tag{8.60}$$

Note that the vertex of the paraboloid (where $R = 0$) is located at a point $z_0 = -H^2/2(L + l)$.

The length of a paraboloid generatrix is calculated in different ways. The simplest one is based on the equation of the parabola, which in our case has the form:

$$z_1 = z + |z_0| = ax^2.$$

At the point of the paraboloid aperture $x = H$. The factor a and the parameter p are accordingly equal to

$$a = (L + |z_0|)/H^2 = [2L(L + l) + H^2]/[2(L + l)H^2]$$

and

$$p = 1/2a = (L + l)H^2/[2L(L + l) + H^2].$$

The generatrix length is calculated by a known formula

$$S = \sqrt{z_1(z_1 + p/2)} + p/2\, sh^{-1}\sqrt{2z_1/p}. \tag{8.61}$$

For the given height H the magnitude L depends on the slope angle θ_0 of the line connecting the focus of paraboloid with the end of the generatrix: $L = H \cot \theta_0$. If, for example, $\theta_0 = 30°$, then $L = 1.732H$, $l = 2H$, $L + l = 3.732H$, $\tau_0 = 1.932\sqrt{H}$, $\sigma_0 = 0.518\sqrt{H}$, $|z_0| = 0.134H$, $a = 1.866/H$, $p = 0.268H$ and the generatrix length is equal to $S = 1.932H + 0.272H = 2.204H$. If $\theta_0 = 15°$, $S = 4.04H$.

It is expedient to compare the lengths of self-complementary radiators of different shapes. The arm length of a vertical flat radiator is $S_1 = H$, the arm of the conic radiator shell for angle θ_0 between the shell and cone axes is $S_2 = H/\sin\theta_0$. In particular, if $\theta_0 = 30°$, then the generatrix length is $S_2 = 2H$. For the paraboloid with $\theta_0 = 30°$ it is equal to $S_3 = 2.204H$. Therefore the arm length of a parabolic radiator is larger than that of a conic radiator (increasing the overall length of the structure by 7.7% due to the segment z_0 leads to 10% increase of the generatrix length). One can expect that the increase of the radiator length increases the SWR in the cable.

In Section 8.4 it is shown that the magnetic V-radiator of parabolic shape creates a field whose force lines coincide with the metal surface of a circular cross section, passing through the axes of this radiator arms. As in the case of a magnetic V-radiator with straight arms, the field of a given radiator in consequence of the surface insertion does not change. That permits to go from V-radiator to the slot antenna, i.e. to obtain the expression (8.38) and to be convinced that the radiators located on the surface of circular paraboloid are complementary.

Calculation results of characteristics for flat and conic self-complementary antennas and also for antennas located a the pyramid are given in Sections 8.5–8.6. The electrical characteristics of the symmetrical self-complementary antenna located along the surface of the circular paraboloid are given in Figures 8.35–8.37 depending on λ/H. The inclination angle of the line between the focal point and the aperture boundary is equal to $\theta_0 = 15°$ and $30°$. Maximal gain and radiation efficiency of these antennas are given in Table 8.3. Besides, in Figure 8.38 the pattern factor of these antennas is shown.

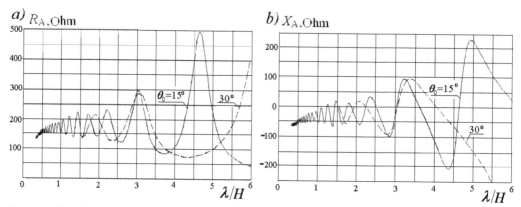

Figure 8.35   Active R_A (a) and reactive X_A (b) impedance of the symmetrical parabolic antennas with the angle $2\theta_0 = 60°$ and $30°$.

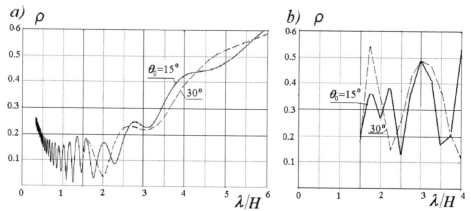

Figure 8.36   Reflectivity of symmetrical parabolic antennas with angle $2\theta_0 = 60°$ and $30°$ in the cable with wave impedance 60π (a) and 100 Ohm (b).

Table 8.3   Maximal Gain and Efficiency of Symmetric Parabolic Antennas

λ/H	Gain at θ_0		Efficiency at θ_0	
	15°	30°	15°	30°
1	12.4	11	0.98	0.98
2	6.81	4.16	0.97	0.96
2.5	5.25	3.51	0.97	0.95
3	4.36	2.78	0.95	0.95

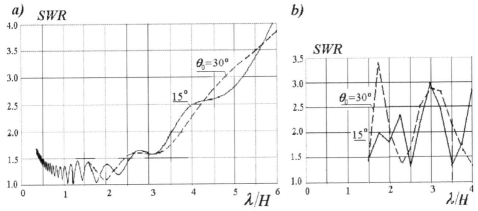

Figure 8.37 SWR of symmetrical parabolic antennas with angle, $2\theta_0 = 60°$ and $30°$ in the cable with wave impedance 60π (a) and 100 Ohm (b).

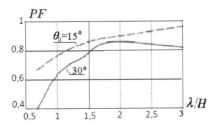

Figure 8.38 Pattern factor of the symmetrical parabolic antennas.

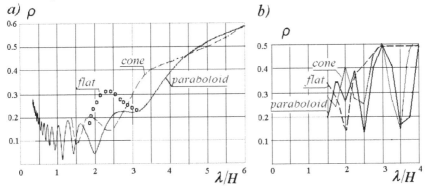

Figure 8.39 Reflectivity of the symmetrical self-complementary antennas located along the paraboloid, cone and vertical plane in the cable with wave impedance 60π (a) and 100 Ohm (b).

As is well known, one of the important requirements for an antenna, especially in the range of long, medium-frequency and short waves, is the requirement to reduce its geometrical dimensions, first of all, its height. If the antenna cost increases linearly with increasing the horizontal dimensions, then increasing the height increases the cost approximately to a third power. When the antenna is mounted on a vehicle, its height affects the air resistance (e.g., to an aircraft), and the stability to the wind. An aspiration of developers to reduce the size of antennas without deterioration of their characteristics is natural. Such developments make possible to improve electrical characteristics of an antenna, while maintaining its size.

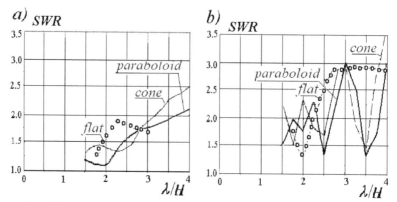

Figure 8.40 *SWR of the symmetrical self-complementary antennas located along the paraboloid, cone and vertical plane in the cable with wave impedance 60π (a) and 100 Ohm (b).*

Therefore, the interest to the three-dimensional antennas is explained by two factors: by a high level of matching and by the opportunity to reduce the antenna height. Besides, it is necessary to take into consideration that at low frequencies it is easier to realize a simpler structure, for example a radiator situated along the pyramid faces is preferable than a radiator in the shape of a cone.

As it is follows from presented results, on the whole the comparison shows that decreasing an inclination angle of the antenna arm with respect to ground (and corresponding increase of the arm length without height increase) improves the electrical characteristics of the antenna. If the inclination angle of the radiator located on the surface of a cone, paraboloid and pyramid is the same, then the radiator on the paraboloid surface has a greater gain, particularly in the lower part of the frequency range. The directional patterns of the radiators situated on the cone and paraboloid in a vertical plane (pattern factor) are approximately the same, and the directivity of the flat vertical antennas and the antennas located on the pyramid is smaller. The radiation efficiency of the different antennas is similar and close to 100%. The level of matching depends on the wave impedance of the cable and may be regulated by the change of angular width of metallic and slot radiators located on the cone and paraboloid and by the ratio of the sides in a rectangular cross section of a pyramid.

8.7 DISTRIBUTION OF ENERGY FLUX THROUGH THE CROSS SECTION OF A LONG LINE

As is shown in Section 7.1, the method of electrostatic analogy and the principle of correspondence allow to use calculating an electrostatic field of charged conductive bodies for calculating constant currents in a homogeneous weakly conducting medium, and also for determining the magnetic field of constant linear currents. In particular one may analyze an energy flux distribution through a cross-section of a long line.

In the case of a plane-parallel electrostatic field, the flux function $V(x, y)$ defines the flux of vector $\vec{E}$ through the cylindrical surface of unit length located between the given and zero surface of the field strength. For two parallel infinitely long charged filaments, the flux function is defined by the expression (8.11). It is proportional to the difference of angles $\varphi_{c2} - \varphi_{c1}$ (see Figure 8.1b), where φ_{ci} is the coordinate of the metal filament i. For any point of the cylindrical surface along the z-axis situated around the left charged filament at distance $\delta \ll b$, angle φ_{c2} is constant and equal to π, i.e. the magnitude of the flux inside the given angle $\varphi_{c2} - \varphi_{c1}$ is proportional to φ_{c1}. It means that each share of the flux is equal to a fraction of a circumference with radius δ. The part of the flux falls into the volume bounded by the surface, passing through both charged filaments. As is seen from Figure 8.1b, the arc length of this fraction is equal to half of the circumference length. This volume is a cylinder, and its cross-section has the form of a circle with the center at the coordinates' origin. That means, in particular, that the part of the flux, which falls into this volume, is equal to a half of the total flux.

Likewise in the case of two convergent infinite charged filaments (see Figure 8.1a) a half of the flux is directed inside a circular cone passing through both filaments. The reason for this is based on the equality of the angles at the transition from the cylindrical problem to conical: in accordance with (8.3) $\varphi_c = \varphi$. Also it is true for two parabolic infinite charged filaments (see Figure 8.15a), since according to (8.46) the angular variables of cylindrical and parabolic coordinates systems are related by equality $\varphi_c = \varphi$.

In accordance with the electrostatic analogy method, the fraction of the constant current inside the volume bounded by a circular cylinder (or cone, or paraboloid) passing through two wires with high conductivity located in a homogeneous weakly conductive medium, will be the same, if a constant voltage is applied to the wires.

In accordance with the conformity principle the magnetic field of constant linear currents coincides with the electric field of linear charges, if the currents and the charges are identically distributed in space. This means, in particular, that a half of the energy flux propagating along a long line, consisting of two parallel (divergent) wires, is concentrated inside the circular cylinder (cone, paraboloid) passing through these wires. Similar postulates are true for the system of two wires with finite radius (see Figure 8.2), if we consider the flux inside the cylindrical (conical, parabolical) surface passing through the filaments, which create the field, whose equipotential surfaces coincide with the outer surfaces of the wires.

If the self-complementary antenna structure is located on the circular cone (or paraboloid of rotation), then the line of two non-closed coaxial shells (see Figure 8.4a) may be imagine as the sum of pairs of the filaments 1-1″, 2-2″, 3-3″ and so on, which are placed diametrically opposite and have opposite in sign charges (see Figure 8.4b). The flux of each pair inside the structure also is equal to a half of the total flux. That means that if symmetrical slotted antennas are located on the circular metal cone (or paraboloid), then the same power is radiated into input and output space independently on the angle at the cone vertex. This conclusion coincides with the inference in Section 8.3. The field symmetry inside the cylinder, cone and paraboloid is the reason of such equality of powers.

One must also remind that (8.22) is true for infinite long line. The input impedance of finite line is closed to 60π, but is not equal to it. This fact is confirmed by calculation results presented in this chapter. Also the magnitude m in the expressions (8.38) and (8.39) for the cone and paraboloid of finite length is closed to 0.5, but is not equal to that.

From the above it follows that only half of the energy flux along the line executed in the shape of wire 1 located over ground 2 (Figure 8.41a) goes between this wire and ground within the marked cross section 3. The second half of the energy, in spite of wide-spread opinion, goes in the surrounding space.

If the cross section of the cylinder (or cone, or paraboloid) has a non-closed form, for example it consists of two arcs 4 (Figure 8.41b) of different radii, then the flux part falling into the structure is not equal to half.

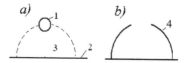

Figure 8.41 Cross section of the line in the form of a wire over ground (a) and a tube (b).

9

Principle of Complementarities

9.1 SELF-COMPLEMENTARY ANTENNAS

The principle of complementarities was briefly formulated in Chapter 8. It proceeds from an interconnection between scattering of metallic and slot radiators of the same shape and dimensions. Two complementary structures are two infinite plates (each plate is aggregate of the metallic and slotted antennas), in which metal sections are replaced by slotted sections and vice versa. Two complementary antennas are metal and slotted radiators of identical shape and dimensions. Their input impedances are related with each other by expression (8.23). When the metallic and slotted radiators of the identical shape and dimensions fill the all plane, they are called self-complementary. In this case their input impedances are the same and correspond to expression (8.24).

The simplest examples of plane self-complementary radiators are demonstrated in Figure 8.8a. As is shown in Chapter 8, these radiators may be placed not only on the plane, but also on the surface of rotation, in particular, conic and parabolic. The main distinctive property of such antennas is the constancy of angles between the limitative lines of the metallic elements of antennas. And thereby the radiator has at all frequencies the same shape (the radiator dimensions are proportional to wave length). Characteristics of these antennas are independent from frequency, i.e. they have the constant and purely resistive input impedance, and hence a high level of matching with the cable in an unlimited frequency range. If the radiator has finite dimensions, the frequency range is limited, but remains a sufficiently wide.

As stated in Chapter 8, the relation (8.23) for a slot antenna in a flat metal plate was obtained by two methods: by means of duality principle and by means of the principle of complementarities. We use the principle of duality. Suppose that the metal sheet consists of two symmetrical radiators (metallic and slot) with different angular widths. In accordance with the principle of duality the slot radiator is a magnetic radiator. Its radiation resistance R_M and input impedance Z_M are related to a radiation resistance R_E

and an input impedance Z_E of the electric radiator, similar in shape and dimensions, by the expressions (8.32) and (8.33) accordingly.

If the each arm of the flat magnetic radiator is to divide along the plane of symmetry by means of a flat metal surface, the radiator field doesn't change, since magnetic field lines coincide with this metal surface. This surface divides the magnetic radiator by two same radiators, and each of them is equivalent to the one-sided slot, magnetic current of which is half as much of magnetic current of the starting magnetic radiator. A magneto motive force in this case does not change. It means that the input admittance of each one-sided slot is twice as much of the input admittance of the starting magnetic radiator. The total admittance of both slots is four times of as much of the input admittance of the starting magnetic radiator, i.e. the input impedance of the slot is

$$Z_S = 1/Y_S = 1/(4Y_M) = Z_M/4, \tag{9.1}$$

and taking (8.33) into account,

$$Z_S = (60\pi)^2/Z_E. \tag{9.2}$$

As shown in Chapter 8, the similar result takes place for two symmetrical radiators (metal and slot), which are located on the surface of rotation, in particular, conic and parabolic. It is also shown that radiators radiate the equal energy into the inner and outer part of space, divided by means of the surface of rotation, regardless of the angle at the vertex of this surface.

The structure of two symmetrical radiators located on a plane or on a surface of rotation can be considered as a two-wire long line. Capacitance per unit length of the long line from two non-closed cylindrical shells with the equal radius (Figure 9.1a) is given in [34] and [71]:

$$C_l = \varepsilon K(\sqrt{1-k^2})/K(k), \tag{9.3}$$

where ε is the dielectric permittivity of the surrounding space (for air it is equal to 1), $K(k)$ and $K(\sqrt{1-k^2})$ are the complete elliptic integrals of the first kind of arguments k and $\sqrt{1-k^2}$, and k is equal to

$$k = \tan^2(\beta/2). \tag{9.4}$$

Here 2β is the angular width of the slot, $2\alpha = \pi - 2\beta$ is the angular width of the metal shell. Accordingly, the wave impedance of this cylindrical line is equal to

$$W = \frac{k}{\omega C} = 120\pi\, K(k)/K(\sqrt{1-k^2}), \tag{9.5}$$

Magnitudes C_l and W are independent on a cylinder radius a. This means that both expressions are correct for a conic (Figure 9.1b) and parabolic (Figure 9.1c) structure with the same cross section. They are true also for a flat antenna, where 2α is an angle at the vertex of each arm of a metal radiator (Figure 9.1d).

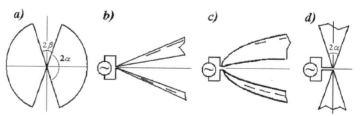

Figure 9.1　Transmission lines on a cylindrical (a), conic (b), parabolic (c) and plane (d) surfaces.

As is shown in Section 8.2, if $2\alpha = 2\beta = \pi/2$, the wave impedance W of each considered long line is equal to 60π. An input impedance of a uniform two-wire line, when its length infinitely increases, tends to its wave impedance. But the length of arm of considered antennas is far away infinity. Their input impedance depends on a frequency, if $2\alpha = 2\beta$, and has a reactive component. This fact is confirmed with calculations presented in Chapter 8.

As already stated, expression (8.24) is written for self-complementary antenna, and expression (8.23)—for complementary metallic and slotted symmetrical radiators. In this case the word "symmetrical" means that axes of both arms of the radiator coincide with each other. It is expedient to compare that antenna with a simple antenna of finite dimensions, shown in Figure 9.2a. In this antenna there is one metallic radiator, consisting of two flat cones, axes of which do not coincide. The method of calculating its input impedance was proposed in [70]. It is based on the conformal transformation of geometric figures and on the property of capacitance to retain its magnitude in this transformation. The transformation on the plane is called conformal, if the angles between any two intersecting lines remain unchanged, and the lengths of all infinitesimal segments, passing through a given point of the plane, are changed the same number of times. In particular, the method of conformal transformations is applied for calculating the capacitance of wires' systems with the help of replacement of the original system by a system with known capacitance.

Before applying this method, it is necessary to reduce the problem to a plane problem. In the case of the structures shown in Figure 9.2a, this requirement is satisfied. Further, the task can be simplified, if to go from a spherical coordinates system to a cylindrical one. When the coordinates system is replaced, Laplace's equation in accordance with the uniqueness theorem must remain true. As shown in [70], this condition is satisfied, if the replaceable variables are related by the equalities (8.3).

As a result of such transition, each radial line becomes a point, and the entire structure turns into two identical "slits" in the shape of two segments of a horizontal straight line (Figure 9.2b). After that the Schwarz-Christoffel transformation maps the structure into the interior area of the rectangle, whose two opposite plates having an

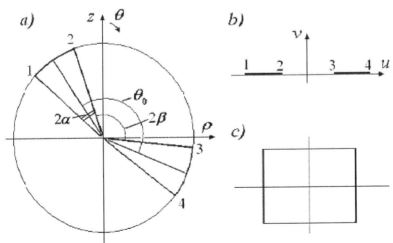

Figure 9.2 Radiator of two flat cones before transformation (*a*), after conversion into two "slits" (*b*) and in the shape of the plane capacitor (*c*).

infinite width create the plane capacitor (Figure 9.2c). The capacitance per unit width of capacitor is calculated in accordance with (9.3), and k is equal to

$$k = \frac{\sin(\theta_0/2) - \cos\beta}{\sin(\theta_0/2) + \cos\beta}. \tag{9.6}$$

Here θ_0 is the angle between the axial lines of the arms of the metal radiator, and β is the half-width of the slot, equal to $\beta = \theta_0/2 - \alpha$ (α is a half-width of a flat metal cone). If the axial lines coincide, (9.6) is reduced to (9.4).

As can be seen from the above, the proposed method permits to solve the problem for a metal radiator of arbitrary angular width with different angles between the axial lines of the arms. But in the case of two dipoles, the described technique leads to four straight segments, and in the case of N dipoles to the $2N$ segments. That offers a method for calculating a few capacitances, but does not permit to bring them together to one capacitance.

In conclusion it is useful to make a few general comments. Existence of self-complementary antennas is founded on the principle of duality, i.e., on the symmetry of Maxwell's equations relative to the magnitudes $\sqrt{\varepsilon_0}\vec{E}$ and $\sqrt{\mu_0}\vec{H}$. It allows to speak about existence of metallic (electric) and slotted (magnetic) radiators, which follows from this symmetry. These radiators radiate identical fields, which after replacement of the corresponding magnitudes may be calculated in accordance with analogous expressions. One can place on a flat metal sheet of infinite dimensions one or more metal radiators and also many slot radiators (it is regarded that the sheet is infinitely thin and has an infinite conductivity). If the radiators fill the entire sheet, the radiators are called complementary. If in addition the radiators are identical in shape and dimensions, they are called self-complementary. The uniqueness of the self-complementary structures is that they have special properties, which are partially retained, if the metal sheet has finite dimensions.

If the metallic and slotted radiators of the identical shape and dimensions occupy the entire surface of rotation, they are called three-dimensional or volumetric self-complementary antennas. But they appeared much later than the flat antennas, and the term itself initially was perceived differently. Details are given in Section 8.3.

In [78] it was shown firstly that a structure consisting of electric and magnetic radiators of the same shape and dimensions must not necessarily be flat, and secondly, the choice of surface for placement of such structures is not random. This is in particular the circular cone, which in the limiting case turns into a plane.

9.2   ANTENNAS WITH ROTATIONAL SYMMETRY

Antenna with rotational symmetry is the most general variant of the self-complementary antennas. It consists of several metal radiators and an equal number of slots. Two antennas with rotational symmetry are shown in Figure 9.3. Both antennas have finite dimensions. Each antenna consists of four symmetrical radiators of the same shape and dimensions: two metal radiators and two slots. In the first antenna (see Figure 9.3a) an arm of each radiator has the shape of a flat cone (this is a sector of a circle whose left and right sides coincide with the radii, and a base coincides with a structure circumference). In the second antenna (see Figure 9.3b) boundaries of radiators do not coincide with the radii, but have a log-periodic shape. In both cases, generator poles can be connected with a different number of elements, forming symmetrical and asymmetrical versions. Also it is possible to excite one antenna by several generators.

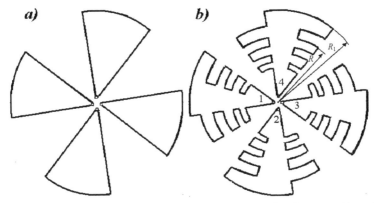

Figure 9.3   Antennas of four metal plates in the shape of a flat cone (*a*) and in the logo-periodic shape (*b*).

Consider a procedure of calculating the wave impedance of an antenna with several metal dipoles at an example of an antenna with two dipoles (Figure 9.4*a*). The antenna consists of four identical flat cones, equally spaced along the circumference, i.e. separated by the same slots. The total angular width of the cone and the adjacent slot is equal to $\theta_0 = 2(\alpha + \beta) = \pi/2$.

First, one must go over in accordance with the conditions (8.3) from the spherical coordinates system to the cylindrical. As a result of the transition, a radial line n (line numbers are shown in Figure 9.4*a*) is transformed to a point (point numbers coincide with the line numbers and are shown in Figure 9.4*b*) with coordinate ρ_n. For values ρ_n the general expression is valid

$$\rho_n = \tan\left\{0,5\left[(2m-1)\theta_0/2 \pm \alpha\right]\right\}, \tag{9.7}$$

where $m = 1, 2, ...$, a sign "minus" refers to an odd value of n and a sign "plus"—to an even n. If the widths of the slot and metal cone are the same, i.e. $\alpha = \beta = \pi/8$, then $\rho_n = \tan\left[(2n-1)\alpha/2\right]$. A general view of the structure, transformed in accordance with the conditions (8.3), is shown in Figure 9.4*b*. This is a horizontal straight line with four segments ("slits"), connecting the points with numbers $2n-1$ and $2n$.

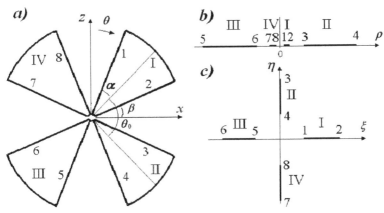

Figure 9.4   Antenna of two metal dipoles before transformation (*a*), in the cylindrical coordinates system (*b*) and after the complex transformation (*c*).

Transforming the structure in accordance with the expression

$$\varsigma = \sqrt{\tan(2\tan^{-1}\rho)} = \sqrt{\tan\left[(2m-1)\theta_0/2\pm\alpha\right]}, \qquad (9.8)$$

i.e., going over from the plane with the cylindrical coordinates ρ and φ_c to the plane with rectangular coordinates ξ and η and considering that $\varsigma = \xi = j\eta$, we obtain Figure 9.4c, where four "slits" of the same length are located on four semi-axes. Point numbers are shown in the figure in Arabic numerals; "slits" numbers are shown in Roman numerals.

As already mentioned, the generator poles can be connected to different elements. Three variants are presented in Figure 9.5. The segments that are connected with each other by solid lines are connected to the same pole of the generator. One pole is marked with a sign "plus", the other pole—with a sign "minus". In the first and second variants (Figure 9.5a, b), each pole is connected with two flat metal cones: in the first case with adjacent cones, in the second case—through a cone. In the third version, one cone is connected with one pole and three cones—with another pole.

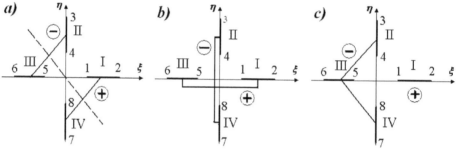

Figure 9.5 Three variants of connecting antenna to poles of generator.

The dashed line in Figure 9.5a is the axis of a system symmetry (it is the line of zero potential). Such a system can be divided into two equal subsystems as shown in Figure 9.6a. In one of them one can calculate the capacitance between the elements, connected to different poles, and then one must double the result. It is easy to see that for a second variant of connecting metal cones to the generator poles, the system is divided into four subsystems that are identical to the structure, shown in Figure 9.6a. So, the antenna input capacitance in this case is twice as many. The magnitude of capacitance between the elements of the structure, shown in Figure 9.6a, may be calculated (see, e.g., [34]) by the method of direct determination of the field strength. It is equal to $2C_l$, where C_l is calculated by the formula (9.3), and k is equal to

$$k = \sqrt{\frac{(a^2+c^2)(b^2+d^2)}{(a^2+d^2)(b^2+c^2)}}. \qquad (9.9)$$

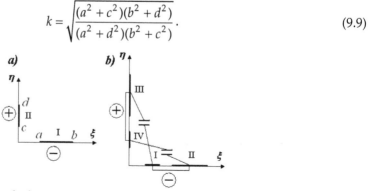

Figure 9.6 Subsystems for calculating capacitances of antennas with two (*a*) and four (*b*) dipoles.

It is easy to see that in our case $a = c$, $b = d$, i.e.

$$k = 2ab/(a^2 + b^2).\tag{9.10}$$

If the angular widths of the cone and the slot are same ($\alpha = \beta = \pi/8$), in accordance with (9.8) and (9.10) $a = 0.6436$, $b = 1.5538$, $k = 0.7071$, and $K(k) = K(\sqrt{1-k^2}) = 1.8541$ This means that the capacitance between the elements of the circuit shown in Figure 9.6a, is equal to $C_0 = 2\varepsilon$. Then the total input capacitance of the first antenna variant is twice as many, and the capacitance of the second variant is four times more than 2ε, so the wave impedances of the first and second variants are equal, respectively, to 30π and 15π. In a third variant of the connecting cones with a generator, the capacitance between the first and second "flit" and between the first and fourth "flit" is 2ε, the capacitance between the first and third "flit" in the first approximation is close to the same value, i.e., $W \approx 20\pi$.

Typical examples of active and reactive components of input impedances for flat antennas with one or two dipoles and identical width of the metal and slot radiators are shown in Figure 9.7. Variant 1 corresponds to the antenna with one metal radiator. Variant 2 corresponds to the antenna with two metal radiators, where each pole of the generator is connected with two adjacent plates. Variant 3 corresponds to the antenna with two metal radiators, where each pole of the generator is connected with two plates through a plate. Using a known method and methods proposed here, one can compare different variants of antennas and come to some general conclusions.

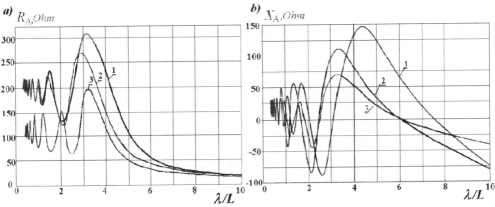

Figure 9.7 Active (a) and reactive (b) components of input impedances for the flat antennas.

Figure 9.8 shows antennas, where one metal dipole is used. Antennas with two metal dipoles are shown in Figure 9.9. They differ from each other by circuit of connection with the generator. Figure 9.10 demonstrates the different circuits of connection with the generator for antennas, with four metal dipoles. Calculation of their wave impedances was made according to the procedure used in the analysis of antennas with two dipoles. It leads to the subsystems in the shape of a quadrant, shown in Figure 9.6b. In the first variant, the total capacitance is the sum of the capacitances in two such quadrants, and in the second and third variants the total capacitance is the sum of the capacitances in four quadrants. The number of summed capacitances in the quadrant is always equal to two, since the two "slits" of four "slits" are connected with each other, and each pair is connected to its pole. Herewith capacitance of the quadrants is equal to the sum of two identical capacitances between the "slits", located on different distances from the coordinate's origin.

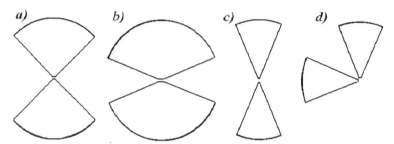

Figure 9.8   Flat antennas with one metal dipole.

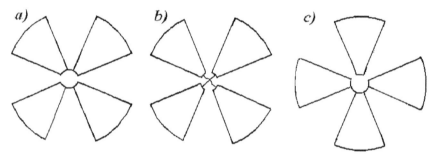

Figure 9.9   Flat antennas with two metal dipoles.

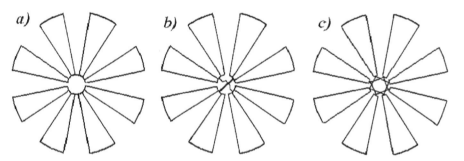

Figure 9.10   Flat antennas with four metal dipoles.

Basic parameters and wave impedances W_1 of considered antennas are given in Table 9.1. The table shows that the wave impedances in it are given for antennas in which the width of the metal plate is equal to the slot width. In this case, the capacitance between the elements of the quadrant in antennas with two dipoles is equal to $C_0 = 2\varepsilon$. If the angular width of the plate in an antenna with two metal dipoles is equal to $60°$ rather than $45°$ (the slot width in this case is $30°$), then $C_0 = 1.28\varepsilon$, i.e. wave impedance of the antenna decreases 1.28 times. In asymmetric antennas, when the symmetrical metal radiators (dipoles) forming the antenna are replaced by asymmetric radiators (monopoles) placed above a conductive surface (ground), the magnitudes of the wave impedances are reduced by half. Figure 9.11 shows the circuits of connecting asymmetrical radiators to a generator, similar to the circuits of connecting for symmetrical radiators, which are shown in Figures 9.9a and 9.9b.

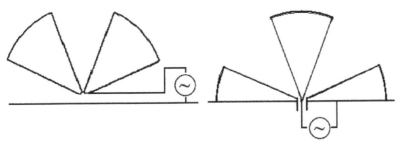

Figure 9.11   Connecting asymmetrical antennas to a generator.

Table 9.1   Parameters and Wave Impedances of Antennas with Rotational Symmetry

Number of metal dipoles	Figure	θ_0	2α	W_1	W_2
1	9.8a	180°	90°	60π	60π
	9.8b	180°	135°	40.8π	40.8π
	9.8c	180°	45°	88.2π	88.2π
	9.8d	90°	45°	84.9π	
2	9.9a	90°	45°	30π	21.2π
	9.9b	90°	45°	15π	10.6π
	9.9c	90°	45°	20π	15π
3	9.10a	45°	22.5°	21.8π	12.8π
	9.10b	45°	22.5°	10.9π	6.4π
	9.10c	45°	22.5°	13.6π	8π

9.3   THREE-DIMENSIONAL ANTENNAS WITH ROTATIONAL SYMMETRY

The creation of flat antennas with rotational symmetry is a step forward compared to the simple antenna in the form of a flat metal dipole with angular width of the arm, equal to the angular width of the slot. Another step forward is the development of three-dimensional (volumetric) antennas. Their appearance became possible, when it has been proved that the self-complementary antenna can be placed not only on a plane surface, but also on the surface of rotation, for example, on the surface of a circular cone or a paraboloid.

Conical and cylindrical problems were compared with each other in [70]. There it is shown that one problem is reduced to another, if the substituted variables satisfy the conditions (8.3). Equivalence of parabolic and cylindrical problems [19] takes place, if variables satisfy the conditions (8.46).

In conical problems (Figure 9.12) the cone axis is located horizontally, the emf excites a structure at the cone vertex, and the radiators' arms are located along the cone surface. The boundaries of each arm coincide with the cone generatrices. The arcs lengths of the cross-section corresponding to a metallic and slotted radiators, are designated as 2α and 2β respectively. A paraboloid has the similar shape of cross section and same designations.

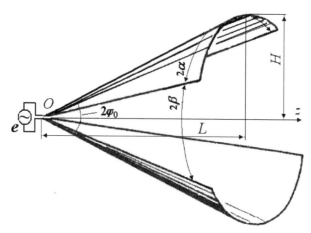

Figure 9.12 Conical antenna with one metal dipole.

The rigorous analysis (see [19]) shows that the input impedances of metallic and slotted radiators, located on a circular cone, are related with each other by the expression (8.23). When $\alpha = \beta$, the expression (8.24) is true. This is natural, since the conical structure can be considered as a two-wire transmission line, and this line is uniform, i.e., does not differ in this respect from the corresponding transmission line, placed on the surface of a circular cylinder with a constant radius of a cross section. That is why, if $\alpha \neq \beta$, for these conical and similar parabolic antennas, the expressions (9.3), (9.5) and (9.6) hold valid. If $\alpha = \beta$, k is calculated in accordance with (9.4).

Three-dimensional antennas as flat antennas may consist of several metal and slot radiators, included in parallel with each other, i.e. they may be named by antennas with rotational symmetry. An example of such antenna is given in Figure 9.13a. When calculating its characteristics we firstly must go over to the plane problem, using the conditions (8.3). Since all points of each cone generatrix are located at the same angle θ to the axis z, then, as a result of transformation, they lie down to one point, and a metal plate becomes a segment of an arc, on so-called "slit" (Figure 9.13b). If the plates and the slots have the same width, the coordinates of the start (x_{m_1}, y_{m_1}) and the end (x_{m_2}, y_{m_2}) of these arc segments are defined by the expressions

$$x_{m1(2)} = \rho \cos \varphi_{m1(2)}, y_{m1(2)} = \rho \sin \varphi_{m1(2)}. \tag{9.11}$$

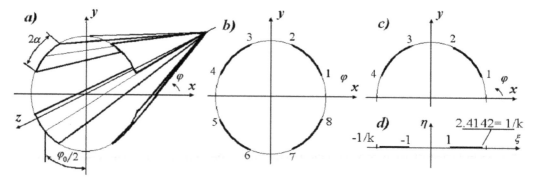

Figure 9.13 Conical antenna with two metal dipoles: general view (a), the transition to a plane problem (b), division along the line of symmetry (c), conformal transformation (d).

Here, a value ρ in accordance with (8.3) is $\rho = \tan(\theta/2)$ (θ is the half-angle at the vertex of the cone), and φ_m is the angle measured in the plane xOy from the axis x, which is equal to

$$\varphi_m = (2m - 1)\varphi_0/2 \pm \alpha,$$

where φ_0 is the angle on the plane xOy between the projections of the axes of two adjacent plates—a sign "minus" corresponds to the arc start and, a sign "plus" corresponds to the arc end, α is half of the arc length.

As in the case of similar flat antenna, there are three possible variants of connecting the generator poles to the different antenna elements (Figure 9.14). The elements that are connected with each other by solid lines are connected to the same pole of the generator. One pole is marked with a sign "plus", the other pole is marked with a sign "minus". In the first (Figure 9.14a) and second (Figure 9.14b) variants, two metal plates are connected to each pole: in the first variant they are the adjacent plates, in the second variant the plates are connected through one cone. In the third variant (Figure 9.14c) one plate is connected to one pole, and three plates—to another pole.

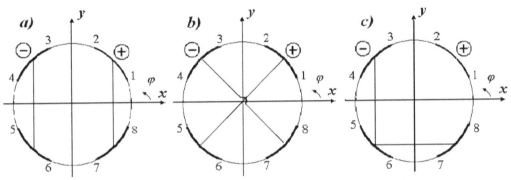

Figure 9.14 Three variants of connecting the antenna to the poles of the generator.

Axis x in Figure 9.14a and b is the axis of the system symmetry (it is the line of zero potential). Such a system can be divided into two identical circuits, shown in Figure 9.13c. In one of them one must calculate the capacitance between the elements, connected to different poles, and then double the result. It is easy to see that for a second variant of connecting metal plates to the generator poles, the system is divided into four subsystems that are identical to the structure, shown in Figure 9.13c, i.e. input capacitance of the antenna in this case is twice as much than in the first variant.

Each arc in Figure 9.13c is a segment of a circumference of radius ρ. If the angular width of the metal plate and the slot is same, then the coordinates of the start and end of the first arc are equal to $x_{11} = y_{12} = \rho \cos \varphi_1$, $x_{12} = y_{11} = \rho \sin \varphi_1$, where $\varphi_1 = 22.5°$. The value z at each point of the arc is equal to $z = \rho(\cos \varphi + j \sin \varphi)$. We use a conformal transformation of the form

$$\varsigma = jz/\rho + \rho/(jz) \qquad (9.12)$$

where $\varsigma = \xi + j\eta$. Then

$$\varsigma = j(\cos\varphi + j\sin\varphi) - j/(\cos\varphi + j\sin\varphi) = -2\sin\varphi. \qquad (9.13)$$

i.e. in the plane ς, the arc segment will transform into the axis segment between points 1 and 2 (Figure 9.13d). A similar transformation will happen with the second arc, connecting points 3 and 4. Dividing the coordinates of points on $2\xi_2$, we get the value given in Figure 9.13d, and in this case $k = 0.4142$. Using the Schwarz-Christoffel transformation, we find, in accordance with (9.3), the capacitance of the structure shown in Figure 9.13c: $C_0 = 1.414\varepsilon$.

As shown before, the total input capacitance of the first variant of the antenna is twice as C_0, and the capacitance of the second variant is greater four times. This means that the wave impedances of the first and second variants are equal, respectively, to 21.2π and 10.6π. In a third variant of the connecting plates to the generator, the capacitance between the first and second "slits" and between the first and fourth "slits" is C_0, the capacitance between the first and third "slits" in the first approximation is close to C_0, i.e. the wave impedance is close to 15π.

The values of wave impedances W_2 of a three-dimensional antenna, located on the conical and parabolic surface of revolution, are shown in Table 9.1 and allow us to compare them with the wave impedances of flat antennas.

The obtained results show that a class of self-complementary antennas is considerably wider than what was regarded previously. This class must be complemented, firstly, at the expense of structures, consisting of several metal and slot radiators and, secondly, by means of three-dimensional structures, located on the surfaces of rotation, in particular on the surfaces of the circular cone or the paraboloid. The technique of calculating the structures, consisting of several radiators is based on the method described by R.L. Carrel [70]. Comparison of cylindrical, conic and parabolic problems allows to determine the relationship between variables, providing a mathematical equivalence of these problems.

Closeness of wave impedances of the self-complementary antennas and cables is a necessary condition for the antenna's effectiveness. Known variants of antennas do not satisfy this condition, since their wave impedances are substantially higher than the wave impedances of standard cables. Antennas which were considered here, have very different including sufficiently smaller magnitudes of the wave impedances. That should greatly facilitate the task of matching and expand the scope of using self-complementary antennas.

9.4   SHAPE AND DIMENSIONS OF A PHANTOM MODEL

Application of parabolic coordinates system allows to solve the difficult problem of evaluating the impact of the phantom shape and dimensions on the measurement results obtained with its help. For this one must to reduce the parabolic problem to the plane. This permits to simplify significantly its decision. The phantom as a model of a human body was developed in order to measure and to study the parameters of fields produced by various radiators, and to determine the level of human body irradiation (SAR). But the excessive aspiration to create a device with the shape and dimensions as close as possible to the shape and size of a human body led to a wide spread of results of measurements in different devices and caused necessity to develop a method for evaluating this difference.

The influence of the volume and shape of the phantom on the accuracy of the measurement is confirmed by the circumstance that there is an interrelation between the shape and dimensions of the phantom, on the one hand, and the measurement results, on the other.

The phantom is a simple model, in which the human body is simulated by a liquid with dielectric permittivity and conductivity equal to their average values for the human tissues at radio frequencies. The phantom is usually constructed as a vessel with thin walls filled with a homogeneous simulation liquid. It is made of fiberglass with low relative permittivity and conductivity and is open from top. A probe is immersed into the liquid to record the electric field strength inside the phantom. The measurement results are processed that to calculate the maximal SAR at the point, the local SAR (the loss per unit mass), and the total SAR (the loss in the head).

Figure 9.15*a* shows the measurement setup of Holon Institute Technology (Israel). As a rule, the setup consists of three vessels in the shape of two heads (left 1 and right 2) and a body 3. The receiving antenna-probe 4 mounted on the rod allows to measure a field at each point of any vessel. An antenna 5 under study is used as the transmitting antenna and is located outside the phantom (Figure 9.15*b*).

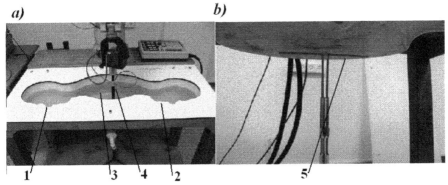

Figure 9.15 SAR measurement setup: a view from the top (*a*) and the bottom (*b*).
1, 2 — the vessels in the shape of heads, 3 — the vessel in the shape of the body, 4 — the rod with the receiving antenna-probe, 5 — the transmitting antenna (dipole).

The field calculation in the phantom is a complicated three-dimensional problem. It is necessary to calculate the near field of the transmitting antenna at the point of location of receiving antenna and take into account the mutual coupling of two radiators. The near region of the transmitting antenna is situated in two different media, and its boundary has an intricate shape. The receiving antenna is placed in the adjacent medium.

As is known from the antenna theory, the near field of an antenna has in a first approximation a quasi-stationary nature, and its structure coincides with the structure of the electrostatic field. Therefore such a problem may be reduced to the electrostatic problem, i.e. to the calculation of electric field of charged conducting bodies.

The signal magnitude in the observation point depends on the capacitance between the transmitting and receiving antennas. The problem of calculating this capacitance has a three-dimensional character. Therefore it is expedient to reduce the three-dimensional problem to a plane (cylindrical) problem. To this end one can use conical or parabolic variants of the three-dimensional problem. But the considered task is not conical, since the vessel shape has little in common with a cone. In considered problem one equipotential surface coincides with the surface of the vessel shell, which is the interface of two different media and, for this reason, this surface is a surface of equal potential. Other equipotential surfaces intersect the vertical axis of the vessel in other points, in particular in point of location of the receiving antenna. In the conical problem the vertexes of the all cones are pllaced in one point, located at the vertical axis and coinciding with the vessel shell.

An approximate procedure of calculating the said capacitance is based on the similarity of the shapes of a phantom and a paraboloid and on reducing a parabolic problem to a cylindrical one. The real phantom structure is more similar to the paraboloid, but this paraboloid is not circular, since the horizontal cross section of the vessel is not circular. If the vessel is not similar to the circular paraboloid, one must approximate it by means of a paraboloid closest to it in shape.

Figure 9.16 shows the arrangement of vessel in the parabolic coordinate system. The antenna under study (transmitting antenna) is denoted here by 1. As seen from Figure 9.15b, this antenna is a dipole, located along the horizontal line tangentially to the bottom of the vessel. The measuring (receiving) antenna is denoted in this figure by 2, and it is a small probe placed inside the vessel along the vertical axis. It is expedient that during the control test of the phantom both antennas must lie in one plane. As is clear from Figure 9.16, in this time the vertical axis of the structure (the z-axis) must pass through the center of the radiator 1 and the measurement point, in which probe 2 is located.

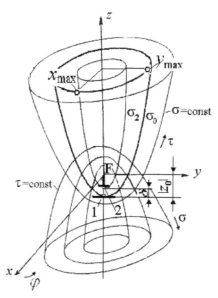

Figure 9.16   The vessel in the system of parabolic coordinates.

As shown in Section 8.4, Laplace's equation holds true at the transition from the parabolic problem to the cylindrical if parabolic (σ, τ, φ) and cylindrical (ρ, φ_c, z) variables are related by the expressions (8.46). The transformation of the parabolic problem to the cylindrical problem results, in particular, in the transition from parabolic surface τ = const to plane (ρ, φ_c). The equivalent cylindrical problem is presented in Figure 9.17. Here the following designations are used: $\rho_{01} = \sigma_{01}$ is the radius of the outer surface of the shell, $\rho_{02} = \sigma_{02}$ is the radius of its inner surface, $\rho_1 = \sigma_1$ and $\rho_2 = \sigma_2$ are radii of the equipotential surfaces crossing the axis z at the distance 1 and 2 cm above the shell. If the paraboloid is circular, the circular cylinder corresponds to it in the cylindrical coordinate system. But usually the vessel boundary is of rather intricate nature. Assume that its cross section perimeter is an ellipse with a major axis of length $2a$ and minor axis of length $2b$, and that the shell thickness is 2 mm. The relative permittivity of the medium is equal to $\varepsilon_1 = 1$ in the range $\rho \geq \rho_{01}$, to 2.7 in the range $\rho_{02} \leq \rho \leq \rho_{01}$, and to permittivity $\varepsilon_2 = 41.5$ of the liquid filling the vessel at $\rho \leq \rho_{02}$.

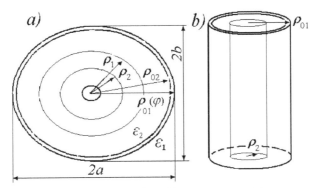

Figure 9.17 The equivalent cylindrical problem: top view (*a*), axonometric view (*b*).

Reducing of the three-dimensional problem to the plane (two-dimensional) simplifies the calculation of capacitance between the equipotential surfaces. The capacitance magnitude determines the coupling level between the surfaces. It can be used for estimating the strength of the field created by the antenna under study at the observation point. As is known, the electric component of the field at a given distance from an antenna is inversely proportional to the medium permittivity:

$$|E_z| = AJ_A F(R)/(\varepsilon_2 R) \tag{9.14}$$

where J_A is the current in the antenna base, and $F(R)$ is a function of the distance R from the antenna axis.

As is seen from Figure 9.17, the capacitance between the antennas is formed of a few capacitances connected in series with each other. The capacitance per unit length between two equipotential cylindrical surfaces of radii ρ_1 and ρ_2 ($\rho_1 > \rho_2$) is equal to (see, for example, [34])

$$C = 2\pi\varepsilon_2 / \ln(\rho_1/\rho_2). \tag{9.15}$$

If the radius of the equipotential surface depends on φ (for example, when the cylindrical cross sections are similar ellipses), it is necessary to take the average radius ρ_{av} of this surface as the cylinder's radius ρ. In order to calculate the magnitude of C for the parabolic structure, we should use (8.46) to replace the parabolic coordinates by cylindrical ones. For this end, we have to determine the shape and dimensions of the phantom in the parabolic coordinate's system in accordance with the drawing and then perform the transition to the cylindrical coordinates. If the vessel has the shape shown in Figure 9.17, and *x*- and *y*-axes are directed along the major and minor axes of the ellipse respectively, the coordinates of the points situated on the vessel walls are

$$x = \sigma\tau f(\varphi)\cos\varphi, \; y = \sigma\tau f(\varphi)\sin\varphi, \; z = 0.5[\tau^2 - \sigma^2 f^2(\varphi)]. \tag{9.16}$$

Here, function $f(\varphi)$, which varies from 1 to b/a, defines the dependence of the ellipse radius on angular coordinate φ.

Comparison of a paraboloid, having the elliptic cross section in the horizontal plane, with the paraboloid of rotation with the same perimeter show that in the latter case, coordinate σ_0 will be constant along the entire parabolic surface and at the lower point of its surface, where $x = y = 0$, coordinate τ in accordance with (9.16) is zero too. Therefore $z_0 = -0.5\sigma_0^2$, i.e. $\sigma_0 = \sqrt{2|z_0|}$. Magnitude z_0 is the coordinate of the lowest point of the vessel.

Point $z = 0$ (where $\tau_0 = \sigma_0$) is the focal point of the paraboloid of rotation. In Figure 9.16 it is designated by the letter F. And the distance between points $z = 0$ and $z = z_0$ is the focal length $|z_0|$, which is equal to a half of the focal parameter. The relationship, similar to $\sigma_0 = \sqrt{2|z_0|}$, is true for each paraboloid, including the paraboloid with the elliptic cross section. In particular, if the probe is located on the z axis at distance δ from the bottom point, then $\sigma = \sqrt{2(|z_0| - \delta)}$ (see Figure 9.16).

As an example, we consider calculation of the field inside the side and central vessels of the phantom shown in Figure 9.15. The approximate dimensions of the set are given in [82]. During the calculation, the following dimensions were taken into account and used:

The left vessel—the length and the width of the entrance (aperture) are respectively 243 and 206 mm, the maximal depth of the vessel, the shape of which follows the shape of the left side of the head, is equal to 79 mm;

The central vessel—the length and the width of the entrance are respectively 370 and 255 mm, the maximal depth of the vessel is 110 mm;

Both vessels—the liquid level is 140 mm.

These dimensions permit to draw the lateral and the longitudinal cross sections of both vessels. The vertical cross sections of the side vessel in the lateral (1) and longitudinal (2) directions are given in Figure 9.18. They allow determining the parameters of parabolic curves. In order to calculate the parameters of the paraboloid, the shape of which is the closest to the vessel shape, we use the parabolic equation $|z_0| + z = ax^2$ and summarize point by point the values of z and x^2. Factor $1/a$ is defined as the ratio of the sums,

$$1/a = \sum_{n=1}^{N} x_n^2 \bigg/ \sum_{n=1}^{N} (|z_0| + z_n). \qquad (9.17)$$

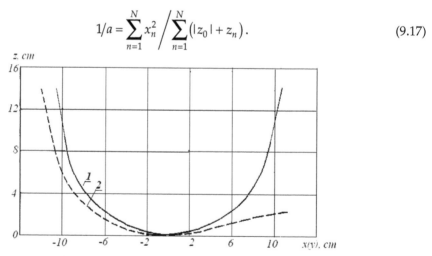

Figure 9.18   The cross sections of the side vessel in the lateral (1) and longitudinal (2) vertical planes.

Applying (9.20) for calculation of the value a, one can largely avoid the computational difficulties associated with distinction between the shape of the vessel and the paraboloid of rotation. The focus of the paraboloid is located at the point $z = 0$. A focal length is the distance from the point $z = 0$ to the bottom point of each paraboloid, i.e. to the point of intersection of axis z with the surface $\sigma = $ const. From (8.40) it follows that at this point

$x = y = 0$, i.e. $\tau = 0$, $z = -\sigma^2/2$. Accordingly $\sigma = \sqrt{2|z|}$, and

$$|z| = 1/(2a) \tag{9.18}$$

Let the antenna-probe 4 (see Figure 9.15) be set in succession at three positions along the z axis: on the inner surface of the shell, and also at a distance 10 and 20 mm from it. In accordance with (9.17) and (9.18) we obtain $1/a = 113$ mm, that is the coordinates of the points on the inner (z_{02}) and the outside (z_{01}) surface of the vessel, and the coordinates of the points inside the vessel (z_1 and z_2) are equal to -56.5, -58.5 (if the thickness of the shell is 2 mm), -46.5 and -36.5 mm respectively. This means that the parabolic and cylindrical coordinates of these points are equal $\sigma_{02} = \rho_{02} = 10.63$, $\sigma_{01} = \rho_{01} = 10.82$, $\sigma_1 = \rho_1 = 9.64$, $\sigma_2 = \rho_2 = 8.54$, respectively. In accordance with (9.15) the capacitances between the adjacent cylindrical surfaces (beginning from the outside) per unit length of the cylinder are equal to $C_{01-02} = 0.86 \times 10^{-8}$ F/m, $C_{02-1} = 2.36 \times 10^{-8}$ F/m, $C_{1-2} = 1.90 \times 10^{-8}$ F/m. Summarizing the capacities, we obtain that the capacity of the shell is $C_{01-02} = 0.86 \times 10^{-8}$ F/m, the capacity of the shell and the first layer of liquid is $C_{01-1} = 0.63 \times 10^{-8}$ F/m, the capacity of the shell and the two layers of liquid is $C_{01-2} = 0.47 \times 10^{-8}$ F/m. The field E_0 created by an external antenna will consistently decrease after passing through each layer: E_0, $0.86E_0$, $0.63E_0$, $0.47E_0$.

Evaluating the capacities per unit length for different vessels, similar in shape, it is expedient to take into account the following regularities, follow from (9.17). If the height of the vessel will increase n times, the value of $|z_0|$ will decrease n times. If the diameter or the vessel perimeter will increase n times, the value of $|z_0|$ will increase n^2 times. If the horizontal dimensions of the vessel in different directions change differently, it is expedient to take the smaller factor as n.

In accordance with the last recommendation we obtain for the central vessel: $1/a = 165_{MM}$, $z_{01} = 84.7$, $z_{02} = 82.7$, $z_1 = 72.7$, $z_2 = 62.7$, $\sigma_{01} = \rho_{01} = 13.0$, $\sigma_{02} = \rho_{02} = 12.9$, $\sigma_1 = \rho_1 = 12.1$, $\sigma_2 = \rho_2 = 11.2$ mm. The capacities per unit length are equal to $C_{01-02} = 1.21 \times 10^{-8}$, $C_{02-1} = 3.59 \times 10^{-8}$, $C_{1-2} = 3.19 \times 10^{-8}$ F/m, i.e. the capacity of the shell is $C_{01-02} = 1.21 \times 10^{-8}$, the capacity of the shell and the first layer of liquid is $C_{01-1} = 0.91 \times 10^{-8}$, the capacity of the shell and the two layers of liquid is $C_{01-2} = 0.71 \times 10^{-8}$ F/m. The field E_0 created by an external antenna will consistently decrease after passing through each layer: E_0, $0.82E_0$, $0.75E_0$, $0.58E_0$. The results of the experimental check are presented in Table 9.2. The measurements of the field were performed at frequency 0.903 GHz. They coincide on the whole with the calculation results. They confirm that the field depends on the shape and dimensions of the vessel and show that the proposed method permits to take this dependence into account.

Table 9.2 Comparison of Calculation and Experiment

Field	E_{02}/E_0	E_1/E_0	E_2/E_0
The side vessel			
Calculation	0.86	0.63	0.47
Experiment	0.86	0.58	0.48
The central vessel			
Calculation	0.82	0.75	0.58
Experiment	0.90	0.54	0.33

9.5 MUTUAL COUPLING BETWEEN SLOT ANTENNAS ON CURVILINEAR METAL SURFACES

The calculation procedure of slot antennas situated on curvilinear surfaces with axial symmetry opens new prospects of rigorous analysis of the electrical characteristics of slots in non-planar screens.

The antenna theory employs the concept of ideal slotted antenna. This is the antenna in an indefinitely large, perfectly conductive, infinitely thin metal screen. In accordance with the duality principle, the characteristics of the ideal slot antenna are easily determined if the characteristics of a metal radiator with the same shape and dimensions are known. As is shown in [83], the outward appearance of integral equation for voltage U between the edges of the slot coincides with the outward appearance of the corresponding equation for the current of an equivalent metal antenna, identical in shape and dimensions. But the linear operator $G(U)$ (integral functional), which is included in the equation, depends on the screen shape, i.e. differs, in the general case, from operator $G(I)$ of a metal antenna in free space.

We will consider two identical magnetic radiators located in parallel in free space and excited in opposite phase (Figure 9.19a). By means of expressions (8.23) and (8.24), one can calculate radiation resistance R_M and input impedance Z_M of each magnetic radiator through the radiation resistance R_E and input impedance Z_E of an electrical radiator, the same in shape and dimensions. In order to perform the transition from magnetic radiator to slot antenna, we shall divide each arm of the magnetic radiator by a metallic surface (of a circular cross section). Since the shape of magnetic field lines coincides with the shape of this surface, the radiator field in consequence of the metallic surface insertion doesn't change. The metal surface divides each magnetic radiator into two radiators situated on the inside and outside of the cylinder. Since magneto motive force e_M, exciting the radiator, and oscillating power of the radiator $P = e_M J_M$ remain unchanged, the fraction of magnetic current J_M in each newly formed radiator is equal to the fraction of the power radiated into each subspace.

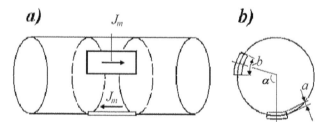

Figure 9.19 Two parallel magnetic radiators (*a*) and the transition to slot antennas (*b*).

As it is shown in Section 8.7, in the case of two parallel, infinitely long filaments with linear charges opposite in sign, the fraction of the energy flux of each filament inside the volume, limited by the circular surface passing through the filaments, is equal to half. Taking into consideration the conformity principle and the quasi-stationary character of an electromagnetic field of a two-wire line, one can make a conclusion that the half energy flux propagating along this line is concentrated inside the circular cylinder passing through the wires. One can also be sure that a similar supposition is correct for wires of finite length (for electrical and also for magnetic radiators), though the magnitude of electrical or magnetic current varies along the wire in these cases.

If we take into account that the half power radiated inside the cylindrical volume, the input admittance of the inside magnetic radiator is

$$Y_1 = e_m/(J_m/2) = 2Y,$$ (9.19)

where $Y = 1/Z_m$ is the total admittance of the original radiator.

We assume that the radiator cross section inside the cylinder has a rectangular (with sides b and a, where $b \gg a$) curved along a cylinder surface (Figure 9.19b). The greater side b represents an arc, which is parallel to the metal surface. Such a radiator is equal to a one-sided slot of width b. The radiator on the outside of the cylinder is equivalent to a similar one-sided slot. Its current and input admittance coincide with the current and the admittance of the radiator inside the cylinder:

$$Y_2 = Y_1 = 2Y.$$ (9.20)

If to replace both one-sided slots by a single two-sided slot and take into account that its admittance is equal to the sum of both slots admittances, then

$$Y_S = Y_1 + Y_2 = 4/Z_M,$$ (9.21)

i.e., the input admittance of the two-sided slot is according to (3.24)

$$Y_S = Z_E/(60\pi)^2.$$ (9.22)

Let us take into account the mutual influence of two parallel magnetic radiators on the basis of mutual communication of equivalent electric radiators. In accordance with this, we assume that Z_E in expression (9.22) is the input impedance of the electrical radiator (of the metal antenna) located near another metal radiator, which is parallel to the first radiator and is excited in anti-phase to it. In this case

$$Z_E = Z_{11} - Z_{12},$$ (9.23)

where Z_{11} is the self-impedance of each radiator, and Z_{12} is their mutual impedance.

The expression (9.22) in view of (9.23) permits to calculate the input impedance of the slot antenna, including in the system of two anti-phase radiators situated along the generatrices of the circular metal cylinder. It is necessary to note that it is two arbitrary generatrices (with an arbitrary length of the arc between them, located along the cylinder circumference). The slot antennas are not obligatorily located at opposite sides of the cylinder (when the arc length is equal to π). Really, the insertion of a cylindrical metal surface dividing each magnetic antenna into two radiators requires to remember that the shape of the magnetic force lines must coincide with the shape of the surface. In given case, this is so, since the field lines created by two charged filaments (see Figure 8.2a) are circumferences passing through the filaments with the centers on the y-axis. The fraction of energy flux inside the volume, which is bounded by the circular metal surface passing through these filaments, is independent of the arc length between the cylinder generatrices, along which the filaments are placed, i.e. it is equal to a half as before.

In accordance with the duality principle, if the moment of current of a magnetic radiator is equal to the moment of current of an electrical radiator, the electric field of a magnetic radiator is equal in magnitude to the magnetic field of an electrical radiator and is opposite to it in sign. The magnetic field of a magnetic radiator is smaller than an electric field of an electrical radiator by a factor of $(120\pi)^2$. It means, in particular, that the directional pattern of a system of two-slot radiators coincides with the pattern of two metal radiators with the same shape and dimensions.

The directional pattern of an array consisting of two identical electrical radiators depends on the single radiator pattern and the array factor, i.e. on the radiator spacing. Accordingly, the pattern of a system consisting of two thin slots depends on their spacing (the pattern of a single slot has the circular shape in the equatorial plane) and is independent of the radius of the circular cylinder in which slots are cut, i.e. on the arc length (in radians) between the slot axes. As follows from (9.19), the input impedance is also independent on the cylinder radius. So, the presence of a cylinder surface has practically no effect on the electrical characteristics of two thin radiators. The cylindrical metal surface is a structure in which the slot is cut, and can be replaced, e.g. with a metal plane passing through both the radiators.

It is easily verified that the expression (9.22) permits to calculate the input impedance of the slot have antenna located on an infinite perfectly conducting plane not far from the other slot antenna (Figure 9.19a). If the second antenna is identical to the first and is excited in anti-phase, magnitude Z_E can be found from (9.23). If several slots are located on a metal plate, then one must substitute it into the expression (9.22), which remains true, the input impedance of a metal radiator Z_E included in the radiators system with allowance for its position and phasing. The directional pattern of a slot antenna coincides with the directional pattern of metal antennas with the same position in space.

If slot antennas have finite width, the electrical radiators, which are compared with them, must have an identical width. They are metal plates with the cross section in the shape of a rectangle curved along the cylindrical surface (see Figure 9.19). Angle α between the perpendiculars to the plates is equal to the arc between their axes, i.e. it is dependent on the cylinder radius. If the arc is π, the plates radiate in the opposite direction. When calculating Z_E, it is necessary to take into account the dimensions of cross section and mutual angle of rotation of metal plates. The directional pattern changes accordingly. It means that the electrical characteristics of slot antennas in the given case will depend on the radius of the cylinder, on which they are set.

All the aforesaid can be extended onto the case of slot radiators situated on the surface of a cone (Figure 9.20a), or a paraboloid (Figure 9.20b), or an arbitrary surface with the axial symmetry. In all cases the shape of the metal radiators, whose directional patterns coincide with the directional patterns of the slot radiators, must coincide with the shape of the corresponding segments of generatrix, the movement of which creates the given surface.

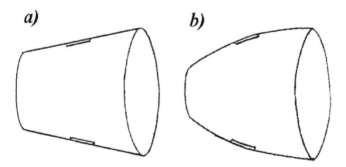

Figure 9.20   Slot antennas at the cone (a) and the paraboloid (b).

Metal bodies of finite length are of particular interest. As the experiments and calculations have shown, the directional pattern and input impedance of a whip antenna located near a circular metal cylinder of finite height, exceeding an antenna height by more than $\lambda/4$, practically coincide with the analogous characteristics of the antenna located near an infinitely long cylinder. Therefore, if the ends of slot antennas are distanced from the end of the metal cylinder with the axial symmetry by more than $\lambda/4$, one may determine the characteristics of two anti-phase slots in the metal body in accordance with the characteristics of two radiators of the same shape and dimensions located in free space.

In substance, in accordance with this conclusion the expression (9.22) for the input admittance of slot antennas placed on the metal cylinder is right. Directional patterns of slot and metal antennas coincide with each other and do not depend on the radius of the metal cylinder. Any restrictions connecting the radius of the cylinder with the antenna length or the wavelength don't imposed. This means that for the transition from a magnetic radiator to a two-sided slot antenna cut in a curvilinear metal surface, and also for the transition from a slot antenna to a magnetic radiator it is enough that the metal surface is smooth, and the antenna surface coincides with the curvilinear surface of the cylinder.

The equivalence of a slot radiator and a magnetic radiator is true, if the field structure at the radiator surface is formed by the extraneous source and is independent of the structure of electromagnetic field in the surrounding space

At present, the mathematical modeling method, in accordance with which a metal surface is replaced by a system of thin wires or metal strips, is used widely to calculate the electrical characteristics of radiators located not far from the metal bodies [33, 84]. In this formulation the problem reduces to the calculation of the current's distribution in a structure of randomly directed segments of wires or stripes. If the currents along the wires (stripes) are known, one can calculate all radiator characteristics.

The considered earlier equivalence of a slot radiator and a magnetic radiator allows to extend the mathematical modeling method to the case of calculation of the electrical characteristics for slot antennas situated on metal bodies of a complex shape. In this case, the magnetic radiator with the magneto motive force is the source of the electromagnetic field rather than the electrical radiator with the electromotive force.

As mentioned above, the coincidence of the electrical characteristics of slot radiators located on an axially symmetric metal surface with the electrical characteristics of metal radiators allows to increase the area of slot antennas, where the rigorous methods of solving the electrodynamics problem are applicable.

10

Problems of
Compensation and Adjustment

10.1 METHOD OF COMPENSATION

Development of radio engineering and the widespread use of the radio devices in the national economy and in everyday people life led to the problem of protecting living organisms and sensitive instruments from strong electromagnetic fields in the near zone of each transmitting antenna. The protection of devices is necessary, since the radiation of nearby devices can disrupt their normal operation, cause spontaneous switching on and switching off of device, change operating regimes, etc. The protection of living organisms is required, firstly, in the vicinity of powerful transmitting centers, where the electrical field strength is great, and, secondly, near mobile equipment in the transmitting mode, which is located next to the user. A cellular phone is, in particular, such a device.

The undoubted advantage of mobile communication consists in the freedom of its use to everyone, regardless of age, wealth and location. If the radio has liberated from the fetters of wire systems, the cellular phone allows replacing the radio station, mounted on a truck or on other vehicle, by a small device that may fit in a child's palm. As a result, the advent of cellular communication looks like a big bang due to the rapid increase in the number of handsets, the widespread proliferation of phone contacts, and also due to the rapid growth of concern about the potential harm of these devices for human health, in particular its carcinogenic effect.

Together with proliferation of mobile communication systems, there has been an increasing concern about possible hazards for the user's body, especially for the user's head, which is irradiated by a handset antenna. During a phone conversation, the personal cellular phone is placed next to the user's head, and its transmitting antenna irradiates sensitive human organs (brain, eyes, etc.). The absorbed power in today's cellular phones can be equal to a half of all radiated power. In order to minimize any possible health risk, it is necessary to reduce the amount of that power.

A correlation between irradiation and parotid tumor is still subject to scientific debate, mostly due to the inherent difficulties of empirical research. Rumors and the truth about the potential harm of irradiation require reducing the Specific Absorption Rate (SAR) in the user's head. This is an extremely complex problem, since one must reduce the near field of the transmitting antenna without decreasing the far field and without deteriorating directional pattern of antenna. In addition, the problem is not limited by the cellular phones, since a man often uses or operates with a low-power transmitter. This transmitter can be placed nearby in a production area or in a vehicle.

The rather obvious idea of head protection by means of screening, i.e. by shading effect, is unrealizable. The near field has no ray structure, and hence the shadow behind a metallic screen can cover only an area approximately equal to the screen size. For example, in order to protect the head of the cellular phone user, the screen must be much larger than the cross-section of the handset housing. For similar reasons, one should discard the idea of using an absorber, i.e. a dielectric shield that absorbs some part of energy. The distortion of the antenna directional pattern is still another obvious disadvantage of using screens and absorbers.

The protective action of the compensation method, proposed by M. Bank [85], is based on a different principle—on the mutual suppression of fields, created by various radiating elements in a certain area. The diverse variants of introducing this principle (mutual compensation of fields of two radiators, use of folded or cavity antennas) possess an opportunity to reduce irradiation of user's organism, especially his head, without distorting the antenna pattern in the horizontal plane. In accordance with the key (main) variant of the compensation method, as shown in Figure 10.1, the main radiator 1 is supplemented with a second (auxiliary) radiator 2, situated in the plane, passing through the head center and the feed point of the main radiator. The second radiator is placed between the head and the main radiator and is excited approximately in anti-phase with it (not exactly in anti-phase, because the phase of the field is changed along the interval between the radiators). So, the fields of the two radiators will compensate each other at a certain point inside the head, and the point will be surrounded by a dark spot, i.e. by a zone of a weak field.

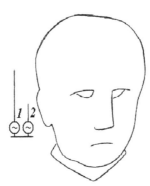

Figure 10.1 Two radiators next to the head.

The dipole moment of the auxiliary radiator must be smaller than that of the main radiator, since the field, close by to an electrical radiator, decreases quickly. In order to get the same magnitude of field at the compensation point, if the currents of radiators differ substantially, it is enough to place the radiators at a distance of a few centimeters from each other (1–2 cm at frequency 1 GHz). Therefore, the field of the auxiliary radiator may be relatively small, i.e. the common field in the far region is little different from the main radiator field, and the directional pattern remains close to the directional pattern in the absence of compensation.

Placement of linear radiators near the human head is shown in Figure 10.2. As seen from the figure, the feed point A_1 of the main radiator, which coincides with the origin of cylindrical coordinate system, and compensation point C are located along the horizontal straight line, passing through a head center. This line is defined as the structure axis. The feed point of the auxiliary radiator is placed on this line, at a distance b from the main radiator. Assume that both radiators are vertical and have equal lengths. Inside the head at compensation point C (at distance ρ_0 from point A_2) the fields of both radiators must be equal in magnitude and opposite in sign:

$$E_{z2}(b + \rho_0, 0, 0) = -E_{z1}(b + \rho_0, 0, 0). \tag{10.1}$$

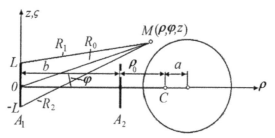

Figure 10.2 Placement of phone radiators close to a human head.

In the capacity of linear radiators with finite lengths, one can employ monopoles with a feed point in their base. A linear radiator creates two electric field components: E_z and E_ρ. If a straight perfectly conducting filament of finite length is used as model of a vertical linear radiator, the electric field components with allowance of a mirror image in a cylindrical coordinate system (ρ, φ, z) are given by expressions (1.31) and (1.32). Similar expressions hold for auxiliary radiators too. As can be seen from them the simultaneous compensation of both E_z and E_ρ field components by adjustment of the current J_{A2} of the auxiliary radiator is impossible. Since E_z is greater than E_ρ, and E_ρ along the ρ-axis (at $z = 0$) is zero, we prefer to compensate the E_z components.

Since both feed points A_1 and A_2 and also the compensation point C lie on the structure axis, components E_z of both radiators in a homogeneous medium with relative permittivity ε_r are determined by (1.31), but for the main radiator $\rho_1 = \rho$, and for the auxiliary radiator $\rho_2 = \sqrt{\rho_1^2 + b^2 - 2\rho_1 b \cos\varphi}$ (Figure 10.3).

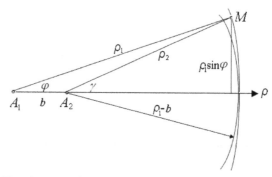

Figure 10.3 Two linear radiators in cylindrical coordinate system (top view).

Suppose we want to find the current and the input impedance of the radiator situated in the near region of the neighboring radiator. If the emf is connected in the input of the first radiator, the circuit looks, as shown in Figure 10.4. The radiators are accomplished as the monopoles of finite lengths, R_{A1} and R_{A2} are the impedances of the first and second generators accordingly. Generally $R_1 = R_2 = R$.

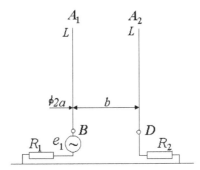

Figure 10.4   Two radiators in the near region of each other.

The calculation method for a system with two linear radiators is based on the folded dipole theory and on the superposition principle. If to connect two voltage generators, equal in magnitude to $e_1/2$ and opposite in direction, in a basis of the right radiator and to divide also the main generator into two generators, the same in magnitude and in direction, then according to the superposition principle, the current at each point is the sum of the currents, created by all generators. Therefore, as shown in Figure 10.5, one can divide the original circuit onto two circuits with two generators in each circuit and then calculate and sum the currents at points B and D, created in each of the circuits. This procedure allows analyzing the antenna system as a superposition of two subsystems with in-phase and anti-phase currents.

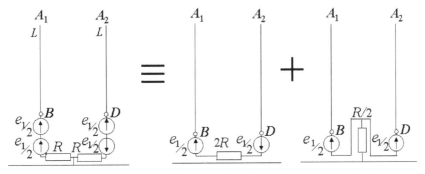

Figure 10.5   Division of the considered circuit onto two circuits.

Let the wires of the first circuit carry only anti-phase currents, e.g. currents equal in magnitude and opposite in direction. The two parallel wires form a long line with the load $2R$ in the basis. The line is open on the end. Let the second circuit carry only in-phase currents, i.e. the potentials of points, situated in both wires at the same height, (including lower points of wires) are identical. With this aim the emf's in the basis of the wires of the second circuit must be equal, if the radii of wires are equal. In this case, the parallel wires form a monopole with resistance $R/2$ between the generator and the ground.

For the two-wire line, we can write

$$e_1 = J_l(Z_l + 2R), \tag{10.2}$$

where J_l is the current in the line basis, $Z_l = -jW_l \cot kL$ is the input impedance of line with length L, $W_l = 120 \ln(b/a)$ is the wave impedance of the line, b is the distance between the wires, and $2a$ is the diameter of each wire. The current at point B is $J_{B1l} = e_1 Y_1$, and the current at point D is $J_{D1l} = -e_1 Y_1$, where $Y_1 = 1/(-jW_l \cot kL + 2R)$.

For the monopole, we can write

$$e_1/2 = J_r(Z_r + R/2), \tag{10.3}$$

where J_r is the current at the monopole basis, and $Z_r = Z_m(L, a_e)$ is the input impedance of the monopole with length L and equivalent radius a_e, equal to $\sqrt{ab}$. The currents at points B and D are the same and equal to

$$J_{B1r} = J_{D1r} = e_1/(4Z_r + 2R) = e_1 Y_2, \tag{10.4}$$

where $Y_2 = 1/[4Z_m(L, a_e) + 2R]$.

So, if emf e_1 is connected in the first radiator input, the currents in the first and the second radiator basis are

$$J_{11} = e_1(Y_1 + Y_2), \quad J_{21} = e_1(-Y_1 + Y_2). \tag{10.5}$$

If emf e_2 is connected in the second radiator input, so, similarly to (10.5), the currents in the first and the second radiator bases are

$$J_{12} = e_2(-Y_1 + Y_2), \quad J_{22} = e_2(Y_1 + Y_2), \tag{10.6}$$

According to the superposition principle, the terminal currents of the radiators are

$$J_{A1} = J_{11} + J_{12} = (e_1 - e_2)Y_1 + (e_1 + e_2)Y_2, \quad J_{A2} = (e_2 - e_1)Y_1 + (e_1 + e_2)Y_2. \tag{10.7}$$

The input admittances of the radiators are

$$Y_{A1} = J_{A1}/e_1 = Y_1 + Y_2 + e_2(Y_2 - Y_1)/e_1, \quad Y_{A2} = J_{A2}/e_2 = Y_1 + Y_2 + e_1(Y_2 - Y_1)/e_2. \tag{10.8}$$

As is known, the current and the input impedance of a radiator depend significantly on the neighboring radiator current. For a system of two radiators one can write

$$e_1 = J_{A1}Z_{11} + J_{A2}Z_{12}, \quad e_2 = J_{A1}Z_{21} + J_{A2}Z_{22}. \tag{10.9}$$

Here, e_1 and e_2 are the driving emf's, connected respectively to the terminals of the first and second radiators, Z_{11} and Z_{22} are their self-impedances, and Z_{12} and Z_{21} are their mutual impedances. Each expression in (10.9) is a Kirchhoff equation for a series connection of circuit elements.

The expression (10.9) can be rewritten as

$$J_{A1} = \frac{e_1 Z_{22} - e_2 Z_{12}}{Z_{11}Z_{22} - Z_{12}Z_{21}}, \quad J_{A2} = \frac{e_2 Z_{11} - e_1 Z_{21}}{Z_{11}Z_{22} - Z_{12}Z_{21}}, \tag{10.10}$$

i.e. the current in each radiator is the sum of the currents produced by its self-generator and the generator of the neighboring radiator (because of the mutual coupling between radiators). The ratio of the currents depends on the level of mutual coupling between the radiators, which is determined by their dimensions and position. Comparing (10.5)–(10.6) with (10.10), one can determine the radiator self- and the mutual impedances (for wires of equal radii). Considering that

$$J_{11} = \frac{e_1 Z_{22}}{Z_{11} Z_{22} - Z_{12}^2} = e_1(Y_1 + Y_2), \quad J_{12} = \frac{-e_2 Z_{12}}{Z_{11} Z_{22} - Z_{12}^2} = e_2(Y_2 - Y_1),$$

$$J_{12} = \frac{-e_2 Z_{12}}{Z_{11} Z_{22} - Z_{12}^2} = e_2(Y_2 - Y_1), \quad J_{21} = \frac{-e_1 Z_{12}}{Z_{11} Z_{22} - Z_{12}^2} = e_1(Y_2 - Y_1),$$

we obtain:

$$Z_{11} = Z_{22}, Y_1 + Y_2 = \frac{Z_{11}}{Z_{11}^2 - Z_{12}^2}, Y_1 - Y_2 = \frac{Z_{12}}{Z_{11}^2 - Z_{12}^2}. \tag{10.11}$$

Summing and subtracting the left-hand and right-hand parts of last two expressions, we find:

$$2Y_1 = 1/(Z_{11} - Z_{12}), 2Y_2 = 1/(Z_{11} + Z_{12}),$$

that is,

$$Z_{11} + Z_{12} = 1/(2Y_2), Z_{11} - Z_{12} = 1/(2Y_1),$$

and consequently

$$Z_{11} = Z_{22} = 0.25(1/Y_1 + 1/Y_2) = Z_m(L, a_e) - j0.25 W_l \cot kL + R,$$
$$Z_{12} = 0.25(1/Y_2 - 1/Y_1) = Z_m(L, a_e) + j0.25 W_l \cot kL. \tag{10.12}$$

One can see from (10.12) that the result of close proximity of the radiators is an additional term of self-impedance of each radiator. A similar approach is true also for the second radiator and for the mutual coupling between them.

The mutual coupling between the radiators affects the current distribution along each radiator and causes the anti-phase currents in addition to the in-phase components. For equal radii of the wires, the in-phase and the anti-phase currents of the first radiator are

$$J_{A1}^{(i)}(z) = (e_1 + e_2)Y_2 \frac{\sin k(L - |z|)}{\sin kL}, \quad J_{A1}^{(a)}(z) = (e_1 - e_2)Y_1 \frac{\sin k(L - |z|)}{\sin kL}. \tag{10.13}$$

In accordance with (1.31), the total fields of the radiators in plane $\varphi = 0$ are given by

$$E_{z1}(z) = -j30 F_1 / \varepsilon_r [(e_1 + e_2)Y_2 + (e_1 - e_2)Y_1].$$
$$E_{z2}(z) = -j30 F_2 / \varepsilon_r [(e_1 + e_2)Y_2 + (e_2 - e_1)Y_1]. \tag{10.14}$$

Here, $F_m = \frac{1}{\sin kL} [\exp(-jkR_{m1})/R_{m1} + \exp(-jkR_{m2})/R_{m2} - 2\cos kL \exp(-jkR_{m0}/R_{m0})]$,

where m is the radiator number, $R_{m1} = \sqrt{(z - L)^2 + \rho_m^2}$, $R_{m2} = \sqrt{(z + L)^2 + \rho_m^2}$, $R_{m0} = \sqrt{z^2 + \rho_m^2}$. In order to bring the total field $E_z = E_{z1} + E_{z2}$ to zero at compensation point $(\rho_0 + b, 0, 0)$, the ratio of emf's, feeding the radiators, must be

$$\frac{e_2}{e_1} = -\frac{(Y_2 + Y_1)F_{10} + (Y_2 - Y_1)F_{20}}{(Y_2 - Y_1)F_{10} + (Y_1 + Y_2)F_{20}}. \tag{10.15}$$

Here, F_{m0} is the value of function F_m at the compensation point. Equation (10.15) allows finding the voltage amplitude and phase at the input of the second radiator, if those amplitude and phase of the first radiator are known. Using (10.9), we find:

$$J_{A1}/e_1 = Y_1 + Y_2 + e_2(Y_2 - Y_1)/e_1, \quad J_{A2}/e_1 = Y_2 - Y_1 + e_2(Y_1 + Y_2)/e_1.$$

Substituting (10.15) in these expressions, we obtain for the driving currents ratio

$$J_{A2}/J_{A1} = -F_{10}/F_{20}.$$

(10.16)

It coincides with the expression in the absence of mutual coupling between the radiators. This occurs when the radiators are of the same length. For $L = \lambda/4$, we obtain in the first approximation $F_{10} = \exp(-jkR_{10})/R_{10}$, $F_{20} = \exp(-jkR_{20})/R_{20}$, i.e.

$$J_{A2}/J_{A1} = -R_{20}\exp[-jk(R_{10}-R_{20})]/R_{10},$$

(10.17)

where $R_{20} = \sqrt{L^2 + \rho_0^2}$, $R_{10} = \sqrt{L^2 + (\rho_0 + b)^2}$. As it follows from the above expressions, the dipole moments of linear radiators must be in inverse proportion to the distances between the radiators and the compensation point.

10.2   DIMENSIONS OF DARK SPOT AND FACTOR OF LOSS REDUCTION

Application of the compensation method requires analyzing its efficiency. To this end, one needs to calculate the field in the space, surrounding radiator, and to estimate the irradiation reduction factor. The problem is complicated due to the heterogeneous nature of the medium. For example, antennas of cellular phone is located in close proximity to the user's head, hand and body, consisting of many tissues with different permittivity, which is much higher than the permittivity of free space. Since the field strength and hence the dissipated power is maximal near the antenna, the capability to calculate correctly the near field of the antenna with due account of the user's body's presence is crucial for results' accuracy.

Consider a linear antenna tangent to the user's head, which is modeled by a vertical prolate ellipsoid (Figure 10.6a). One may interpret the structure in the capacity of a linear radiator situated along the flat boundary between two half-spaces—between air with $\varepsilon_r = 1$ and the head with $\varepsilon_r \neq 1$, as shown in Figure 10.6b. It is a crude approximation, because the head dimensions are magnitudes of the order of the wavelength.

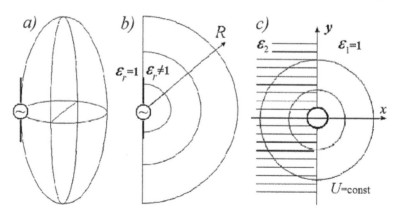

Figure 10.6   Model of the head and the linear antenna (a), the antenna at the interface (b), the charge in a heterogeneous medium (c).

The problem of finding the electromagnetic field of a linear radiator, located along the boundary between two media, reduces to calculation of the electrostatic field in a heterogeneous (or more exactly, piecewise-homogenous) medium [86]. Such problem arises, in particular, if an isolated wire is located at the interface of two media, e.g., air and a dielectric medium with $\varepsilon_r \neq 1$, as shown in Figure 10.6c.

As already mentioned in Chapter 7, a structure of the quasi-stationary electrical field of alternating linear currents is similar to a structure of the magnetic field, created by constant linear currents. The magnetic field structure coincides with the structure of the electric field of a line charge (principle of correspondence). The near field of an antenna in the first order approximation has a quasi-stationary nature. This analogy allows to reduce the calculation of the antenna near field in a piecewise-homogenous medium to a calculation of a field in a homogenous medium. But this analogy is true only in the approximation of the first order and requires checking.

The solution for the electrostatic field in a heterogeneous medium is demonstrated by Figure 10.6c, where an isolated charged straight wire is located at the interface of two homogeneous media. As shown in Chapter 7, the field near an interface of a piecewise-homogenous medium coincides with the field in a homogenous medium, if to assume that equivalent permittivity is equal to

$$\varepsilon = (1 + \varepsilon_r)/2. \tag{10.18}$$

Thus, the field of a linear radiator located at the boundary of two media differs only in the magnitude of relative permittivity. Analysis of a heterogeneous environment that uses a homogeneous medium as an equivalent replacement substantially simplifies the problem of calculating the irradiation power and the factor of loss reduction. The obtained results are verified with the program CST, based on the Moment Method, which allows to take into account detailed characteristics, including heterogeneity of a medium.

The relative permittivity of the brain, muscles and skin has similar values and on the average is about 40–50. Only the permittivity of bones differs substantially from this value. Accordingly, equivalent permittivity ε_e of the human tissues and the air is close to 20–25, and the field magnitudes in the vicinity of an antenna, located near the head, can be calculated in a homogenous medium with ε_{er}.

The dark spot dimensions for dipoles of finite length can be calculated, using equations (10.14), written for the field of such radiators in the cylindrical coordinates. The dark spot boundary is determined by radius ρ_n, at which $n = |E_z/E_{z10}|$ is equal to a given value, where E_z is the common field and E_{z10} is the field of the main radiator.

In order to simplify calculation of the factor of loss reduction, assume that the field strength increases linearly from the compensation point with the coordinate $t = 0$ to the dark spot boundary, where $t = t_0$, so that $|E_z/E_{z10}| = nt/t_0$, where $n = \text{const}$, and hence the loss power grows in proportion with the square of a distance. If $s = t/t_0$, the total power of losses inside the dark spot is given by

$$P = P_0 \int_0^1 (ns)^2 \, ds = P_0 n^2/3 \tag{10.19}$$

where P_0 is the loss power within the boundaries of dark spot caused by the main radiator in the absence of the auxiliary radiator. For $n_0 = 0.2$, the loss of power is smaller by a factor of $3/(0.2)^2 = 75$ (or 18.8 dB). For $n_0 = 0.1$ the factor of loss reduction is 300 (24.8 dB), and for $n_0 = 0.04$ the factor is 1875 (32.7 dB). In practice this factor is equal to roughly a half of the calculated value, since the field within the dark spot has a more complicated structure, i.e. the factor of reduction is approximately equal to

$$P_0/P = 1.5/n_0^2. \tag{10.20}$$

In order to find the boundary of the dark spot, one must determine the radius ρ_n as a function of z and φ. In Figure 10.7 ratio $n = |E_z/E_{z10}|$ is plotted as a function of ρ for $\lambda = 30$, $L = 7.5$, $\rho_0 = 1$ and different values of b (all dimensions are in centimeters). The magnitudes of fields E_{z10} and E_z are calculated at $z = 0$, $\varphi = 0$ by the above formulas. The boundaries of the dark spot in the ρ-direction, denoted as points ρ_1 and ρ_2, are found as the intersection points of curve n with given level n_0. Along the segment $\rho_1\rho_2$ (between the points ρ_1 and ρ_2), n is smaller than the required value of n_0. Length $\Delta\rho = \rho_2 - \rho_1$ of the segment, i.e., the dark spot length, is presented in Table 10.1 for different values of b and n_0, i.e. for different levels of field reduction at the boundary of the dark spot. One can see from Table 10.1 that the length of dark spot for given level n_0 decreases, when distance b between the radiators increases. But at large distances, when n_0 at the boundary of dark spot increases, the length of dark spot increases also. Therefore, for small values of b, the dip of curve n becomes narrower and deeper. For great values of b, the dip of curve n is wider.

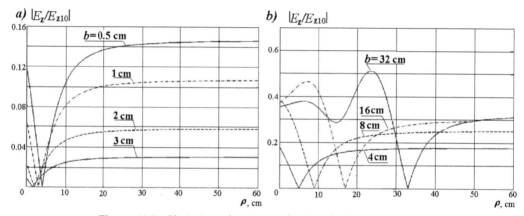

Figure 10.7 Variation of ratio $n = |E_z/E_{z10}|$ along the ρ-axis.

Table 10.1 The Length of Dark Spot for Different b and n_0

b	$n_0 = 0.01$	0.02	0.04	0.07
0.5	3.53	6.80		
1	1.89	3.89	7.67	
2	1.00	2.00	4.17	8.12
3	0.71	1.42	2.91	5.43
b	$n_0 = 0.07$	0.10	0.14	0.25
4	4.28	6.58	11.38	
8	2.89	4.29	6.43	49.60
16	2.44	3.56	5.24	23.63
32	2.29	3.35	4.88	11.73

Figure 10.8 shows a few examples of the dark spot boundaries in the horizontal plane ($z = 0$) for $f = 1$ GHz, $\rho_0 = 1$ cm and different values of b and n_0. In the figure, fields E_{z10} and E_z are calculated, using equations (1.31) at $z = 0$ and given angles φ, in the manner similar to the preceding case. The difference between coordinates ρ_1 and ρ_2,

denoting the start and end of the corresponding segment with $n \leq n_0$, determines the
length of dark spot in the given direction.

From Figure 10.8 it is seen that the dark spot width is, as a rule, greater than its
length. The spot height is close to its length. This result is particularly important, as it
shows, in which direction movements of the head are more dangerous in terms of the
growth of the absorbed power.

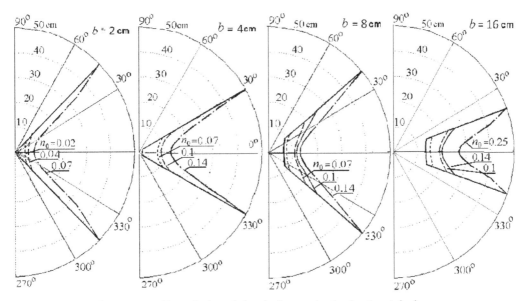

Figure 10.8 Boundaries of the dark spot in the horizontal plane.

The calculations also show that at the optimal selection of the structure in terms of
the relative positions of the radiators and the compensation point, the volume of the
dark spot increases, and the field inside it decreases substantially. One can choose the
structure parameters so that the spot dimensions would coincide with the human head
dimensions. That will allow to diminish sharply loss power in the head, and gives high
tolerance with respect to movements of the user's head.

Employing several auxiliary radiators allows increasing the dark spot volume further.

The use of an auxiliary radiator in the compensation method on the one hand allows
to reduce the field in the near region and on the other hand must not distort the circular
pattern in the horizontal plane. Retention of the far field strength and directional pattern
is a crucial issue for any method of SAR reduction. In order to retain the shape of
directional pattern and, in particular, to avoid a deep dip in it along the structure axis,
the phases of the far fields created by the two radiators should not differ by more than
a few degrees. The small distance between the antennas and the closeness of phases of
driving currents ensure the preservation of the circular shape of the directional pattern.
Yet, in order to reduce the common field at the compensation point to zero, the fields
of radiators at this point must have opposite phases.

If the driving current of the main radiator is J_{A1}, the driving current of the auxiliary
radiator can be written as $J_{A2} = J_{A1}De^{-j\psi}$, where D is the ratio of the dipole moments of
the radiators, and ψ is the phase difference between the driving currents. In this case,
the field of the auxiliary radiator is $E_{z2} = E_{z1}De^{ju}$, where E_{z1} is the field of the main

radiator, and $u = kl - \psi$. Here l is the path difference between rays from the radiators to a point located at angle φ (see Figure 10.3). The common directional pattern in the horizontal plane can be written as

$$|F| = \left|1 + D\exp(ju)\right| \Big/ \sqrt{(1 + D\cos u_m)^2 + D^2 \sin^2 u_m}. \qquad (10.21)$$

Here u_m is the value of u at the maximum of the denominator. Since D and u_m are constant, the denominator of (10.21) is independent of φ. If the directional pattern of the main radiator is circular, the ratio of the maximum of directional pattern to its minimum, which is a measure of the distortion of directional pattern, is equal to

$$\frac{|F_{max}|}{|F_{min}|} = \frac{\left|1 + De^{ju}\right|_{max}}{\left|1 + De^{ju}\right|_{min}} = \frac{1 \pm D}{1 \mp D}, \qquad (10.22)$$

where the upper sign applies when D is positive, and the lower sign applies if D is negative. It is easy to show that when the maximum of directional pattern differs from the minimum by 6 dB (deviation of 3 dB from the average level), we obtain $|D| = 0.33$, and when the difference between the maximum and the minimum is 3 dB, we have $|D| = 0.17$.

It is necessary here to explain that cellular communication is not the only area of application where one must create a weak field region near the transmitting antenna. The task is vital, if, for example, a mobile transmitter is located in a vehicle close to users and other passengers. Creating a weak field region in such cases is an efficient technique of protecting against irradiation. A problem is often complicated because of the operation of the transmitter in a broad frequency band. The problem is considered in Section 10.4. The analysis of directional pattern is essential for this problem, since there is habitually some degree of freedom to choose the antenna location. This helps to keep the undistorted directional pattern.

From (10.17), taking into consideration that $R_1 = \rho_0 + b$, $R_2 = \rho_0$ (see Figure 10.2), we obtain the expression $D = R_2/R_1 = \rho_0/(\rho_0 + b)$, which reduces to $\rho_0 = b/\beta$, where $\beta = 1/D - 1$. Let the main radiator is situated on the structure axis, at a distance 3 cm from the head, i.e. $\rho_0 + b = 3$, and the compensation point is placed on the head surface. In this case, if $D = 0.33$, then $\beta = 2$, $\rho_0 = b/2 = 1$, $b = 2$; for $D = 0.17$ we have $\beta = 4.88$, $\rho_0 = 0.5$, $b = 2.5$. The results point out the necessity for a tradeoff between the level of irradiation reduction and the directional pattern of the two-antenna structure, especially in cases when the antenna platform is subject to spatial restrictions. However, the changes of directional pattern are predictable and small.

Figure 10.9 shows three variants of position of the dark spot relative to the head, where boundaries of the dark spot are given by a dotted line and the head boundary are given by the solid curve. Feed points A_1 and A_2 of the main and the auxiliary radiators and compensation point C in Figure 10.9a and 10.9b lie on the structure axis, and so the dark spot is symmetric relative to the axis. Figure 10.9a has the compensation point placed outside the head, and Figure 10.9b has it inside the head, close to its surface. In the latter case, the whole dark spot lies inside the head, resulting in a significant reduction of loss inside the head. Figure 10.9c corresponds to the case, where the auxiliary radiator is located out axis, so that the axis of dark spot would not coincide with the structure axis. This may result in such displacement of the dark spot from the head center that the loss power in the head increases.

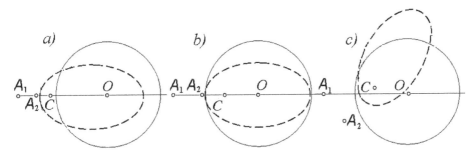

Figure 10.9   The dark spot: compensation point is located outside the head (*a*), inside the head (*b*) and outside the symmetry axis (*c*).

Calculation of the dimensions of dark spot were accomplished by the Matlab program and verified both experimentally and by means of the CST program. The simulations were carried out with allowance for the user's head and without that. The results of calculations, simulations and measurements with allowance for the user's head are given in Figure 10.10. The main radiator was located at point $\rho = 0$, the auxiliary radiator was located at point $\rho = 2$ cm. The head was situated at a distance 4 cm from the main radiator. The point C was located at $\rho = 5$ cm.

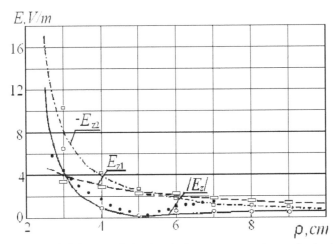

Figure 10.10   Field strength in the dark spot: comparison of calculation, simulation and measurements.

As mentioned earlier, the calculations were performed by the Matlab program. The calculation results are presented in the form of dashed, dash-dotted and solid curves plotted for absolute values of fields E_{z1}, E_{z2} and $E_z = E_{z1} + E_{z2}$, respectively. The presence of the head was taken into account by replacing $\varepsilon_r = 1$ with ε_e. It is easy to be convinced that in the interval from $\rho = 4$ to $\rho = 10$ cm, the field E_z is substantially less than the field E_{z1}.

The simulation was performed by means of CST program. The rectangles, squares and empty circles are the simulation results (model of the head is included in the CST program). The measurements of the common field E_z were carried out in laboratory conditions; they are presented with full circles. The presence of the user's head was taken into account. If the user's head is absent, the results of calculations, simulations and measurements are in good agreement. However, as one might expect from the approximate estimation of ε_e, the coincidence between the results with allowance for the head is not too good, but a qualitative agreement exist.

Table 10.2 shows the total and maximum local SAR (in W/kg) with the auxiliary radiator and without it. he compensation points are located in accordance with Figure 10.9a and Figure 10.9b. The maximal absorbed power and the maximal local SAR exist near the main radiator. And the maximal reduction of the loss power occurs here, too. The obtained results show that under realistic conditions the compensation method allows to reduce the power, absorbed by the head, of three to four times as well as to reduce the maximal local power of five to ten times. The distortion of directional pattern due to the auxiliary antenna is relatively small provided that this antenna is properly located.

Table 10.2 Level of *SAR*

Number of radiators	Figure	Total SAR	Maximal local SAR (in 1 g)
1		0.00793	0.166
2	10.9a	0.00257	0.0235
2	10.9b	0.00225	0.0162

10.3 STRUCTURES FOR A STRUGGLE WITH EXTERNAL INFLUENCES

Application of the compensation method under realistic conditions often meets with difficulties because of the changeable environment, since the operation of a complicated radiating structure is often disrupted by various external actions. The system disturbances are to be counteracted. In particular, approaching of the metallic objects to the antenna system or relative displacement of the antennas as a consequence of user's movement can result in a tuning disturbance. In such cases, the compensation point may be displaced and the field inside the dark spot can grow significantly. Consider the impact of metallic objects approaching to an antenna as an example of the external action.

In order to eliminate the consequences of these influences, one can try to retain the amplitudes and phases of the driving voltages and currents (the first method) or prevent the change of the radiator fields (the second method). It should be noted, first, that the appearance of a metal body near a radiator changes the current at all points along the wire of the radiator, whereas the feedback circuits are capable of adjusting the current only at a single point in each radiator, e.g., at its input. And, second, one must say that a metal body causes different changes of the radiator fields in the entire space, whereas the feedback circuits are capable of adjusting the field only at one or two points. Therefore, the efficiency of both methods, especially of the first one, is inherently limited. This Section is devoted to a comparative analysis of efficiency of both methods. It should be emphasized that the analysis regards cases of severe distortion in antenna systems.

The circuit of the antenna system is given in Figure 10.11. The circuit consists of two monopoles (*A* is the main radiator, *B* is the auxiliary radiator) mounted on a metal plate near the model of a human head. The compensation point is located inside the head, near to its front boundary. Dimensions in the figure are indicated in centimeters. Figure 10.12 shows the same circuit for the case, when a vertical metal sheet in a shape of a square is placed not far from the antenna system (0.5 m from the radiators). Presence of a metal sheet in the proximity to a cellular phone may occur, e.g., when a phone's user enters an elevator or a car. The metal sheet affects the antenna system and causes the growth of fields in the dark spot.

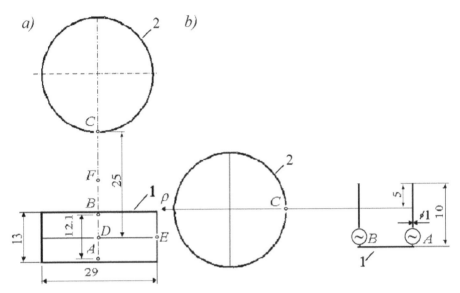

Figure 10.11 Circuit of antenna system near the model of a human head: top view (*a*) and side view (*b*).

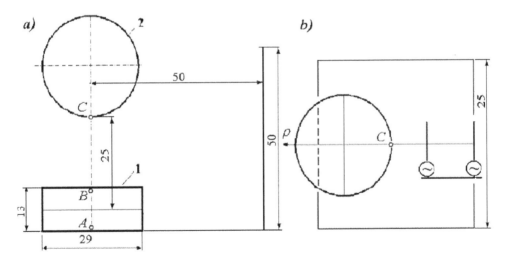

Figure 10.12 A metal sheet placed not far from the antenna: top view (*a*) and side view (*b*).

Let us start with the relatively simple case of a single radiator. The input current of a single radiator in the absence of any metal sheet is

$$J_{A1} = e_A/Z_A,$$

where e_A is electromotive force, and Z_A is its input impedance. The metal sheet changes an input impedance of the radiator and makes it equal to Z'_A. To avoid the current change in the radiator base, one must change the emf at its input to

$$e'_A = J_A Z'_A = J_A Z'_A / Z_A. \qquad (10.23)$$

For a single radiator of height 10 cm and diameter 1 cm we have: $e_A = 1$, $Z_A = 38.6$ + $j12.8$, $Z'_A = 41.7 + j14.3$, $e'_A = 1.09 \exp(j0.0145)$. Figure 10.13 shows the field of single radiator A along the horizontal line, passing through the radiator and the center of the head in the absence (1) and presence (2) of a metal sheet and after adjustment of the emf (3). Figure 10.13 and several others figures were divided into two parts in order to use the different scales.

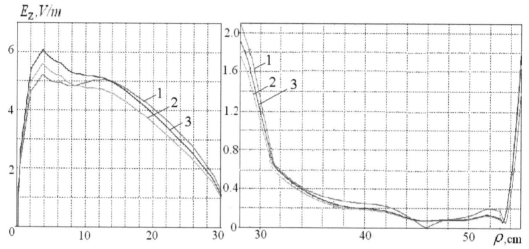

Figure 10.13 The field of the single radiator A in the absence (1) and presence (2) of the metal sheet and after adjustment of the emf (3).

In the case of two radiators one must first calculate the amplitude and the phase of the emf of the second radiator, which ensures the field compensation at a given point. For this purpose one can in turn excite both radiators by emf e_1, calculate the fields E_1 and E_2 of each radiator at the compensation point and take emf $e_2 = -e_1 E_1 / E_2$ as emf for the second radiator. It should be noted that the field and the input impedance of each radiator are calculated in the presence of the other radiator, when its input is grounded.

Calculations were performed with the CST program, which permits to simulate the total circuit with both generators and find the self-impedances Z_{11} and Z_{22} of each radiator and their mutual impedance Z_{12}. The solution of the set of two equations $e_1 = J_1 Z_{11} + J_2 Z_{12}$, $e_2 = J_2 Z_{22} + J_1 Z_{12}$ allows to find currents J_1 and J_2. The metal sheet changes the self- and mutual impedances of the radiators. To avoid the change of the currents at radiators bases, the emf's must be modified to

$$e'_1 = J_1 Z'_{11} + J_2 Z'_{12}, \; e'_2 = J_2 Z'_{22} + J_1 Z'_{12}, \qquad (10.24)$$

where the new impedances are marked by primes. Calculating the new emf's, we may find the new fields and ascertain the adjustment results.

For two radiators of the same dimensions, we initially obtain: $e_1 = 1$, $E_1 = 0.48 \exp(j1.11)$, $E_2 = 1.09 \exp(j2.54)$, i.e. $e_2 = 0.44 \exp(j1.71)$. Accordingly in the compensation mode we obtain for impedances: $Z_{11} = 37.3 + j16.2$, $Z_{22} = 39.5 + j16.8$,

$Z_{12} = 0.67 - j28.9$, and for currents: $J_1 = 0.013 - j0.0008$, $J_2 = 0.0064 + j0.018$. The adjustment results for impedances are as follows: $Z'_{11} = 40.6 + j17.1$, $Z'_{22} = 42.0 + j18.9$, $Z'_{12} = 2.04 - j28.02$, and for emf's: $e'_1 = 1.04 \exp(j0.04)$, $e'_2 = 0.51 \exp(j1.70)$. The calculated curves for the fields of the two-radiator structure are presented in Figure 10.14.

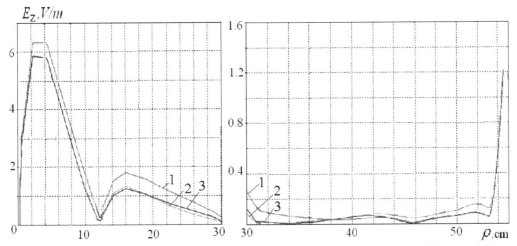

Figure 10.14   The field of two radiators in the absence (1) and presence (2) of the metal sheet and after adjustment of the emf (3).

The verification of adjustment results involves computation of fields along the horizontal line, passing through the radiators and the center of the head, as well as calculation of the total SAR and the maximal local SAR (in 1 g). The results obtained for the single radiator and for the structure from two radiators in the absence and presence of a metal sheet and after adjustment of the emf's in accordance with expressions (10.23) and (10.24), are compared with each other. The total SAR and the maximal local SAR for the corresponding cases are given in Table 10.3. The calculated amplitudes of field at the compensation point are also given in Table 10.3. The SAR level and the fields are presented in W/kg and in V/m, respectively. As one can see from the table and figures, the results indicate the rather low efficiency of the correction method, based on retaining the driving currents of radiators.

Table 10.3   *SAR and Field at Adjustment of emf in Accordance with Currents*

Characteristic	Total SAR	Local SAR	Field	Total SAR	Local SAR	Field
	One radiator			Two radiators		
Sheet is absent	$4.5 \cdot 10^{-5}$	$1.54 \cdot 10^{-3}$	0.79	$1.4 \cdot 10^{-5}$	$0.3 \cdot 10^{-3}$	0.0029
Sheet is present	$2.8 \cdot 10^{-5}$	$0.29 \cdot 10^{-3}$	0.68	$1.7 \cdot 10^{-5}$	$0.11 \cdot 10^{-3}$	0.0044
With adjustment	$3.4 \cdot 10^{-5}$	$0.34 \cdot 10^{-3}$	0.74	$2.1 \cdot 10^{-5}$	$0.18 \cdot 10^{-3}$	0.1214

The implementation of the second method calls to use one radiator as a measuring antenna and the second radiator as a local transmitting antenna (i.e., as a field source), or to use in turn both radiators as a measuring and a transmitting antenna, or to use the third radiator as a measuring antenna. The third radiator may be mounted at any suitable and convenient place. The adjustment is performed in the following way. The field at the receiving point is measured in the presence of a metal sheet and is compared with its magnitude in the absence of the metal sheet (i.e. in the compensation mode), and afterwards the emf of transmitting antenna is varied until it is obtained the initial value of the field.

The results of such adjustment are given in Figure 10.15 and Table 10.4 for the following variants: (1) when the main antenna is used as the transmitting and the auxiliary antenna is used as the receiving, (2) the opposite case: the main antenna is used as the receiving, whereas the auxiliary antenna is used as the transmitting, (3) both emf's are changed. Figure 10.16 and Table 10.4 present the results of field adjustment, using the third antenna located in the center of a metal plate, i.e., at point D (see Figure 10.11) for the following variants: (4) when the emf and the field of the main antenna is changed so that its field at point D becomes equal to its original field (before distortion), (5) when the field of the auxiliary antenna is changed for this purpose, (6) when the fields of both antennas are changed.

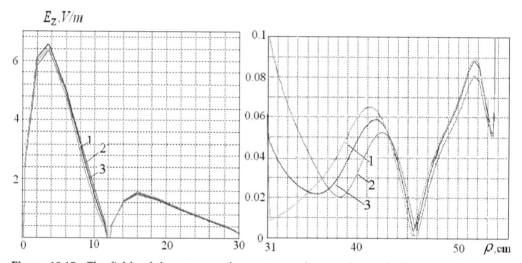

Figure 10.15 The fields of the structure from a two radiators after emf adjustment, based on the fields, received from by these radiators.

Table 10.4 *SAR* and Field at Adjustment of emf in Accordance with Fields

Variant	Total SAR	Max. Local SAR	Field
1	$1.99 \cdot 10^{-5}$	$0.137 \cdot 10^{-3}$	0.035
2	$1.83 \cdot 10^{-5}$	$0.151 \cdot 10^{-3}$	0.070
3	$2.02 \cdot 10^{-5}$	$0.152 \cdot 10^{-3}$	0.052
4	$1.57 \cdot 10^{-5}$	$0.102 \cdot 10^{-3}$	0.070
5	$1.45 \cdot 10^{-5}$	$0.125 \cdot 10^{-3}$	0.130
6	$1.18 \cdot 10^{-5}$	$0.090 \cdot 10^{-3}$	0.047

As one can see from Table 10.4 and Figures 10.15 and 10.16, the method, based on the measurement of the fields, demonstrates a higher efficiency. But acceptable results are obtained only at application of variant 6. Other variants, including placement of the third antenna at points E and F (see Figure 10.11), do not give satisfactory results.

And it should be pointed out that the monitoring signal is a weak signal that serves as the signal of feedback for both the main and auxiliary radiators. Therefore, proper adjustment by both methods requires that no signal from the transmitter impinges on the measuring antenna during the measurements of field.

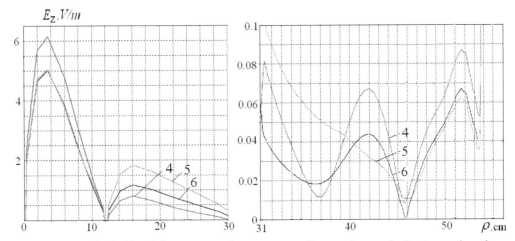

Figure 10.16   The fields of the structure form a two radiators after emf adjustment, based on the fields, received by a third radiator.

In order to prevent the changes of field under external actions (e.g., at the approach of a metallic object), one may use a manual or an automatic adjustment. Figure 10.17 gives the block diagram of an automatic adjustment. It contains transmitter 1, main radiator A_1 and auxiliary radiator A_2, connected to the transmitter through the power divider 2, the amplitude controller 3 and the phase shifter 4. The amplitude controller and the phase shifter provide the initial tuning and the field compensation at a given point. The amplitude controller is usually implemented by means the potentiometer, and the phase shifter is implemented by means the delay line, the low-pass filter or the high-pass filter. Two circuits, consisting of the amplitude controllers 13 and 15 and the phase shifters 14 and 16, provide two reference signals for the radiators A_2 and A_1, respectively.

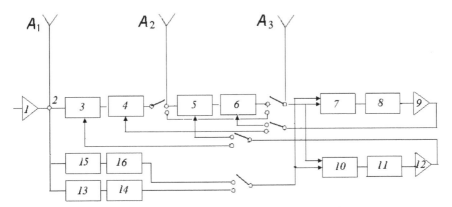

Figure 10.17   Block diagram of the proposed automatic adjustment circuit, based on the constancy of fields.

Antenna A_3, used for adjustment of the antenna system, receives, in turn, signals from the main and auxiliary radiators and changes (switches) the radiators. This procedure permits to determine the amplitude and the phase of the radiated and received signals.

The external action (e.g., an approach of a metal body) changes the phases of the signals of the both radiators, received by antenna The phase detector 7 compares in turn these phases with phases of the reference signals and produces an error signal, proportional to the difference of the said phases.. Low-pass filter 8 removes short-term fluctuations of the error signal. The error signal passes through amplifier 9, controls the phase shifters 4 and 6 and brings up the optimal phase differences. As it can be seen from Figure 10.17, a feedback circuit is constructed, and it provides a phase self-tuning action, similar to action of a phase locked loop (PLL).

The second feedback circuit is used for predicting the optimal signal amplitudes. It is similar to an automatic gain control (AGC) circuit. The input signal of radiator A_2 is compared in amplitude by comparator 10 (operational amplifier) with the reference signal. Low-pass filter 11 removes short-term fluctuations of the signal at the comparator output. The error signal passes through the amplifier 12, controls the amplitude controller 3 and brings up the amplitudes relationship to the optimal ratio.

As a result, two feedback circuits allow optimizing the amplitude relationship and the phases of the emf, feeding main radiator A_1 and auxiliary radiator A_2. In contrast to the conventional automatic gain control circuits, the amplitude difference and the phase difference of the two different radiator signals are not zero in this case.

The obtained results show that it is possible to automatically adjust a complicated antenna system under conditions of intensive disturbances. The proposed method, based on measurement of the fields, demonstrates a higher efficiency. It allows obtaining acceptable practical results under severe disturbance of the antenna system operation even in cases, when (as in our example) the zero-field point is not achievable.

10.4   WIDEBAND FIELD COMPENSATION

The compensation method was proposed to protect the user's head from the electromagnetic field of a cellular phone antenna. But the problem is not limited by the cellular phone, since cellular communication is not the sole scope of application where it is necessary to create a region of weak field near the transmitting antenna. For example, the passengers of a vehicle may be placed nearby a mobile radio station. Secondly, one must protect from a strong electromagnetic radiation not only living organisms, but also devices. The protection of devices is necessary, since the irradiation of devices can disrupt their normal operation, cause spontaneous switching on and switching off, change operating regimes, etc.

Requirements to the weak field may be changed partially. On the one hand, they often get more complicated due to the necessity of operation in a broad frequency range. On the other hand, the requirements may become less stringent. For example, large dimensions of the transmitter allow increasing the room for mounting auxiliary radiators. The requirement to the shape of the horizontal directional pattern is somewhat weakened.

Operation in a broad frequency band gives rise to additional difficulties, since the fields of the main and the auxiliary radiators at the compensation point must have the same magnitude and opposite sign at each frequency. The phase shift can be easily implemented by using an output stage of a transmitter with balanced output. But, firstly, two specimens of a given radiator type must have identical characteristics at all frequencies of the operation range. Secondly, since the main and auxiliary radiators are located at different distances from the compensation point, and signal velocities in the air and in transmission lines, connecting the radiators and transmitter, are different, it is necessary to place a phase shifter in one of a channel. This phase shifter must provide the required delay time throughout the frequency range.

It is not easy to satisfy all these requirements. In order to provide the same signal amplitudes, it is necessary to secure identity of both radiators or to install a controlled attenuator in one channel. Wideband antennas often have a complicated structure, which uses different conducting and dielectric materials. That is why antenna characteristics are unstable, i.e., different antenna specimens may have different frequency responses, and so their fields are not equal. Attenuators in a wide frequency range must be tunable.

Placement of both radiators at equal distance from the compensation point yields better results. A simple structure, consisting of two radiators placed at identical distance from the compensation point, is shown in Figure 10.18a (top view). Here the following notation is used: 1 is the transmitter top cover, 2 is the user's body, B_1 is the main radiator, B_2 is the auxiliary radiator, and C is the compensation point. The radiators are chosen in the form of two asymmetrical specimens of the same antenna type, which have a circular pattern in the horizontal plane and analogous amplitude-frequency characteristics. They are located at the same distance $B_1C = \sqrt{a^2 + b^2}$ from point C, where $a = AC$ is the distance from point C to the middle line of the transmitter cover and $b = AB_{1(2)}$.

Let the phase of an auxiliary radiator differ from the phase of the main radiator by 180°, and the signal amplitude of the auxiliary radiator is equal to $E_2 = DE_1$, where E_1 is the signal amplitude of the main radiator, and $D \le 1$. If $D = 1$, i.e., the amplitudes of the auxiliary signal and the main signal coincide, the sum of the signals in the plane of the structure symmetry is zero, since the signals in this plane are the same in amplitude and opposite in phase. If $D \ne 1$, the summary signal $E = E_1 + E_2 = (1 - D)E_1$ is equal to a fixed small fraction of the main signal and weakly dependent on the frequency, since the distance from the radiators to the compensation point is the same, and the radiators themselves are similar. A phase shifter in this case is not needed.

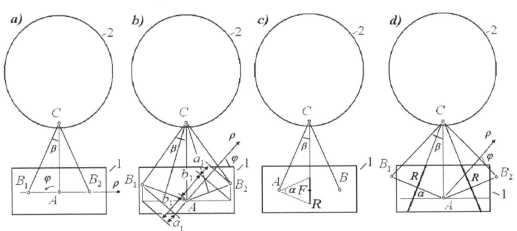

Figure 10.18 Compensation structure created by two (*a*) and three (*b*) radiators, with one (*c*) and two (*d*) reflectors.

One disadvantage of the proposed circuit is the absence of signal along the symmetry plane in the far region (in both directions), when $E_2 = -E_1$. But, first, is not crucial in many applications. Second, the angle (gap width), in which the radiation signal is nearly zero, is very small. And third, the influence of the ground and neighboring metal bodies leads to a smoothing of the directional pattern, i.e., the depth and the width of the gap in the horizontal directional pattern will be reduced in real conditions.

It is useful to estimate the gap width by calculating an angle $\Delta\varphi$ between the boundaries of the gap, i.e. between the points corresponding to the given small value f_1 of the directional pattern. Let $a = 25$ cm and $\beta = 30°$. The total signal consists of two signals with equal amplitude E_1 and opposite phase. It is created by two radiators located at distance $d = 2a \tan \beta = 28.9$ cm from each other and is equal to $E = 2E_1 \sin [(kd/2) \sin \varphi]$, i.e. $E_{max} = 2E_1$. If the angle φ is small, then

$$f_1 = \sin[(kd/2)\sin\varphi] \approx kd\varphi/2, \tag{10.25}$$

i.e. $\Delta\varphi = 2\varphi = 4f_1/(kd)$. In particular, if $f_1 = 0.1$ and the wavelength λ is 30 cm, then $\Delta\varphi = 3.8°$.

Another disadvantage is the fact that when $D \neq 1$, the factor of loss reduction decreases, since the total field even at the compensation point is not equal to zero. But this disadvantage is also inherent in the case of antennas placement at different distances from the compensation point, due to the limited dimensions of the transmitter top cover. Experimental results for the signal magnitude at the compensation point are given in Figure 10.19. Curve 1 corresponds to the signal of the main antenna, placed at point A (see Figure 10.18a), and curve 2 corresponds to the signal of two antennas, placed at points B_1 and B_2. As seen from the results, mounting both radiators at the same distance from point C permits to ensure compensation in a wider frequency band. The auxiliary radiator decreases the signal magnitude at the compensation point by 10–15 dB (3–6 times) in the band from 1.7 to 2.7 GHz.

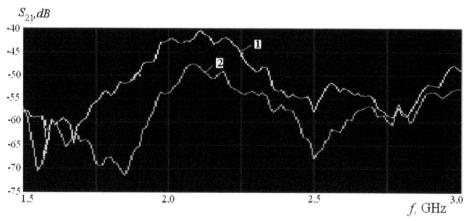

Figure 10.19 The frequency dependence of the field at the compensation point, created by the one (1) and two (2) radiators.

Using of two auxiliary radiators gives additional advantages. In this case, the amplitude of each auxiliary radiator signal may be smaller than that of the main radiator. One variant of this compensation structure is presented in Figure 10.18b (top view) where the same notation is used. If E_0 is the amplitude of the main radiator signal at the compensation point, the amplitude of each auxiliary radiator signal must be $E_1 = E_2 = E_0/2$. In the cylindrical coordinate system, the origin of which coincides with point A, the ratio of the auxiliary signal to the main signal in direction φ is

$$E_1(\rho)/E_0(\rho) = 0.5 \exp[jk(a_1 - b_1)], \quad E_2(\rho)/E_0(\rho) = 0.5 \exp[jk(a_1 + b_1)], \qquad (10.26)$$

where $E_0(\rho)$ is the main radiator field at distance ρ, $E_i(\rho)$ is the field of the auxiliary radiator, located at point B_i, and $a_1 \pm b_1$ are the path-length differences of the signals from the main and auxiliary radiators to the observation point. Here, $a_1 = a(1 - \cos \beta)$ $\sin \varphi$, $b_1 = a \sin \beta \cos \varphi$ (these segments are marked in Figure 10.18b). The total signal is

$$E(\rho) = E_0(\rho)\left[1 - \cos(kb_1)\exp(jka_1)\right]. \qquad (10.27)$$

The ratio of the total to the main signal is equal to

$$|E(\rho)|/|E_0(\rho)| = \sqrt{\left[1 - \cos(kb_1)\cos(ka_1)\right]^2 + \cos^2(kb_1)\sin^2(ka_1)}. \qquad (10.28)$$

In the symmetry plane, the total signal in the far region is

$$E(\rho)/E_0(\rho) = 1 - \exp\left[jka(1 - \cos \beta)\right] = -2j\exp\left[jka(1 - \cos \beta)/2\right]\sin\left[ka(1 - \cos \beta)/2\right],$$

i.e.

$$|E(\rho)/E_0(\rho)| = 2\sin\left[ka(1 - \cos \beta)/2\right]. \qquad (10.29)$$

If, for example, $a = 25$ cm and $\beta = 30°$ or $60°$, we obtain at 1 GHz $|E(\rho)/E_0(\rho)| = 0.69$ and 1.22, respectively. The total far field in the symmetry plane in both directions is other than zero. But the total signal in the far region at any φ is other than zero too. This is an inherent advantage of such a structure.

The calculated horizontal patterns at frequency 1 GHz of one radiator, placed at point A, of two radiators (see Figure 10.18a) and of three radiators (see Figure 10.18b) located on the cover of infinite dimensions, are given in Figure 10.20 and marked with symbols 1, 2 and 3 respectively.

The considered compensation structures comprise radiators, placed at the same distances from the compensation point. The structures require no use of any phase shifter. The compensation structures, in which flat metal reflectors (mirrors) are used instead of auxiliary radiators, permit to create a weak field in the subspace on the side of the reflectors, where the antenna is located, which is similar to the weak field generated by three radiators.

An example of such a structure is presented in Figure 10.18c (top view). Here, the following notation is used in addition to the notation used above: R is the metal reflector, and B is the location of the equivalent radiator. As seen from the figure, the flat metal plate, in which the main radiator A is reflected, is mounted on the transmitter top cover along the axis of symmetry (in the vertical plane passing through compensation point C). The reflection (or imaging) of the main radiator signal by the metal plate is equivalent to the presence in the space before the reflector the second signal created by the equivalent radiator located at the same distance behind the plate.

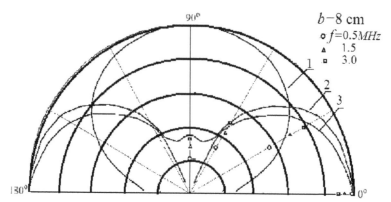

Figure 10.20 The horizontal patterns of different structures at 1 GHz.

The phase of reflected signal differs from the phase of main radiator by 180°, since vector $\vec{E}_1$ of the incident signal is located along the tangent to the reflector surface. The amplitude of the reflected signal $\vec{E}_2$ is $E_2 = DE_1$, where D depends on dimensions of the reflector and on a distance from it. If the plate dimensions are infinite (or if they are in all directions much larger, than the main radiator dimensions), then $D = 1$, i.e. the amplitudes of reflected and main signals coincide, and the total signal in the reflector plane is zero. If $D \neq 1$, total signal $E = E_1 + E_2 = (1 - D)E_1$ is a fixed fraction of the main signal, and the share is independent of frequency, since the distances from both radiators to the compensation point are the same, and the radiators themselves are identical.

On the whole, the characteristics of this structure are similar to the characteristics of the two-radiator structure, presented in Figure 10.18a. A phase shifter is not needed in this structure. An additional and very important advantage of the described structure is that the external actions, e.g., an approaching of the metallic objects to the antenna system, exert practically no effect on the structure operation, since the signal of the main radiator and the reflected signals undergo the same changes upon any external actions. Only the appearance of a metal body between a signal source and the compensation point is an exception and a very rare one at that.

The influence of a mirror (a metal plate) on the antenna input impedance decreases the matching between the antenna and the transmitter. In order to offset the decreased matching, it is necessary to increase the distance between the antenna and the mirror as well as the mirror width so that the second signal would not decrease. The reflector width has only a weak effect on the antenna input impedance, but essentially increases the signal. Moreover, the mirror influence on the antenna input impedance is constant and can be allowed for a priori. The mirror height must exceed the height of the antenna.

Flat reflectors can be manufactured in the form of a light-weight, strong and collapsible construction (if necessary). In such case, each reflector is implemented as a set of vertical parallel wires, located at a distance of $0.06\lambda_{min}$, where λ_{min} is the minimal wavelength in the frequency range. Use of two reflectors, i.e., presence of two equivalent radiators, gives additional advantages. In this case, the amplitude of each reflected signal may be smaller than that of the main radiator signal, and so the requirements to the reflector become weak. A version of such compensation structure is presented in Figure 10.18d (top view), where, as before, B_1 and B_2 denote the locations of equivalent radiators. The characteristics of the structure given in Figure 10.18d is similar to the characteristics of the three-radiator structure presented in Figure 10.18b. The distance from all radiators to the compensation point is the same, i.e., field variation due to frequency dependence has no effect on the compensation quality.

As stated above, the amplitude of the reflected signal (vertically polarized) is $E_2 = DE_1$. Here, D depends on the mirror dimensions. An essential drawback of the structure with two reflectors is that the reflectors creating the dark spot in the near region between them at the same time interfere with the propagation of electromagnetic waves in the angular sector, shadowed by them. The signal in this sector increases approximately twice. This drawback limits the application of such structures.

The structure with two reflectors was tested experimentally. A general view of the experimental setup is presented in Figure 10.21. The following notation is used in the figure: 1 is the radiator, 2 is the receiving antenna, placed at the compensation point, 3 and 4 are flat reflectors, located symmetrically on either sides of the straight line passing through the compensation point and the radiator. The distance between the radiator and the receiving antenna is 25 cm. The reflector is manufactured in the shape of a vertical rectangular plate. Contact with the ground is maintained along the whole width of the reflector by means of a metallic segment. In the course of the measurements, each reflector was separately moved along a straight line to the point where the received signal at 2 GHz (close to the center of frequency range) will decrease by 6 dB. As a result the field of antenna at the compensation point in the presence of both reflectors is substantially smaller than the field of the antenna without reflectors. The measurements are performed at two different values of reflectors width—8 and 15 cm, and at different angles (10, 15 and 20°) between the axis of symmetry and the lines, along which the reflectors were moved to attain the zero field at the compensation point. In all variants, when the distance between the compensation point and the near edge of the reflector was 9 cm, the signal was half as strong as its value without the reflector.

Figure 10.21   General experimental setup.

The measurement results are given in Table 10.5. The frequency dependence of the signal magnitude at the compensation point without reflectors (1) and with reflectors (2) is presented in Figure 10.22, for reflector widths of 8 (*a*) and 15 cm (*b*). As is clear from the table and figures, the reflectors with width 8 cm permit to decrease the signal magnitude without tuning by 10–15 dB for the whole frequency range, and the reflectors with width 15 cm perform even better.

The boundaries of the dark spot created by three radiator (see Figure 10.18*b*) in the horizontal plane ($z = 0$) are determined by the procedure described above. They are given in Figure 10.23 for different values n_0. They show that these spots have sufficiently large dimensions.

Table 10.5 Measurement Results for Signal Magnitudes at the Compensation Point, in dB

f, GHz	Without reflector	With reflectors of width (in cm)			Difference		
		8		15	8		15
		β/2 = 10°	15	15	10	15	15
0.5	−17	−32.1	−33.3	−35.8	15.1	16.3	18.8
1.0	−19.5	−33.5	−35.8	−34.3	14	16.3	14.8
1.5	−39	−46.4	−43.4	−54.9	7.4	4.4	15.9
2.0	−28.8	−37.3	−40.4	−62.5	8.5	11.6	33.7
2.5	−27.7	−47.1	−42.6	−41.3	19.4	14.9	13.6
3.0	−36.3	−43.9	−47	−42.6	7.6	10.7	6.3

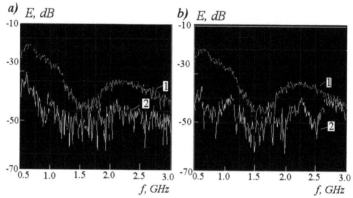

Figure 10.22 Frequency dependence of signal at the compensation point without (1) and with reflectors (2) for width 8 cm (*a*) and 15 cm (*b*).

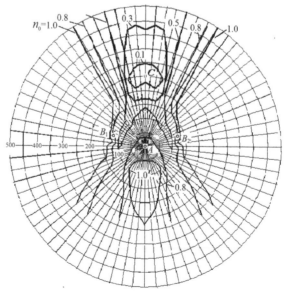

Figure 10.23 The boundaries of the dark spot created by three radiators.

10.5   NEW ANTENNA FOR A PERSONAL CELLULAR PHONE

Cellular communication imposes high requirement for antenna of personal phone and at the same time limits the possibility of its installation. An increase in the number of subscribers requires using several high frequency bands, and accordingly installation of separate antennas (one for each band). One can use a single antenna, but it must be very wideband, and that complicates greatly the task. In addition, cell phones support the diversity of applications, which require placing many components and devices into the phone housing. These circumstances lead to space constraints and leave a small area for the antennas.

Extensive use of mobile phones may lead to potential health risk as a result of the impact of radiation (or more correctly irradiation) on human organism.

A new antenna, which is presented in this section, offers a possible solution to the described problems. The structure of the offered antenna is shown in Figure 10.24. It is an asymmetrical dipole, which is excited by a generator 1 at a feed point 2. The lower arm of the dipole is a metal plate 3, to bottom edge of which a small plate 4 is attached at an acute angle. The upper arm is a multi-folded structure 5, which is fed at its mid-point and open at the ends (at points 6). A dielectric plate 7 is inserted inside this structure.

The first distinction of the offered antenna from the known antennas is its use of the phone's ground (chassis) as a radiator. This suggestion was made in [87]. The phone's ground is a rectangular metal plate, whose length is close to a quarter of a wave length at the main operating frequency. This distinction permits to integrate the antenna with ground, removing the need for an installation of a separate antenna and a special metal counterpoise. Components of radio transmitter and receiver can be mounted on the metal antenna, which replaces the ground for those elements. Filters, placed on this plate, may provide a short circuit for direct currents and insulation (interruption) of circuit at high frequency.

The second distinction of the offered antenna is implementation of the upper arm in the shape of a multi-folded radiator. This arm has a small height. Its contribution to the antenna's radiation is small, but it allows to match the antenna to a cable or a generator. The reactive component of its impedance compensates the reactive component of the lower arm, providing serial resonances at operating frequencies. Its complex structure has many degrees of freedom, including number of sections, their dimensions, the width and thickness of the wires, types and magnitudes of concentrated loads. That permits to change the input reactance of the antenna in wide limits and to provide operation at few bands of frequencies.

The third distinction of the offered antenna is the use of special measures for decreasing the field in its near region and hence reducing the irradiation of the user's head (reducing SAR). These special measures are firstly the multi-folded radiator 5 with the dielectric insert 7 and secondly a small metal plate 4. The multi-folded radiator is fabricated in the shape of a three-dimensional structure that protrudes to the direction of the user's head. The dielectric plate is inserted between the wires of this structure. The short metal plate is attached to the bottom edge of the lower arm. These structures create anti-phase currents in the antenna and hence small auxiliary fields, which compensate the main field at a certain point (compensation point) and form around this point an area of a weak field (a dark spot).

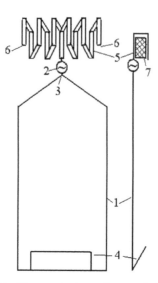

Figure 10.24 The equivalent circuit of the antenna.

The multi-folded structure 5 is more sensitive to external objects and actions than the plate 3. In order to reduce the impact of the user's hand, it is expedient to apply the plate as the lower arm of the antenna, and the multi-folded structure as the upper arm. The user's hand in this case will lie on the handset housing not far from the plate, but far from the multi-folded structure. It is the fourth distinction of the offered antenna.

The input impedance of an asymmetrical dipole, consisting of two different arms, in the first approximation, is equal to half the sum of input impedances of two symmetrical radiators: one with arms identical to the lower arm of the asymmetrical dipole, and other radiator with arms identical to its upper arm [27]. This relation is exact for the input reactance of the antenna and has an approximate nature for its resistance. From the above it follows that

$$X_A = X_{A1} + X_{A2}, \quad R_A \approx R_{A1} + R_{A2}, \tag{10.30}$$

where X_{A1} and R_{A1} are respectively the reactive and resistive components of the input impedance of the radiator, shown in Figure 10.25a, X_{A2} and R_{A2} are corresponding components of the radiator, shown in Figure 10.25b.

The lower arm of the offered antenna is a monopole, realized as a wide metal plate with length L_1 and width d. As is said in Section 2.7, such an antenna is called a strip monopole. Its characteristics are similar to the characteristics of a linear cylindrical radiator with a circular cross section of radius a_e. In [16] it is shown that a_e is equal to $d/4$. It is also considered that it is a trade-off between input impedance and scattering characteristics of the radiator [13]. The input impedance of the monopole in the first approximation is equal to

$$Z_{A1} \cong R_{A1} - jW_1 \cot kL_1. \tag{10.31}$$

Here $W_1 = 30$ Ohm is the wave impedance of the monopole, $\Omega = 2\ln(2L_1/a_{e1})$ is the parameter of the theory of linear antennas, R_{A1} at frequencies near the first series resonance is close to 40 ohm. If $d \approx L_1/2$, then $a_{e1} = L_1/(2\pi)$, and $\Omega = 2\ln(4\pi) \approx 5$.

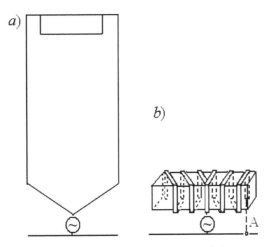

Figure 10.25 The lower (a) and the upper (b) arms of the asymmetrical dipole.

For good matching, it is expedient to realize a region of the monopole's excitation as a tapered line in the shape of a flat triangle (see Figure 10.25a). It is the transition from the wide plate of antenna to the inner conductor of a coaxial cable. Calculations show that the wave impedance of the tapered line is close to the standard cable's wave impedance, when the angle at the vertex of the tapered plate is close to $80°$.

The upper arm of the antenna in the first approximation may be considered as a multi-folded radiator, mounted on a metal plate (see Chapter 4). The current in each wire can be divided into in-phase and anti-phase components. This system is reduced to an aggregate of a radiating linear antenna (monopole) and few non-radiating long lines. The multi-folded radiator can be either with shorting to the ground or with a gap (see Figure 10.25b). In the latter case the input impedance of the multi-folded radiator is a serial connection of the monopole and the two-wire long lines, shorted at their ends. The expressions for the input impedances $Z_{A2}^{(n)}$ of these antennas with a different number n of wires are presented in Chapter 4.

As it can be seen from these expressions, the radiation resistance of this antenna is equal to the radiation resistance of a monopole with the same height. The reactive component of its input impedance has additional resonances, and the first parallel resonance is defined by a line of length $SL/2$, i.e., its frequency is smaller approximately by a factor of $S/2$ in comparison with the frequency of the first serial resonance of an ordinary monopole with the same height. The frequency of the first serial resonance of the multi-folded antenna is still smaller, approximately twice.

Such type of the input impedance simplifies selecting dimensions of a structure, whose serial resonances coincide with the given operating frequencies. Thus, in this case the monopole's effective length is

$$h_e = (1/k) \tan (kL_1/2) + L_2/2, \tag{10.32}$$

where L_2 is the length of the upper arm. If, for example, $kL_1 = \pi/2$, and $L_2 = L_1/4$, then $h_e = 1/k + L_1/8$. Accordingly, the radiation resistance of the antenna is

$$R_\Sigma = 20k^2h_e^2 = 20(1 + \pi/16)^2 \approx 29. \tag{10.33}$$

From the above it is clear that the radiation resistance of the monopole (of the metal plate) determines mainly the antenna's resistance on the whole, i.e. the latter is close to

30 Ohm, if the operating frequency is close to the frequency of the first series resonance. Contribution of the upper arm to the radiation resistance is small as compared with the contribution of the lower arm. But the reactance of the monopole is close to zero, if the plate length is equal to a resonant length, i.e. if it is close to $\lambda/4$. Therefore, to ensure matching with a cable or generator, the upper arm should have a series resonance at the first operating frequency. Accordingly, the reactance of the upper arm of a multi-band antenna should compensate the monopole reactance at other operating frequencies.

The directional pattern of the antenna in the horizontal and vertical planes do not differ from a typical directional pattern of monopole, since the radiation of the long lines can be neglected, provided the distances between the wires are small.

The analysis of the considered antenna demonstrates that its structure firstly allows to obtain the serial resonances at required frequencies. Secondly, the considered antenna contains a radiating element of large length, close to $\lambda/4$, and that permits providing an effective radiation and reception of signals.

If the multi-folded antenna is shorted to ground at points A, its input admittance is equal to

$$Y_{A2}^{(S)} = 1/\left[4Z_{A2}^{(S/2)} \right] + 1/\left[j120 \ln (b/a) \tan (SkL/2) \right], \tag{10.34}$$

where $Z_{A2}^{(S/2)}$ is the input impedance of an antenna with a gap, whose conductors consist of two wires. Therefore, in this case the input impedance is a parallel connection of a $S/2$-folded radiator with a gap and a close-end line of length $SL/2$. Such type of the input impedance complicates selecting dimensions of a structure.

In order to produce a small auxiliary field with the aim to compensate the main field in the user's head without changing the far field, it is expedient to use the multi-folded structure, which protrudes toward the user's head. This structure allows to create an anti-phase field in the near region of the antenna and to nullify the total field at a compensation point, located not far from the neighboring edge of the user's head. Around the mentioned point an area of a weak field (a dark spot) is created. We shall demonstrate this effect by means of a folded radiator with a gap in a point A, which is located along a plane passing through the compensation point and placed on the side of the phone housing near to the head.

A folded radiator consists of two parallel wires. If to connect two current generators of equal magnitude $J/2$ and opposite directions to the bottom of the right wire in parallel with each other and to divide also the main generator into two parallel generators, equal in magnitude (to $J/2$) and coinciding in direction, then as a result, voltages and currents of this circuit do not change. According to the superposition principle, the voltage and the currents at each point are equal to the sum of the voltages and the currents, produced by all generators. Therefore, as shown in Figure 10.26, one can divide the considered circuit onto two circuits with two generators in each and then calculate

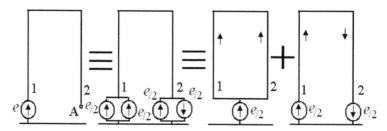

Figure 10.26 Division of the folded radiator into two circuits.

and sum up the currents in any wire, created in each circuit. The left circuit is a linear radiator (monopole), the right circuit is a two-wire long line.

The current in each wire consists of an anti-phase current of the line and the fraction of the in-phase current of the linear radiator. In Figure 10.27 the currents' distribution along the wires is presented: the distribution of the in-phase currents $J_1^{(in)} = pJ$ and $J_2^{(in)} = mJ$ along wires 1 and 2 of the monopole (a), the distribution of the anti-phase current $J_1^{(an)} = mJ$ and $j_2^{(an)} = mJ$ along the wires of closed at the end line (b) and the distribution of the total currents (c).

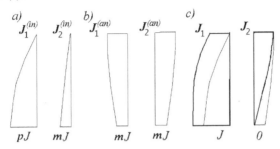

Figure 10.27   The distribution of in-phase (a), anti-phase (b) and total (c) currents.

The in-phase currents in both wires are distributed by sinusoidal law; the ratio of their magnitudes depends on the wire capacitances, and for identical wires the magnitudes are equal. The anti-phase currents are the same in magnitude, but opposite in sign and are distributed by cosine law. In the first wire the currents are added, since they have the same sign. In the second wire they are opposite in sign, and under these conditions the dipole moment of negative current is greater due to cosine distribution. This gives a possibility for compensation of the fields, created by the wires. The total current of the second wire is less than the total current of the first wire. But since the second wire is located nearer to the compensation point, then, although the wires' currents are different, their fields at the compensation point are the same in magnitude.

This effect is more pronounced if, firstly, the folded radiator is replaced by a multi-folded structure, and, secondly, if a dielectric plate is placed between the wires. In this case the field of the first wire is attenuated quicker with the wires spacing. As a result, firstly, one can decrease this distance, i.e. decrease the thickness of phone housing. Secondly, the electric lengths of transmission lines increase, i.e. one can decrease geometric lengths of lines, for example, to excite the multi-folded structure in the middle of its width and not at a side point.

The additional anti-phase current is produced by the small plate 4 (see Figure 10.24).

It should be emphasized that the structure without a connecting bridge between the upper ends of the wires is divided into a monopole and an open-end two-wire line. In this case, the in-phase and anti-phase currents of the second wire are equal in magnitude, i.e. the total current of the second wire is zero, and hence there is no possibility of compensating the field of the first wire.

It is useful to compare the characteristics of the proposed antenna with the characteristics of a symmetrical dipole. As previously indicated, if the lengths L_1 and L_2 of the lower and upper arm of the proposed antenna are equal, respectively, to $\lambda/4$ and $\lambda/16$, the active component of the input impedance is close to 30 Ohm, and the reactive component is close to zero. On the same length of the phone housing one may place a symmetrical dipole with arm length $\lambda/8$, i.e. smaller twice. Its resistance is

$$R_\Sigma = R_{\Sigma 0}/n^2, \tag{10.35}$$

where $R_{\Sigma 0} \approx 80$ Ohm is the active component of the resonant dipole's input impedance. From (10.35) it follows that in this case $R_\Sigma \approx 20$ Ohm.

It is more substantial that the dipole impedance has a reactive component, which increases the losses in the cable much stronger than the low value of R_Σ. If the length of a dipole arm is $\lambda/8$, its input reactance is equal to the wave impedance. Let us take a relatively small wave impedance of 100 Ohm. It is easy to be convinced that in this case the travelling wave ratio (TWR) for the dipole with $R_\Sigma = 20$ Ohm and reactance 100 Ohm in the cable with $W = 50$ Ohm is equal to 0.08, while TWR for a monopole with resistance 30 Ohm and zero reactance in the same cable is 0.6, i.e. 7.5 times greater. This example shows the obvious advantage of the proposed antenna, the resonant dimensions of which fit into the dimensions of the phone chassis.

Simulation of the proposed antenna was carried out using the CST program, and the results are compared with the results of the planar inverted F antenna (PIFA). The results of tuning are given in Figure 10.28, where the standing wave ratio (SWR) of the proposed antenna is shown. The calculated results of reducing field in a near zone in the presence of user's head along a line perpendicular to the antenna plane are given in Figure 10.29 (solid curve—in the direction of user, dotted curve—in the opposite

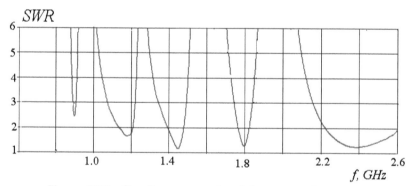

Figure 10.28 Standing wave ratio of the proposed antenna.

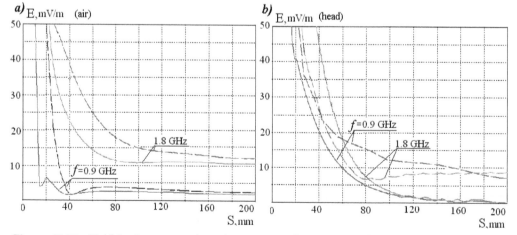

Figure 10.29 Field in the near region of antenna in the presence of user's head as a function of distance S from the plane of the antenna.

direction). The calculation uses the model of the head as part of the program CST. The fields are presented at frequencies 0.9 and 1.8 GHz as the functions of distance S from the plane of the antenna.

A full-scale model of the new antenna was fabricated in accordance with results of the calculation. Photo of this model is presented in Figure 10.30. The measurement setup is demonstrated in Figure 10.31a. In Figure 10.31b the calculated curves and

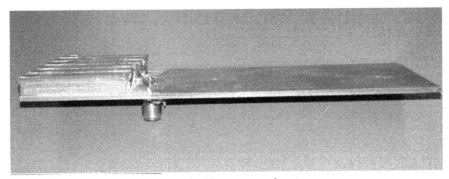

Figure 10.30   The proposed antenna.

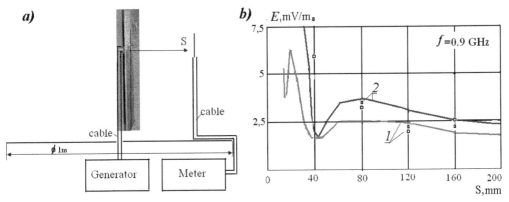

Figure 10.31   The measurement setup (a) and calculated curves and experimental values of the near field at the frequency 0.9 GHz (b) in the direction of the head (1) and in the opposite direction (2).

experimental values of the antenna near field in the head at the frequency 0.9 GHz are compared. Fields in the direction of the head are denoted by the number 1 and in the opposite direction—by the number 2. As can be seen from the Figure, the field of the antenna in the direction of the head is substantially smaller. The experimental values were determined by means of measurement in the phantom model.

The directivity of the offered antenna is presented in Table 10.6. In Table 10.7 the level of SAR in the user's head is shown for the proposed antenna and antenna $PIFA$ at the frequency 0.9 GHz.

Table 10.6   Directivity of the Proposed Antenna

Frequency, GHz	Directivity, dB
0.9	2.59
1.8	4.48

Table 10.7 SAR of Antennas

Antenna	f, GHz	Total	Max local in 10 g	Max local in 1 g	Max point
PIFA	0.9	0.061	3.62	5.8	345.8
Proposed	0.9	0.013	0.762	1.41	21.6

The proposed asymmetrical antenna with the long arm and zero reactance on the operating frequencies has high electrical characteristics and creates in the far region the electromagnetic field, which exceeds significantly the field of the other antennas of the same dimensions. In particular, the field of the new antenna at the same transmitter power is much greater than the field of antenna PIFA. The results of researching the new antenna, including simulations and experimental tests, corroborate that this antenna is promising for use in modern cellular phones. In contrast to known antennas this antenna is multi-frequency, i.e. it can operate on multiple frequencies simultaneously and has high electrical characteristics, since it provides without switching frequency, good match with the cable and correspondingly high efficiency.

Additional advantage of this antenna is reducing the user's head irradiation, since this antenna uses the new method for its reducing. It should be emphasized that this method may be used for reducing the irradiation produced by other antennas, in particular by the antenna PIFA.

10.6 DIVERSITY RECEPTION

Cellular communications, as well as other forms of communication, suffer from various types of noises and interferences. Diversity reception, i.e. mounting two receiving antennas on the finite (in wavelengths) distance from each other and the creation of the new signal from two received signals, is effective measure of a struggle against interferences. As a result, the interference or is absent completely or is relaxed substantially. It is obvious that at the user's phone such separation is not possible. But it can be used at the base station, where the interference signal can be drastically reduced. In order to filter the interference, one can use the bridge circuit and the modulator (Figure 10.32). Receiving antennas A_1 and A_2 are mounted at points 1 and 2 of the bridge.

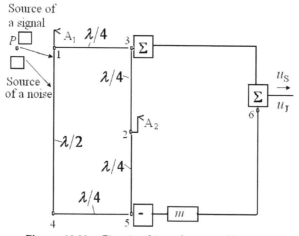

Figure 10.32 Circuit of interferences filtration.

Let the source of the useful signal be placed at some point P. The signals from this point come to antennas A_1 and A_2 with different phases, caused by the distances difference. The magnitudes of the received signals with allowance for their changes in receiving circuits can be written as $u_1 = S_1 \exp(j\varphi_1)$ and $u_2 = S_2 \exp(j\varphi_2)$ respectively. If the distance between the antennas is small, one can assume that the amplitudes of the received signals in the first approximation are the same.

These signals arrive to the point 3 of the bridge circuit in phase and summed in it:

$$u_1 + u_2 = S \exp(j\varphi_1)[1 + \exp(j\Delta\varphi)], \tag{10.36}$$

where $\Delta\varphi = \varphi_2 - \varphi_1$. The resulting signal comes to an adder, located at the point 6. To the point 5 of the bridge circuit the signals come in anti-phase, i.e. one signal is subtracted from the other:

$$u_2 - u_1 = S \exp(j\varphi_1)[\exp(j\Delta\varphi) - 1]. \tag{10.37}$$

Further, the signal difference passes through the modulator, i.e. is multiplied by its gain $m = M \exp(j\varphi_m)$, and also comes to the adder. The useful signal at the output of the adder is

$$u_S = u_1 + u_2 + m(u_2 - u_1) = S \exp(j\varphi_1)[1 - m + (1 + m)\exp)(j\Delta\varphi)]. \tag{10.38}$$

A similar expression can be written for the interference. If the distance between the antennas A_1 and A_2 is small, the amplitudes of the interfering signals in these antennas are respectively $u_3 = J \exp(j\psi_3)$, $u_4 = J \exp(j\psi_4)$. Interference at the output of the adder is

$$u_J = J \exp(j\psi_3)[1 - m + (1 + m)\exp(j\Delta\psi)], \tag{10.39}$$

where $\Delta\psi = \psi_4 - \psi_3$. In order that interference is equal to zero at the output of the adder, it is necessary that the gain of the modulator is equal to

$$m = \frac{1 + \exp(j\Delta\psi)}{1 - \exp(j\Delta\psi)} = j \cot(\Delta\psi/2). \tag{10.40}$$

From this expression it follows that the modulator has to change the phase of the transmitted signal. If the amplitudes of an interfering signals at the antennas A_1 and A_2 are not the same, i.e. $u_3 = J_3 \exp(j\psi_3)$, $u_4 = J_4 \exp(j\psi_4)$, then assuming $J_4/J_3 = \exp(\beta)$, we find

$$u_J = J_3 \exp(j\psi_3)[1 - m + (1 + m)\exp(\beta + j\Delta\psi)],$$

whence

$$m = -\coth\frac{\beta + j\Delta\psi}{2} = \frac{1}{\coth^2(\beta/2) + \cot^2(\psi/2)} \times$$
$$\left[\coth(\beta/2)\csc^2(\psi/2) - j\cot(\psi/2)\operatorname{csch}^2(\beta/2)\right] \tag{10.41}$$

In this case, the modulator must change not only the phase but also the amplitude of the signal, passing through it.

Accordingly, for the useful signal at the output of the structure we obtain:

$$u_S = S \exp(j\varphi_1)\left[1 + \coth\frac{\beta + j\Delta\psi}{2} + \left(1 - \coth\frac{\beta + j\Delta\psi}{2}\right)\exp(\alpha + j\Delta\varphi)\right], \tag{10.42}$$

where

$$\exp(\alpha) = J_2/J_1 = S_2/S_1.$$

Since

$$\coth x = \left[1 + \exp(-2x)\right]/\left[1 - \exp(-2x)\right],$$

then

$$u_S = \frac{2S \exp(j\varphi_1)}{1 - \exp\left[-(\beta + j\Delta\psi)\right]}\left\{1 - \exp[\alpha - \beta + j(\Delta\varphi - \Delta\psi)]\right\}. \tag{10.43}$$

As one can be seen from expression (10.43), unlike the interference, which is equal to zero at the structure output, the useful signal is different from zero, if the value $\exp[\alpha - \beta + j(\Delta\varphi - \Delta\psi)]$ is not equal to 1.

The value $\Delta\varphi$ in the expression (10.43) is the phase difference of useful signals received by different antennas. It is

$$\Delta\varphi = kd \cos\gamma,$$

where $k = 2\pi/\lambda$, d is the distance between the antennas, γ is the angle between the normal to the segment d and the direction of the useful signal source. Similarly, the value $\Delta\psi$ in this expression is a phase difference $\Delta\psi = kd \cos\delta$ between interfering signals, received by different antennas (δ is the angle between the normal to the segment d and the direction to the interference source). If the arrival angles of the signal and the interference are close to each other ($\delta \approx \gamma$), then

$$\Delta\varphi - \Delta\psi = kd (\cos\gamma - \cos\delta) \approx kd(\delta - \gamma)\sin\gamma.$$

Obviously, α and β coincide also. In this case,

$$u_S = \frac{2S \exp(j\varphi_1)}{1 - \exp\left[-(\beta + j\Delta\psi)\right]}\left\{1 - \exp\left[jkd(\delta - \gamma)\sin\gamma\right]\right\} = \frac{2jkd(\delta - \gamma)\sin\gamma\, S\exp(j\varphi_1)}{1 - \exp\left[-(\beta + j\Delta\psi)\right]}. \tag{10.44}$$

From (10.44) it follows that the useful signal at the output of the system at similar arrival angles of the signal and the interference is proportional to the difference between the angles of its arrival.

If the signal and the interference come from one and the same azimuth, it is necessary to separate the antennas for height. Otherwise, if the angle in the vertical plane between the directions of arrival of the signal and the noise does not exceed a few degrees with the same polarization, the reception is not possible.

Thus, application of this circuit together with the interference attenuation leads to a weakening of the useful signal. Also, the areas of weakened reception may appear. Indeed, suppose that the antennas A_1 and A_2 are the same and have a circular directional pattern in the horizontal plane. The modulator gain in accordance with (10.40) is equal to $m = j \cot(\Delta\psi/2)$. If the phase difference between the interferences, received by the antenna, is equal to $\Delta\psi$, and the phase difference between the useful signals is

$$\Delta\varphi = \Delta\psi \pm n\pi,$$

where n is the natural number, then in accordance with (10.43) $u_S = 0$. However, if d is not too great, this does not take place. Suppose, for example, that $\Delta\psi = 0.3\pi$, $d \approx 0.35\lambda$, i.e. $kd = 0.7\pi$, $\delta = \cos^{-1} 0.43 = 0.36\pi$, $\gamma = \cos^{-1}(\pm 0.86n\pi + 0.36\pi)$. The absence of real γ testifies about the absence of these zones.

It is necessary to say a few words about the impact of inaccurate adjustment of the modulator gain. If the gain differs from the optimum value m in magnitude and phase, for example, it is equal to

$$m_1 = m(1 + \varepsilon_1) \exp(j\varepsilon_2), \tag{10.45}$$

where $\varepsilon_1, \varepsilon_2 \ll 1$, i.e. $m_1 \approx m(1 + \varepsilon_1 + j\varepsilon_2)$, then the interference signal at the output will be different from the zero signal:

$$u_J = -\frac{2J \exp(j\psi_1)}{1 - \exp\left[-(\beta + j\Delta\psi)\right]} \times$$
$$\left\{1 + \exp\left[-(\beta + j\Delta\psi)\right] - \exp(\beta + j\Delta\psi)\left[1 + \exp(-\beta - j\Delta\psi)\right]\right\}(\varepsilon_1 + j\varepsilon_2). \tag{10.46}$$

But, as calculations show, it will be significantly less than the initial interference in each channel. If, for example, $\varepsilon_1 = 0.01$, $\varepsilon_2 = 1° = 0.017$, then, neglecting the differences of the interferences amplitudes at two antennas, we obtain

$$u_J = -\frac{2J \exp(j\psi_1)}{1 - \exp\left[-(\beta + j\Delta\psi)\right]}\sqrt{0.01^2 + 0.017^2} = -0.0197\frac{2J \exp(j\psi_1)}{1 - \exp\left[-(\beta + j\Delta\psi)\right]}$$

Application of diversity reception as an effective measure of a struggle against interferences is known since long. But its effectiveness is essentially enhanced if in order to obtain optimum results we use the method of mathematical programming. Advantages of this approach are clearly manifested in the struggle against the constant interferences. In this case, the optimization problem is the correct choice of amplitude and phase modulator characteristics, and the variable parameters x are the modulus and phase of the modulator gain.

As already it was said, the method of mathematical programming allows to obtain the minimum of an objective function $\Phi(\vec{x}_m)$, and is based on finding the gradient. This method is an iterative procedure, which step by step moves from one set of the parameters to another set in the direction of the greatest decrease of the objective function. An iteration m is the movement along the surface $\Phi(\vec{x}_m)$ in the space of the vector $\vec{x}_m$. Figure 10.33 dipicts a particular case of such movement, when the vector $\vec{x}_m$ has two parameters: x and y.

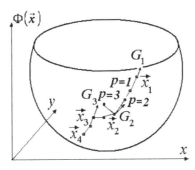

Figure 10.33   Iteration as a movement along the surface $\Phi(\vec{x}_n)$.

The movement is performed in a predetermined direction (along the selected curve). At each iteration the objective function minimum and the values of the parameters, under which it takes place, are determined. Each subsequent iteration is a descent from the point, reached in the previous step, in a new direction, opposite to the gradient direction (it is determined by calculating the gradient at a reached point). The optimization process ends, when the objective function decrease as the result of the next iteration becomes less than the predetermined decrease.

An algorithm of the iterative procedure is shown in Figure 10.34.

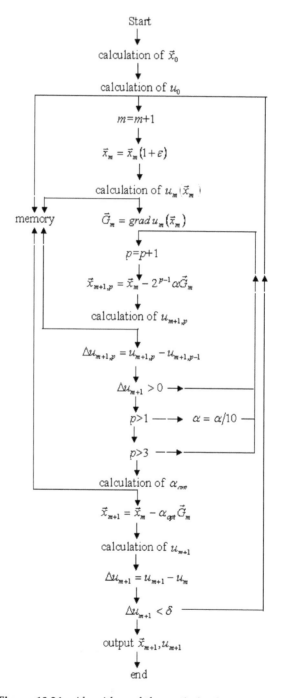

Figure 10.34   Algorithm of the optimization program.

Solution of the problem would be accelerated, if the set of initial values is adequate for the problem. That dramatically reduces the probability of an error, caused by the fact that due to an arbitrary choice of the initial parameters the optimization process may lead to a local, but not to the true extremum of the objective function. Therefore, as the initial values of the parameters x the values, obtained from the expressions (10.40) or (10.42), must be taken. Due to the inaccuracy of the quantities in these expressions, in the real conditions it is advisable to directly measure the magnitude of interference at different parameters x.

Another option is the use of multiple cycles, starting with the jump of the phase (arg m) of the modulator gain on a given section of the period. Each jump leads to a new set of initial parameters and starts a new search of the minimum of objective function. If found minima will differ from each other, one must take the smallest value as a true extremum.

11

Arrays

11.1 CHARACTERISTICS OF DIRECTIVITY

As noted in Section 1.1, if sources of electromagnetic field are distributed continuously in some volume V, and a medium surrounding volume V is a homogeneous isotropic dielectric, the solution of equation (1.19) for a harmonic field is given by (1.21). In this expression, R is distance from the integration point to the observation point, which in the far region in the first approximation corresponds to Figure 1.1. If radiators inside the volume are parallel to each other, i.e. have the same direction of radiation, then

$$A_p = \frac{A_p(0)}{j_p(0)} \iiint\limits_{(V)} j_p \exp(jkz\cos\theta)dV, \qquad (11.1)$$

where $A_p(0)dV$ is the vector-potential of the field, created by an elementary volume with current density $J_p(0)$ located at the coordinate origin. If $E_p(0)dV$ is the field of this volume, then the total field is

$$E_p = \frac{E_p(0)}{j_p(0)} \iiint\limits_{(V)} j_p \exp(jkz\cos\theta)dV. \qquad (11.2)$$

As an example, we shall consider an elementary linear radiator (Hertz dipole) located along the z-axis. It is a filament of length b with current amplitude I, constant along the filament. In the far region, the filament field is defined by expression (1.7). Substituting it in (11.2), we obtain expression (1.9) for the field of a symmetrical radiator (dipole) located along the z-axis with the center at the coordinate origin (see Figures 1.4a). If, according to (1.8), we consider that the current distribution along the radiator arm is sinusoidal one, we come to expression (1.10), the last factor of which gives the vertical directional pattern of solitary radiator.

Another example is a linear array situated along the x-axis (Figure 11.1a). In this case, one must replace (11.2) with the sum

$$E = \frac{E_1}{J_1(0)} \sum_{n=1}^{N} J_n(0) \exp \left[jk(n-1)d_1 \right], \tag{11.3}$$

where E_1 is the field of the first radiator, $d_1 = d \cos \varphi \sin \theta$ is the path difference of beams from the adjacent radiators to the observation point with arbitrary coordinates φ and θ, n is the radiator number, and the antenna arrays are excited by currents with equal amplitudes and linearly growing phase displacement:

$$J_n(0) = J_1(0) \exp[-j(n-1)\psi]. \tag{11.4}$$

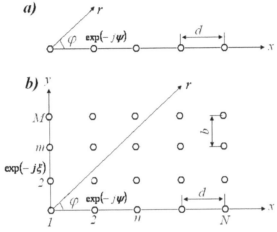

Figure 11.1 Uniform antenna array: linear (*a*), rectangular (*b*).

Here, ψ is the phase shift between currents of adjacent radiators. Using the formula for a sum of N terms of geometric progression with ratio $\exp[j(kd \cos \varphi \sin \theta - \psi)$, omitting factor $\exp[j(N-1)(kd \cos \varphi \sin \theta - \psi)/2]$ defining phase characteristic of array, and normalizing the result to 1, we obtain expression for the amplitude characteristic of an array

$$F_N(\theta, \varphi) = \frac{\sin \left[N (kd \cos \varphi \sin \theta - \psi)/2 \right]}{N \sin \left[(kd \cos \varphi \sin \theta - \psi)/2 \right]}. \tag{11.5}$$

As seen from (11.3), the directional pattern of a system of identical, equally oriented directional radiators is the product

$$F(\theta, \varphi) = F_1(\theta, \varphi) F_N(\theta, \varphi), \tag{11.6}$$

where $F_1(\theta, \varphi)$ is the directional pattern of a solitary radiator. Equality (11.6) is called the theorem of multiplication of directional patterns, and $F_N(\theta, \varphi)$ is the array factor.

If a system of radiators consists of no direction in horizontal plane antennas, e.g., of vertical monopoles, its horizontal pattern coincides with the array factor, and the vertical one depends on the corresponding directional pattern of a solitary radiator. One should emphasize that the array factor also depends on angle θ.

A rectangular array (Figure 11.1*b*) can be considered as a linear structure consisting of M linear arrays. Therefore, the array factor of the rectangular array is

$$F_{MN} = F_M F_N, \tag{11.7}$$

where F_M is the array factor of a linear array of M radiators situated along the y-axis:

$$F_M(\theta, \varphi) = \frac{\sin[M(kb\sin\varphi \sin\theta - \xi)/2]}{M\sin[(kb\sin\varphi \sin\theta - \xi)/2]}. \tag{11.8}$$

The electrical characteristics of radiators system at given frequency are conditioned by its phasing mode, i.e. by choice of the phase shift between the currents in radiators. Antenna array can have two modes of phasing: the maximal radiation forward and the minimal radiation backward. In the first mode, the phases of radiators fields are identical at the observation point with a given azimuth; in the second mode, the field is minimal or zero in the direction opposite the direction towards the correspondents. The first mode can be set up in any array, while the second one is not feasible always. For example, it is unfeasible in a linear array.

The phase difference of signals coming from the adjacent radiators of one row of the rectangular array to the observation point, located in the same horizontal plane as the array, is equal to the value $\Phi_{12} - \Phi_{11} = kd \cos\varphi - \psi$, and the phase difference from radiators of adjacent rows is $\Phi_{21} - \Phi_{11} = kb \sin\varphi - \xi$. Here, Φ_{mn} is the phase of signal, coming from the radiator m of the row n. At the observation point with azimuth φ_m, phases of all signals will be the same, if

$$\psi_m = kd\cos\varphi_m, \; \xi_m = kb\sin\varphi_m = (b\psi_m/d)\tan\varphi_m. \tag{11.9}$$

These conditions are necessary and sufficient for implementation of the first phasing mode.

The second mode can be applied, e.g., in a two-row array, where the fields of radiators of each row are summed together in phase, and then the fields of different rows are summed together in anti-phase. The condition of no signal in direction φ_0 has the form

$$\psi_0 = kd\cos\varphi_0, \; \xi_0 = kb\sin\varphi_0 + \pi = \frac{b\psi_0}{d}\tan\varphi_0 + \pi. \tag{11.10}$$

Calculation of the phase shifts in accordance with equalities (11.9) and (11.10) for different angles φ of the maximum radiation (in the first case $\varphi = \varphi_m$, in the second case $\varphi = (\varphi_0 + \pi)$ clearly demonstrates the difference between the modes (Figure 11.2).

Apart from a rectangular array, on object, e.g., aboard a ship, the other variant can be implemented. The variant uses the existing set of ship antennas. As a result, an array of arbitrarily located radiators is formed. Let the current in the base of the radiator n be $J_n(0) = |J_n \exp(-j\psi_n)|$. Figure 11.3$a$ presents phase shift φ_m, ensuring coinciding of the field phase of the radiator n in the far region with the field phase of the radiator located at the coordinate origin. If the condition is met for all radiators, their fields are summed together in the observation point (in the mode of radiation forward).

The phase of the field of radiator n in the far region is

$$\Phi_n = -kr_n - \psi_n,$$

where $r_n = r - x_n \cos\varphi - y_n \sin\varphi$ is the distance from the radiator to the observation point. If $\Phi_n = \text{const}(n) = -kr$, then $\psi_n = k(x_n \cos\varphi + y_n \sin\varphi)$. Let $x_n/\lambda = D + d/\lambda$, $y_n/\lambda = B + b/\lambda$, where D and B are integers, d/λ and b/λ are proper fractions. Then

$$\psi_n = D\psi_1 + B\xi_1 + \psi_m = \xi_m. \tag{11.11}$$

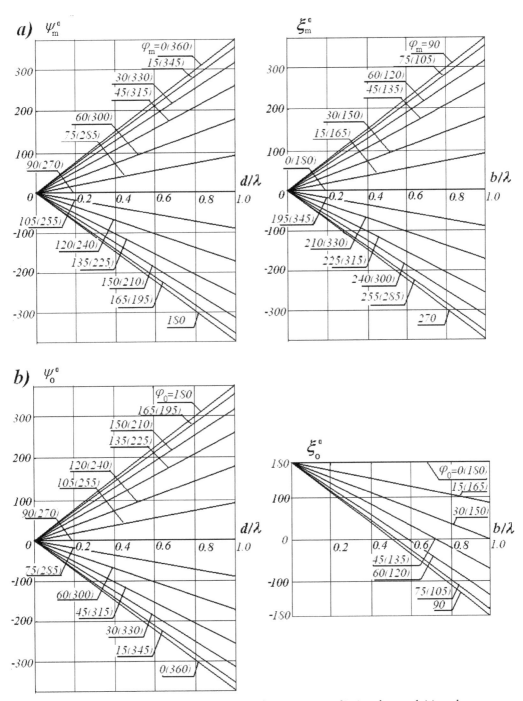

Figure 11.2 Phase shifts in the modes of maximum radiation forward (*a*) and zero radiations backwards (*b*).

Here $\psi_1 = 2\pi \cos\varphi$ and $\xi_1 = 2\pi \sin\varphi$ are found from Figure 11.3*b* (depending on φ), ψ_m and ξ_m—from Figure 11.2*a* or expression (11.9).

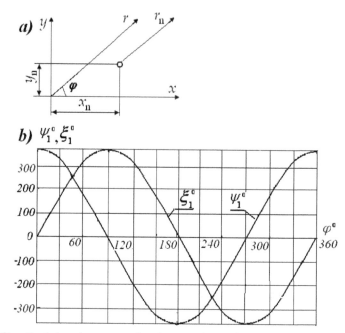

Figure 11.3 Circuit of the placement of radiator n (a) and dependence of ψ_1 and ξ_1 on φ (b).

As is well known, the directional characteristics of a radiator are described by the directivity, which is given by the ratio of the maximal intensity of radiation to its average intensity in the radiation sphere. This parameter represents the factor, into which one must increase the power in going from a directional antenna to isotropic antenna under condition of field preservation at the receiving point.

Since the field strength of a directional antenna is $E = E_m F(\theta, \varphi)$, where E_m is the field in the direction of the maximal radiation, then the power of such antenna is equal to an integral of the Poynting vector over the surface of a sphere with great radius r (in the far region)

$$P_\Sigma = \int\limits_{(S)} \frac{E_m^2 F^2(\theta, \varphi)}{Z_0}\, dS. \tag{11.12}$$

Here the surface element is equal to $dS = r^2 \sin\theta d\theta d\varphi$. The radiation power for no directional antenna is

$$P_{\Sigma 1} = \int\limits_{(S)} \frac{E_1^2}{Z_0}\, dS = \frac{4\pi r^2 E_1^2}{Z_0}, \tag{11.13}$$

where E_1 is the field of no directional antenna, which in accordance with the condition of the field's equality at the observation point must be equal to $E_1 = E_m(\theta_1, \varphi_1)$. Therefore we obtain for directivity in an arbitrary direction

$$D_1 = \frac{P_{\Sigma 1}}{P_\Sigma} = \frac{4\pi F^2(\theta_1, \varphi_1)}{\int\limits_0^{2\pi} d\varphi \int\limits_0^\pi F^2(\theta, \varphi) \sin\theta d\theta}. \tag{11.14}$$

In particular, in direction of the maximal radiation, when $F(\theta_1, \varphi_1) = 1$, the directivity is

$$D_m = \frac{4\pi}{\int\limits_0^{2\pi} d\varphi \int\limits_0^{\pi} F^2(\theta, \varphi) \sin\theta \, d\theta} \tag{11.15}$$

As one can see from (11.14) and (11.15), the magnitude D_1 is determined easily, if D_m and $F(\theta_1, \varphi_1)$ are known: $D_1 = D_m F^2(\theta_1, \varphi_1)$.

Figures 11.4–11.6 present the maximum directivity of linear and two-row arrays consisting of isotropic radiators. Directivity is calculated in accordance with (11.15). For the two-row array $F(\theta, \varphi)$ in conformity with (11.7) is replaced by $F_{MN}(\theta, \varphi)$. The calculated directivity of several arrays at three frequencies at different azimuths of a main lobe is given in Table 11.1 in the mode of the maximal radiation forward. Processing calculation results allows to build other characteristics of directivity for the system of radiators: the width of main lobe, the level of side lobes, and the level of radiation in the opposite direction. In particular, the half-power width of the main lobe is presented in Figure 11.7.

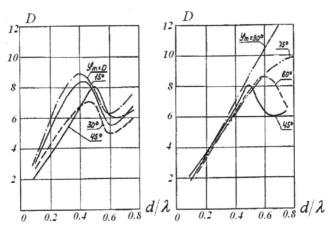

Figure 11.4   Directivity of a linear array of 8 radiators.

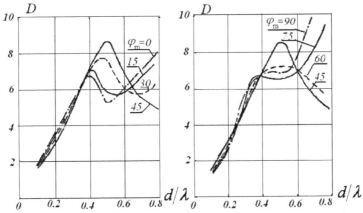

Figure 11.5   Directivity of a two-row array of 8 radiators in the mode of the maximal radiation forward.

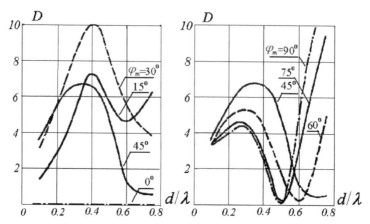

Figure 11.6 Directivity of a two-row array of 8 radiators in the mode of the zero radiation backward.

Table 11.1 Directivity of Antenna Arrays in the Mode of Maximal Radiation Forward

Array type	$d/\lambda = 0/2$	0.4	0.6	0.2	0.4	0.6
	in units			in decibels		
Linear, 4 radiators	1.8–3.2	3.1–5.5	2.9–4.9	2.5–5.0	4.9–7.4	4.6–6.9
Linear, 8 radiators	3.3–6.0	6.2–8.9	5.0–10.3	5.2–7.8	7.9–9.5	7.0–10.1
Square, 4 radiators	1.8	3.5–4.0	3.5–3.8	2.5	5.4–6.0	5.0–5.8
Two-row, 8 radiators	2.5–3.5	6.7–7.2	5.7–7.9	4.0–5.4	8.2–8.6	7.5–9.0

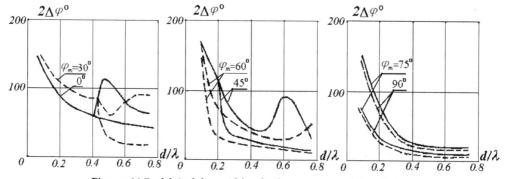

Figure 11.7 Main lobe width of a linear array of 8 radiators.

If there are several main lobes, two curves are presented: the lower curve—for width of one lobe, the upper curve—for the summary width of two lobes.

The analysis of results presented in table and in figures allows to compare the characteristics of different arrays depending on the number of radiators.

11.2 PRINCIPLE OF SIMILARITY AND RECIPROCITY THEOREM

This section is devoted to the principle of electrodynamics similarity and to the reciprocity theorem.

As is known, antennas with the shape defined by angular dimensions, e.g., conic radiators of infinite length, satisfy the principle of electrodynamics similarity. In this case, a change of the scale does not lead to a change of antenna characteristics, i.e. the radiator shape and its dimensions in wavelengths are the same at different frequencies.

Relationships of similarity follow from Maxwell's equations. For the harmonic field created by an antenna in surrounding space, we can write, accordingly (1.1),

$$curl \ \vec{H} = (\sigma + j\omega\varepsilon)\vec{E}, \ curl \ \vec{E} = -j\omega\mu\vec{H}. \tag{11.16}$$

Similarly, at another frequency

$$curl \ \vec{H}' = (\sigma' + j\omega'\varepsilon')\vec{E}', \ curl \ \vec{E}' = -j\omega'\mu'\vec{H}', \tag{11.17}$$

where $\vec{E}' = k_E\vec{E}, \vec{H}' = k_H\vec{H}, \omega' = k_\omega\omega, \varepsilon' = k_\varepsilon\varepsilon, \mu' = k_\mu\mu, \sigma' = k_\sigma\sigma, l' = k_l l'$, here, $k_E, k_H, k_\omega, k_\varepsilon, k_\mu, k_\sigma, k_l$ are the coefficients, interrelating magnitudes at different frequencies in the equation, and l is the distance (arbitrary linear coordinate). The said coefficients are called coefficients of modeling, since the use of the principle of similarity allows studying antennas characteristics by means of experiment on a model.

If to substitute the coefficients into (11.17) and take into account the linearity of magnitudes and of an operator $curl \ \vec{A}$, where $\vec{A}$ is an arbitrary vector, then the resulting equations are to coincide with equations (11.16), i.e. $k_H = k_\sigma k_E k_l, k_H = k_\omega k_\varepsilon k_E k_L, k_E = k_\omega k_\mu k_H k_l$. If $k_\varepsilon = k_\mu = 1$, then $k_H/k_E = k_\sigma k_l = k_\omega k_l = 1/(k_\omega k_l)$, i.e.

$$k_\omega k_l = 1, k_H = k_E, k_\sigma k_l = 1. \tag{11.18}$$

From the expression it follows that the electromagnetic field at different frequencies will be the same, if the electrical dimensions of an antenna (the ratio of linear dimensions to the wavelength) at different frequencies coincide and the material conductivity is in inverse proportion to magnitude k_l, i.e. increases with a frequency.

A similar conclusion is true for studying an antenna's model. When the geometrical dimensions of a model are smaller than the dimensions of the original in N times, it is necessary to increase the signal frequency and the conductivity of model material in N times. Since $k_H = k_E$, i.e. relationship of the currents and voltages remains the same, the resistances of resistors should remain the same, whereas the capacitances of capacitors and the inductances of inductors connected in the model should be smaller in N times. As a rule, the conductivity of model material could not be increased in N times. So, the resistances of an antenna and of a model as well as the characteristics depending on them (e.g., efficiency, Q-factor, gain) differ substantially from each other.

Self-complementary and log-periodic antennas belong to the class of frequency-independent antennas. Their characteristics change weakly, if an operation frequency changes. The log-periodic antenna is an antenna array. It consists of a few dipoles of the same shape and different dimensions, which are connected in a single structure. The self-complementary antenna of several dipoles of the same shape and dimensions is also an antenna array, although its elements are not parallel to each other.

Another issue, which as it will be shown is concerned with antenna array, is the reciprocity theorem. Reciprocity theorem argues that the current in the receiving antenna and all characteristics of the antenna can be found, if the characteristics of the antenna used in transmission mode are known. Suppose there are any two antennas, which are remote from each other by such a distance that their mutual impedance is zero (Figure 11.8). This condition is accepted for the sake of simplicity of a proof and has no principled significance. In the middle of each antenna with the input impedance Z_A the load impedance Z_L is connected. If emf e_I creates a current $J_I = \dfrac{e_I}{Z_{AI} + Z_{LI}}$ at the input of the first antenna, then field $E_I = \dfrac{-j30J_IF_I}{\varepsilon_r}$ arises near the antenna II and creates in this antenna the current J_{II} (see Figure 11.8a). Here the designations adopted in Chapter 10 are used. In a similar case, when emf e_{II} is connected in the second antenna and creates in it the current J_{II}, the current J_I arises in the first antenna (see Figure 11.8b).

In accordance with the electrodynamics principle of reciprocity

$$e_I/J_{II} = e_{II}/J_I. \tag{11.19}$$

From these expressions, which are given for the first antenna, it follows that

$$\frac{e_I}{J_{II}} = \frac{J_I\left(Z_{AI} + Z_{LI}\right)}{J_{II}} = j\frac{\varepsilon_r E_{II}\left(Z_{AI} + Z_{LI}\right)}{30F_IJ_{II}}$$

Similarly, for the second antenna

$$\frac{e_{II}}{J_I} = j\frac{\varepsilon_r E_I\left(Z_{AII} + Z_{LII}\right)}{30F_{II}J_I},$$

i.e.

$$\frac{E_{II}\left(Z_{AI} + Z_{LI}\right)}{F_IJ_{II}} = \frac{E_I\left(Z_{AII} + Z_{LII}\right)}{F_{II}J_I}.$$

If the magnitudes relating to each antenna are transferred into one part of the equation and it is assumed that each side of the equation is a constant, which does not depend on the properties of the antenna, we get:

$$\frac{J_i\left(Z_{Ai} + Z_{Li}\right)}{E_iF_i} = \text{const.} \tag{11.20}$$

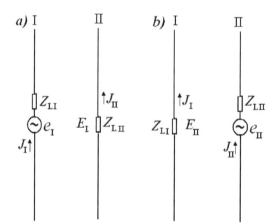

Figure 11.8   The reciprocity theorem: antenna I radiates (a), antenna II radiates (b).

If to assume that this expression refers to the receiving antenna, then J_i is the current in the middle of this antenna, E_i is the field strength near it, Z_{Ai} and Z_{Li} are its input impedance and load impedance, F_i is a function characterizing the effective length and the shape of the directional pattern of antenna in the transmit mode.

The constant on the right side of expression (11.20) can be determined by considering the simple electrical dipole (Hertz' dipole) as an antenna. If the axis of receiving dipole lies in the plane of a wave incidence, the constant is equal to 1. In a general case, it is necessary to take into account the polarization of the field and the azimuth, if the horizontal antenna pattern differs from circular. As a result we obtain in accordance with (11.20) that the current in the middle of the receiving antenna is equal to

$$J_i = \frac{E_i F_i(\varphi)}{Z_{Ai} + Z_{Li}} = \frac{e_i}{Z_{Ai} + Z_{Li}}. \tag{11.21}$$

The reciprocity theorem shows that the main characteristics of the antenna (input impedance, effective length, directional pattern) coincide for transmit and receive modes.

This theorem allows also to analyze a relationship between the field incident onto an array element (single radiator) and reflected from that element. It is known that when a wave is incident on a flat perfectly conducting metallic surface it is reflected at an angle equal to an angle of incidence. Amplitudes of the incident and reflected fields are identical, and the wave phase changes in a stepwise fashion by π. However, when a metal surface is replaced with a system of radiators, e.g., with a linear equally spaced array, direction and phase of the reflected field could be essentially changed, because they depend on parameters and electric characteristics of a separate radiator.

An example of such structure is the in-phase reflector array (Figure 11.9a). It is a flat equivalent of a parabolic reflector. The structure consists of primary exciter 1 of antenna array (e.g., a horn) and an equally spaced array of secondary micro strip radiators 2, situated in one plane along surface 3. In order to sum the signals of secondary radiators in the direction, perpendicular to the array plane, their phases should be identical. Since distances r_i (i is the number of reradiator) between the primary exciter and an arbitrary reradiator are not identical, this results in a phase path difference, which should be compensated with a phase step in the signal reradiating.

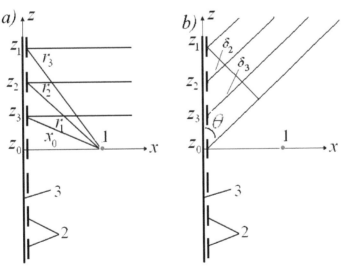

Figure 11.9 Reflector array with the radiation direction perpendicular to the array plane (*a*) and in any desired direction (*b*).

The method of calculating the phase step of the reflected field in comparison with the incident field phase in the signal rer-adiating can be constructed on the basis of the reciprocity theorem. The reciprocity theorem for two antennas is described above in the form of (11.19). In [14] the theorem is formulated as follows: if emf e_I applied to the terminals of antenna I establishes current J_{II} at the input of an antenna II, then equal emf e_{II} applied to the terminals of an antenna II will create at the input of an antenna I the same current J_I, i.e.

$$J_{II}/e_I = Y_{I,II} = Y_{II,I} = J_I/e_{II}, \qquad (11.22)$$

where $Y_{I,II} = |Y_{I,II}| \exp(j\varphi_{I,II})$ and $Y_{ba} = |Y_{II,I}| \exp(j\varphi_{II,I})$ are the mutual admittances between antennas. It follows that

$$\varphi_{I,II} = \varphi_{II,I}, \qquad (11.23)$$

i.e. the difference of phases between the exciting emf and the current excited in an adjacent antenna is the same in both cases (in these expressions I is used instead of a and II instead of b).

In our example, incident field E_I acts as a signal source instead of the antenna I with emf e_I. If into the circuit to introduce a source of field – an antenna I, which is excited by the generator e_I with infinitely high output resistance and to regard as usually that a linear antenna is the aggregate of elementary dipoles with appropriate currents, then, since the phase of radiation field outstrips the current phases by $\pi/2$, the phases difference between the current J_{II} and emf e_I is (see Figure 11.8a)

$$\varphi_{I,II} = \varphi_{11} + \varphi_{12} + \varphi_{13} + \varphi_{14}, \qquad (11.24)$$

where φ_{11} is the phase shift between the current in the antenna I and emf e_I (in this case it is absent), φ_{12} is the phase shift between the radiated field E_I and the current in the antenna I (it is $\pi/2$), φ_{13} is the phase shift due to the distance between antennas, φ_{14} is the phase difference between current J_{II} and field E_I that is

$$\varphi_{14} = \varphi_{I,II} - \pi/2 - \varphi_{13}. \qquad (11.25)$$

The other source of a signal is the current in antenna II (rather than emf e_{II}), which creates the reflected field E_{II}. A current distribution along the receiving antenna differs from distribution along a transmitting antenna. Antenna II is the aggregate of elementary dipoles, each of which is excited by its generator. The currents of the dipoles create in-phase fields. Let the current in the middle dipole (at the antenna center) be excited by the generator e_{II}; it is equal to the product of emf e_{II} and the dipole admittance. Since the dipole impedance is capacitive, the current phase outstrips the emf phase by $\pi/2$. That assertion holds true for other dipoles of antenna II. Accordingly, the phase difference between current J_I and emf e_{II} is equal (see Figure 11.8b)

$$\varphi_{II,I} = \varphi_{21} + \varphi_{22} + \varphi_{23} + \varphi_{24}, \qquad (11.26)$$

where φ_{21} is the phase shift between the current in antenna II and emf e_{II} (in this case it is $\pi/2$), φ_{22} is the phase difference between field E_{II} and the current in antenna II, φ_{23} is the phase shift due to the distance between the antennas (it is φ_{13}), φ_{24} is the phase shift between the reflected field E_{II} and current J_I (it is zero, since the input impedance of antenna I is infinitely large), that is

$$\varphi_{22} = \varphi_{II,I} - \pi/2 - \varphi_{13}. \qquad (11.27)$$

In accordance with the equation (11.23), this implies that the increment of the phase of the receiving antenna current compared with that of the incident field and the increment of the phase of the reflected field compared with that of the receiving antenna current are identical in magnitude.

As to the amplitudes of incident $|E_1|$ and reflected $|E_2|$ fields, since the total tangential component of both fields on a perfectly conducting metal surface is zero, then at the reflection point

$$|E_2| = |E_1|\cos\gamma\cos\delta, \tag{11.28}$$

where γ is the angle of ray incidence onto an antenna, δ is angle of reflection.

The realistic variants of reflect arrays are considered in the follow section.

11.3   REFLECTOR ARRAYS

Lately, reflector arrays became widespread in the capacity of a flat equivalent of a parabolic reflector. The calculation method of the array is based on the reciprocity theorem described in the previous section. For example, an array of micro strip radiators is such an array.

The simplest micro strip antenna is a rectangular metal plate of length L and width b situated on a dielectric substrate above a metal plane (Figure 11.10a). Length L of the plate is about $\lambda_1/2$, where λ_1 is the wavelength in the substrate material. A simplified model of a micro strip antenna is a planar dipole with a sinusoidal current distribution (Figure 11.10b) and dimensions coinciding with those of the micro strip antenna. The propagation constant of the current is close to $k_1 = k\sqrt{\varepsilon_r}$, the propagation constant of a wave traveling in the substrate material (ε_r is the relative permittivity of the substrate).

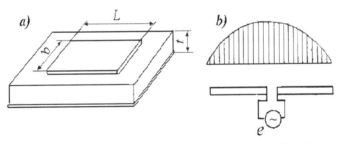

Figure 11.10   General view of a rectangular micro strip antenna (a) and it's excitation at the center of the metal plate (b).

The equivalent circuit of the antenna operating in the receive mode is shown in Figure 11.11a. Here Z_A is the antenna impedance in the transmit mode. In this mode, the current distribution along a planar radiator is similar to distribution along an impedance electric dipole excited at its center. In the first approximation, the input impedance of the radiator when $L/2 < 0.3\lambda_1$ is

$$Z_A = R_\Sigma - jW_A \cot(k_1 L/2), \tag{11.29}$$

where R_Σ is the radiation resistance, W_A is the wave impedance of the planar dipole, the value of which is equal to the doubled wave impedance of a strip line. The wave impedance of a strip line is $W = 120\left\lfloor \pi t/(b\sqrt{\varepsilon_r}) \right\rfloor$, where t is the thickness of the substrate [88]. If $L/2 < 0.3\lambda_1$, the radiation resistance can be calculated by the formula $R_\Sigma = 20k^2 l_e^2$, where $l_e = (2/k_1)\tan(k_1 L/4)$ is the effective length of the antenna. The magnitudes of the input impedance, corresponding to $L/2 < 0.3\lambda_1$, can be calculated with the use of the Moment method.

The amplitude and phase of the current created in a reradiator depend on the load impedance. For the circuit shown in Figure 11.11b,

$$I_A = e/Z_A = e/|Z_A|\exp\{j\cot^{-1}[(W_A/R_\Sigma)\cot(kL/2)]\}. \qquad (11.30)$$

Magnitude $\cot^{-1}[(W_A/R_\Sigma)\cot(kL/2)]$ is the phase increment of the current running in the antenna relative to the phase of the incident field. In accordance with the reciprocity theorem, the increment is equal to the phase increment of the reflected field relative to the phase of the current running in the antenna. Hence, the phase step during reradiation is

$$\varphi_1 = 2\tan^{-1}[(W_A/R_\Sigma)\cot(kL/2)]. \qquad (11.31)$$

The value of the step is zero for a tuned antenna, negative for an elongated antenna, and positive for a shortened antenna. Increase of dipole radius a lowers its wave impedance W_A and decreases the phase step (Figure 11.12).

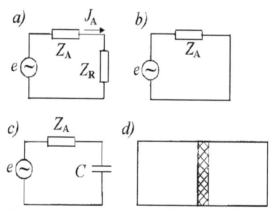

Figure 11.11   Equivalent circuits of a receive antenna with load Z_R (a), zero load (b), and a capacitive load (c), which is formed by the slot filled with a dielectric (d).

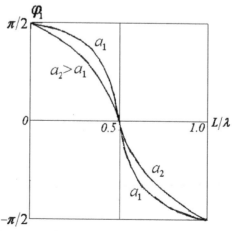

Figure 11.12   Field phase step during re-radiating as a function of the antenna length.

The approximate method for calculating the phase step of the field reradiated by an element of a reflection array is based on the reciprocity theorem and the theory of dipoles. The method is simple and efficient. After light clarifications, this method can be used to analyze loaded antennas. For the circuit shown in Figure 11.11c,

$$I_A = \frac{e}{Z_A - (j/\omega C)} = e\bigg/ \sqrt{|Z_A| + (1/\omega C)^2} \, \exp\left\{ j \tan^{-1}\left[\frac{1}{R_\Sigma}\left(\frac{1}{\omega C} + W_A \cot\frac{kL}{2}\right)\right]\right\},$$

i.e., the phase step during re-radiating is

$$\varphi_2 = 2\tan^{-1}\left[\frac{1}{R_\Sigma}\left(\frac{1}{\omega C} + W_A \cot\frac{kL}{2}\right)\right]. \qquad (11.32)$$

Here ω is the circular frequency of the signal.

One can apply the above results to calculation of the phase step of a micro strip antenna during the re-radiating. In order to sum up the signals from secondary radiators in the direction perpendicular to the array plane, the phases of the signals should be equal. If the coordinates of the primary feed are x_0, y_0 and z_0 (see Figure 11.9a), then the phase step in the re-radiator i, needed to compensate the phase difference in the direction of the x-axis, must be equal to $\xi_i = k\left[\sqrt{x_0^2 + (y_i - y_0)^2 + (z_i - z_0)^2} - x_0\right]$. The choice of geometric dimensions of the re-radiator i allows it to receive phase step $\varphi_{1i} = \xi_i$.

Figure 11.13 plots the field phase step φ_1 created at 60 GHz by a micro strip antenna situated on a substrate with a thickness of 0.254 mm and a relative permittivity of 2.22 as a function of the antenna length. The results were obtained by the proposed technique. Curves 1 and 2, respectively, correspond to antenna widths of 0.3 and 2.3 mm. The rigorous method for calculating the step value was described in [89]. It relies on the analysis of an infinite periodic array of identical elements illuminated by a plane wave, i.e., on solution of the analysis problem in the spectral domain and on the Floquet's theorem. The open and closed circles in Figure 11.13 correspond to the results presented in [89]. As seen from the figure, the correspondence is satisfactory.

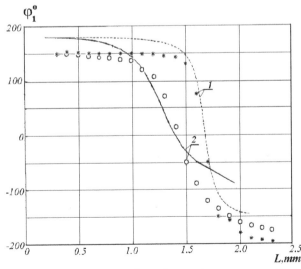

Figure 11.13 Phase steps φ_1 for micro strip antennas with lengths of 0.3 (curve 1) and 2.3 mm (curve 2).

Along with simple micro strip antennas, in reflection arrays, multilayer (multiple-stack) micro strip antennas can be used. If the field phase step arising during reradiating in a single-layer antenna is less than 360° (the phase increment arising with a sufficiently thick substrate, which ensures smooth phase variation during plate length changing and a wider frequency band, does not exceed 300°), the maximum phase step, attainable, e.g., in a two-stack antenna, is 540°.

The design of a two-stack micro strip antenna is shown in Figure 11.14. In this antenna, two rectangular metal plates (of lengths L_1 and L_2 and widths b_1 and b_2, respectively) are separated from each other and from the metal plane by a dielectric substrate. The upper plate is smaller than the lower one. Plate length L_2 is larger than $\lambda_1/2$, where λ_1 is the wavelength in the substrate material.

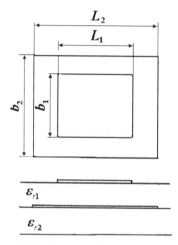

Figure 11.14   Two-stack micro strip antenna.

Characteristics of a multiple-stack antenna can be calculated by the method described above. The phase step of the field generated at 12.5 GHz by a two-stack micro strip antenna containing substrates with a thickness of 3 mm and a relative permittivity of 1.03 is shown in Figure 11.15 as a function of antenna length L_2. The antenna is a square. The calculating curves were obtained with the use of the proposed technique. The open and closed circles in the figure indicate the results of [90].

In order to set the maximum of the radiation pattern of an antenna array in the prescribed direction, the phases of the radiator fields must be equal in this direction. Hence, if this angle varies in plane xOz, the phases must vary linearly as functions of coordinate z_i (see Figure 11.9b): $\psi_i = k(z_1 - z_i) \cos \theta$. To simplify the control procedure, the field phases should be controlled by an electric signal.

Among the circuits of micro strip antennas, circuits with loads are most suitable for continuous control of the phase of the reradiated field. In order to connect the load impedance in series into the circuit of a receiving antenna, a slot can be cut in the central part of the planar radiator across its long sides (Figure 11.11d). In this case, a slot filled with a dielectric forms the simplest load: a capacitor with capacitance C (see Figure 11.11c) and reactance $1/j\omega C$. The capacitance of this load can be controlled readily via variation of the permittivity of a special material placed between the capacitor plates during application of voltage between the plates.

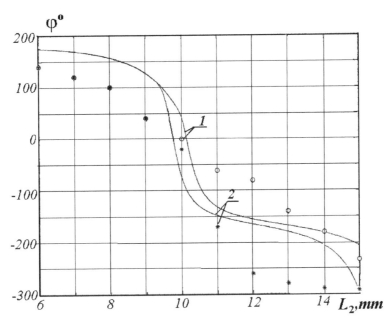

Figure 11.15 Phase steps φ for a two-stack micro strip antennas with $L_1/L_2 = 0.6$ (1) and 0.8 (2).

For the circuit presented in Figure 11.11c the phase step during reradiation is determined by (11.32). The tangent of angle φ_2 depends on two terms. The first term, $\alpha = 1/(R_\Sigma\omega C)$ is due to the presence of the capacitor in the radiator center while the second term, $\beta = (W_A/R_\Sigma)\cot(k_1L/2)$ is related to the deviation of the radiator length from the resonant value. The second term can be used to compensate of the phase difference caused by the differences in distances r_i between primary feed 1 and reradiators 2 (see Figure 11.9a), while the first term can be used for turning the radiation pattern. The total phase step in the reradiator i should be $\varphi_{2i} = \xi_i + \psi_i$. Here, if reactance $1/(j\omega C_i)$ of capacitors C_i, corresponding to $\varepsilon_r = 1$ varies linearly with coordinate z_i, then, applying equal voltages to the capacitors filled with the same dielectric, we find that the reactance of these capacitor retain a linear dependence on this coordinate. This statement is equally true for the first term in the expression for the above tangent.

Note, however, that the second term takes different values for different reradiators because it depends on distance r_i. Therefore, phase φ_2 will vary nonlinearly after application of the same voltage to all capacitors. Note that even if the second term is absent, i.e. $\tan\psi_i = 1/(R_\Sigma\omega C_i)$ and the angular turning the radiation pattern requires a linear dependence of phase ψ_i on coordinate z_i, the value $\tan\psi_i$ and, accordingly, capacitance C_i do not possess this property. Therefore, in the general case, the angular displacement of the radiation pattern requires application of an individual voltage to each capacitor.

The case when angle θ of the maximum radiation of the antenna array is close to $\pi/2$, i.e. $\alpha \ll \beta$, requires a special analysis. In this case, by expanding the function $\varphi_2 = \tan^{-1}(\alpha + \beta)$ into the Taylor series, we obtain

$$\varphi_2(\alpha + \beta) = \tan^{-1}\beta + \alpha/(1 + \beta^2).$$

Here, the term $\dfrac{\alpha}{1+\beta^2} = \left[R_\Sigma\omega C\left(1 + \dfrac{1}{R_\Sigma^2}W_A^2\cot^2\dfrac{k_1L}{2}\right)\right]^{-1}$ is proportional to reactance

$1/(j\omega C)$ of capacitor C. Hence, phases of the fields created by secondary radiators will vary linearly with the coordinate upon application of equal voltages to the capacitors filled with the same dielectric.

If the direction of the maximum radiation differs substantially from the perpendicular to the array plane, different voltages can be applied to several groups of antennas in order to bring the law of variation of phase φ_2 along the antenna closer to a linear function. Note that the number of these voltages can be substantially less than the total number of radiators.

During calculation of the capacitance formed by a slot cut in the plate of a microstrip antenna, it should be taken into consideration that this capacitance consists of two terms: capacitance C_1 between thin planar plates and capacitance C_2 of the planar capacitor formed between the surfaces of plate edges.

11.4 INDEPENDENT CONTROLLING THE DIRECTION OF THE MAIN BEAM AND OF THE NOTCH

Phased antenna array found a wide use in antenna technology. These are a system of radiators, which allow the dramatically increase the directivity of the structure in a given direction. They are used both to transmit and to receive signals. Receiving arrays have their features, since often, except for receiving a weak signal, they must suppress interfering signals, coming from other directions. To do this, the main lobe of the array pattern should be directed to the source of the useful signal, and the zeros of the pattern—to the sources of interference. In particular, zero in the pattern may be utilized for suppression of silencer's signal.

Beam forming technique for solving this problem is described in [91]. As the authors point out, in the simplest embodiment, the structure is divided into two arrays, one of which provides a maximum directivity in the arrival direction of the desired signal, and the second array receives the interference signal and compensates it with the help of an interference signal at the output of the first array. In contrast to this approach side lobe compensation can be realized by providing in array the required amplitude-phase distribution of current. In order to simplify its search, one can use the expansion of the pattern in a series in terms of so-called sinc-functions, forming an orthogonal system of functions [92, 93]:

$$\sin c(x) = \sin x / x . \tag{11.33}$$

The pattern of a linear equidistant array of N elements is usually written as

$$F(u) = \sum_{q=0}^{N-1} |I_q| \exp\left[j\varphi_q + 2ju\left(q - \frac{N-1}{2}\right)\right], \tag{11.34}$$

where $|I_q|$ and φ_q are the amplitude and the phase of q-radiator current, $u = (kd/2)^*$ $\sin\varphi$, $k = 2\pi/\lambda$ is the propagation constant, λ is the wave length in free space, d is the distance between the radiators, φ is azimuth. This expression can be written through sinc-functions as

$$F(u) = \sum_{p=0}^{N-1} J_p \sin c(Nu - p\pi). \tag{11.35}$$

Magnitude J_p is called the selective factor and can be found as

$$J_p = \frac{N}{\pi} \int_{-R}^{R} F(u) \sin c(Nu - p\pi) du. \tag{11.36}$$

Number of selective factors J_p is equal to the number of current amplitudes I_q. Equate the directivity pattern in the form (11.34) and (11.35). The result is a system of linear algebraic equations related I_q and J_p. Its solution gives

$$I_q = \sum_{p=-(N-1)/2}^{(N-1)/2} M_{q,p} J_p, \tag{11.37}$$

where $M_{q,p} = \exp\left[jp\pi\left(1 - \frac{1}{N} - \frac{2q}{N}\right)\right]$.

Let, for example, the initial current distribution (Figure 11.16a) be created, and the desired pattern (Figure 11.16b) constructed. The pattern is calculated in accordance with (11.34). In order to ensure zero reception in the pattern in a given direction, the following form is used:

$$F_1(u) = F_0(u - u_0) \cdot Z(u, u_2, \delta), \tag{11.38}$$

where

$$Z(u, u_2, \delta) = \begin{cases} a, & \text{if } u_2 - \delta < u < u_2 + \delta \\ 1, & \text{otherwise} \end{cases}.$$

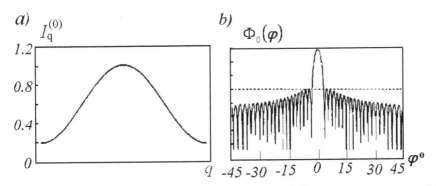

Figure 11.16   The current amplitudes in the radiators (a) and the array pattern, created by the initial currents (b).

Position of the main lobe is determined by parameter u_0, the zero position in the pattern—by parameter u_2, the magnitude δ defines the width of the zero zone and does not affect its placement point. The notch depth ξ, measured from a zero level, is characterized by the parameter a. If $a = 1$, the notch in the pattern is absent. If $a < 0$, then the phase is modulated too. This phase change in the given pattern results in the corresponding change of the current distribution along the antenna array. It is important to emphasize that the zero position and the position of the main lobe are not dependent on each other and are determined separately by parameters u_2 and u_0. The amplitude

$|I_q^{(I)}|$ and phase $\varphi_q^{(I)}$ distribution in the linear array may be found by using (11.36) and (11.37) (Figure 11.17). Substituting the calculated values into (11.34) gives the directivity pattern $\Phi_1(\varphi)$ shown in Figure 11.18.

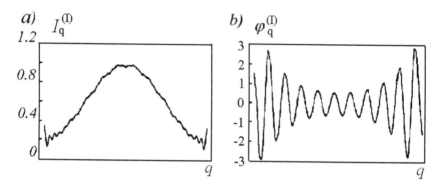

Figure 11.17   Distribution of the current amplitude (*a*) and phase (*b*) in the linear array with zero in the directional pattern.

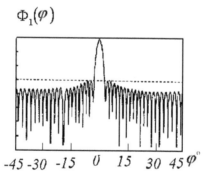

Figure 11.18   The array pattern for the fixed positions of the main lobe and zero in the directions $\varphi_0 = 0°$ and $\varphi_2 = -20°$.

A simple circuit of the receiving antenna array with zero is shown in Figure 11.19. It consists of the radiators 1, a set of attenuators/amplifiers 2, the phase-shifters 3, and the power divider 4 too. Power divider defines the initial current distribution. Attenuators/amplifiers and phase shifters are used to adjust the amplitude and to form the desired

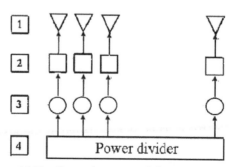

Figure 11.19   Control system of the antenna array.

phase shifts respectively. Both the phase shifters and attenuators/amplifiers operate in an analog mode, providing the correct linear coupling of the main parameters and the real phases and amplitudes.

The described procedure of forming antenna array pattern considers the ideal case where the mutual influence of radiators on each other is absent. In fact, the mutual influence of the array elements can be quite strong. In this case, the pattern may be substantially destroyed. Therefore, the effect of mutual coupling must be taken into account during the pattern formation. The first step is forming the array impedance matrix. For simplicity, the relationship between each radiator and its two nearest neighbors, located at a distance d and $2d$ respectively, is considered.

As previously mentioned, the divider forms an initial current distribution $I_q^{(0)}$. Mutual influence of the radiators can be accounted by means of impedance matrix related radiators currents with applied voltages:

$$U_q^{(0)} = [Z] \cdot \left| I_q^0 \right|,\qquad (11.39)$$

where $[Z]$ is the impedance matrix, $U_q^{(0)}$ are the applied voltages, and $I_q^{(0)}$ are complex amplitudes of currents. Directivity pattern, formed by voltage $U_q^{(0)}$ is equivalent to the directivity pattern, formed by the current distribution $I_q^{(0m)} = U_q^{(0)} \cdot Y_{11}$, where Y_{11} is the admittance of a single radiator. Thus, one may say that a mutual coupling of the radiators results in the transformation of the distribution $I_q^{(0)}$ into the new distribution, which is equal to

$$I_q^{(0m)} = \left([Z] \cdot I_q^{(0)} \right) \cdot Y_{11}.\qquad (11.40)$$

Here the letter "m" of superscript means taking mutual coupling between the radiators into account.

This account allows to change the synthesis procedure. If to replace $I_q^{(0)}$ by $I_q^{(0m)}$ in all subsequent calculations, then instead of pattern $\Phi_0(\varphi)$ the pattern $\Phi_2(\varphi)$ appears (Figure 11.20a). Figures 11.20a and 11.16b differ from each other in that the presented in the figures pattern are constructed with and without taking into account the mutual coupling between the radiators.

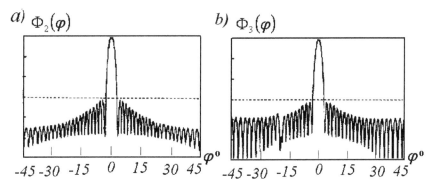

Figure 11.20   Array pattern with taking into account mutual coupling (a) for the fixed positions of the main lobe and the zero (b).

The next step relates to the synthesis of the distribution $I_q^{(m)}$ that compensates the side lobe in a given direction. Taking into account the mutual coupling between radiators we obtain a current distribution $I_q^{(1m)}$ created by attenuators/amplifiers and phase shifters

$$I_q^{(1m)} = \left([Z] \cdot I_q^{(1)}\right) \cdot Y_{11}. \tag{11.41}$$

The current distribution $I_q^{(1m)}$ obtained in this manner forms a directivity pattern $\Phi_3(\varphi)$ (Figure 11.20b). One can find after that the transfer factors $K_q^{(m)}$ (Figure 11.21a) and phases $\varphi_q^{(m)}$ (Figure 11.21b):

$$K_q^{(m)} = |I_q^{(1m)}| \,/\, |I_q^{(0)}|, \; \varphi_q^{(m)} = \arg(I_q^{(1m)}). \tag{11.42}$$

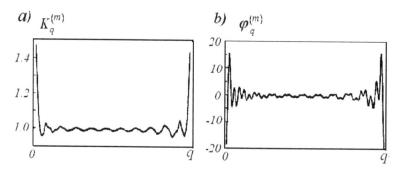

Figure 11.21   Transfer factors (*a*) and phases of phase-shifters (*b*) with allowance for the mutual coupling and the zero.

As already mentioned, the mutual coupling of the array elements requires a substantial correction of the amplitude and phase current distribution in the radiators. After that the shape of the zero zone has little effect on array performance, i.e. one can assume that the procedure of forming a pattern without side lobe is more or less stable with respect to the effect of mutual coupling.

Considered procedure of pattern synthesis with zero in a given direction is based on the changing phase and amplitude current distribution along the array. Its use requires additional attenuators/amplifiers, which leads to complication of the entire system. Therefore it is expedient to check what will happen if we use the initial amplitude distribution and change only the phase distribution in accordance with the described procedure. Then the final directivity pattern is calculated as

$$F_2(u) = \sum_{q=0}^{N-1} |I_q^{(0)}| \exp\left[j\varphi_q^{(l)} + 2ju\left(q - \frac{N-1}{2}\right) \right], \tag{11.43}$$

where $I_q^{(0)}$ is the initial current distribution along the antenna array, and $\varphi_q^{(l)}$ is the phases distribution calculated in accordance with the described procedure.

Formed in this manner the directivity pattern is shown in Figure 11.22. The position and depth of zero is not changed, but the pattern level rose, and the depth of the null node decreased by 6 dB. This example allows to compare two procedures: the amplitude and the phase one. The first one allows us to obtain a more uniform pattern, without rise, but with a more complicated control system because of requirement in additional attenuators-amplifiers. In the second case, their installation is not needed.

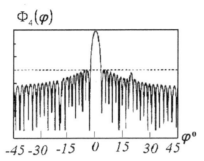

Figure 11.22 The array pattern with unchanged position of the main lobe and zero, based on the change of phase distribution.

11.5 ADAPTIVE ARRAY

During receiving a radio signal, possible filtration of interference, i.e. separation of useful signal and simultaneous suppression of interfering signals, is of important significance. The spatial filtration is one of the most efficient methods of fighting interference. In order to apply it, one must know the direction of the useful signal arrival so that the main lobe of the reception pattern could be oriented on the source of the signal, and nulls of the pattern—on the sources of interferences and disturbances.

The array of receive antennas forming a unified system with receiving equipment can act as an efficient spatial filter [94]. Adjustment of such a system is performed with the help of special weighing devices (attenuators), forming the required pattern, and adaptive controlling circuit (feedback loop), which uses an iterative procedure to automatically choose optimal parameters of the system and then automatically adapts to the changing conditions. For this reason, the described antenna system is called an adaptive one.

An advantage of adaptive processing is the fact that the suppression of interference, as a rule, involves no decrease of the useful signal. The automatic control of parameters is of special importance where constant factors, degrading the antenna performance, act often together with variable factors that exist, for example, aboard ships: running rigging, motion of various steel ropes under the action of wind or pitching and rolling, rotation or tuning of nearby antennas, the weather effects, etc.

An adaptive antenna system operates in a situation when the spectrum of the useful signal and the direction of its arrival is known, whereas the field structure of the source incorporating noise and interference and the direction towards the source are not. The system uses an artificially introduced reference signal that is produced in the receiver and has the spectral characteristics and azimuth coinciding with those of the useful signal approximately known.

The principle of beam forming in the adaptive antenna system with the help of weighting devices is clear from Figure 11.23. When multiplying output signals of array elements by weighting coefficients, the latter can be selected to secure that the main lobe undergoes almost no change (i.e. that the magnitude of the received useful signal remains the same), and the direction of zero reception coincides with that towards the interference source. A possible variant to implement the required weighting coefficient W is using a circuit two parallel channels at each element output with system adjustment in amplitude and phase delay by $\pi/2$ in one channel only. Such an element is called a circuit with quadrature channels. Introduction of the phase delay equal to $\pi/2$ is unnecessary, yet useful, since it allows obtaining a close magnitude of the weighting coefficient in the adjacent channel.

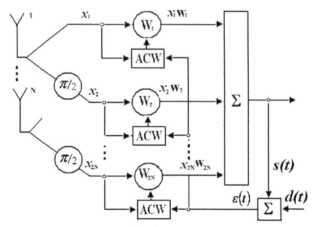

Figure 11.23   Antennas array with adaptive control of weighting coefficients.

The adaptive control of weighting coefficients (ACW) is performed with the help of a controlling circuit–adaptive processor (AP). It automatically adjusts the weights by the iterative procedure in accordance with the chosen algorithm. Error signal $\varepsilon(t)$ is used as a controlling one in adjustment circuits for weighting coefficients. The error signal is equal to the difference between reference signal $d(t)$ (close to required output signal) and actual signal $s(t)$ at the adder output:

$$\varepsilon(t) = d(t) - s(t), \tag{11.44}$$

where the output signal $s(t)$ is the sum of signals $x_n(t)$ with weighting coefficients W_n:

$$s(t) = \sum_{n=1}^{2N} x_n(t)W_n. \tag{11.45}$$

Here, N is the number of antennas, $2N$ is the number of weighting coefficients. Quantity $x_n(t)$ is to take into account the phase delay, equal to $\pi/2$ for even values of n.

Three adaptation algorithms are applied: (1) differential or greatest steepness, (2) least mean squared error, (3) random search [95]. The first two are based on the steepest descent method. The adaptation process starts with a set of several arbitrary coefficients. Then, if the steepest descent method is used, the gradient of error function is measured, and the weighting coefficients are set such that the error function will change in the direction opposite to the gradient. The procedure is repeated to ensure that the error decreases and the weighting coefficients approach the optimal values. If the differential method is used, the gradient is evaluated directly according to the error function derivatives. In the least mean squared error method, the error value is squared, and the derivatives of the square are used to calculate the gradient. The random search method includes the measurement of the mean squared error before and after an arbitrary change of the weighting coefficients and the comparison of the results to decide whether to accept the change, if the error has decreased, or to discard it otherwise.

The least mean squared error algorithm ensures either the most rapid convergence to the same value of error in all cases or the least error in the same operation time.

Implementing it in practice is easier than other one, the algorithm bases on using of feedback and adjusting each weighing coefficient following the law

$$dW_n/dt = \mu \overline{\partial \varepsilon^2(t)/\partial W_n}, \qquad (11.46)$$

where μ is a negative constant governing the convergence rate and the system stability, and the overline denotes the mathematical expectation. In this case, the number of arithmetical operations is linear in the number of weighting coefficients, that is, far less than in the direct calculation of the coefficients with the help of a covariance matrix, where the number of operations is proportional to the third power of the number of weighting coefficients.

Rather than calculate the gradient of the mean squared error, which requires a great number of statistical samples, it is expedient to use the gradient of a single sample of the squared error (the gradient estimate), i.e. to replace the derivative $\overline{\partial \varepsilon^2(t)/\partial W_n}$ with the derivative $\partial \varepsilon^2(t)/\partial W_n$. Then the law of feedback takes the shape

$$dW_n/dt = -2\mu x_n(t)\varepsilon(t). \qquad (11.47)$$

One can show expected value of the gradient estimate to be the gradient, i.e. the gradient estimation is unbiased.

If the number of iterations increases without limit, the mathematical expectations of the weighting coefficients converge to the Wiener solution, for which gradient

$$\overline{\nabla \varepsilon^2(t)} = \sum_{n=1}^{2n} \overline{\partial \varepsilon^2/\partial W_n}$$ vanishes. But the convergence is secured only in the case, when

constant μ lies within certain limits. A practically convenient restriction (although stricter than necessary) is inequality

$$-1/P < \mu < 0, \qquad (11.48)$$

where P is the total power of input signals.

In accordance with (11.47), input signal $x_n(t)$ and error signal $\varepsilon(t)$ i.e. the difference between reference signal $d(t)$ and actual signal $s(t)$ at the adder output, are fed into the processor. The error signal would have performed best, if the error signal had rather included the required output signal instead the reference one. But the latter is unavailable in the receive antenna, and we have to resort to the reference signal, close to the required one. For this reason, the main lobe of array in the process of adaptation orients in the direction specified by the reference signal. The amplitude response of the antenna system in the frequency band of the reference signal becomes uniform, and the phase response becomes linear.

Lest the reference signal distort the useful signal, two manners of adaptation, single-mode and dual-mode, are developed and used. In the dual-mode adaptation (Figure 11.24) only one processor is used, i.e. it is more economical. As seen from the figure, the reference generator signals (RG) outputs have two signals. One signal goes as the reference signal $d(t)$ to the circuit for the processing. The second (control) signal imitates the useful signal arrival from the given direction. It goes through the circuits of delay δ_n to inputs of array channels. Delays δ_n are chosen so that the received input signals are identical to the signals coming from the given direction.

In the first mode, with the switch set to position I, the control signals are fed to the inputs of the adaptive processor channels, and the processor adjusts the weighting coefficients so that the output signal does not differ from the reference signal, i.e. turns the pattern main lobe in the given direction. In the second mode, with the switch set to position II, the signals from array elements (i.e. from the surrounding space) are fed to the inputs of adaptive processor channels, and the reference and control signals are removed, lest they distort the external signal. Since there is no reference signal, i.e. $d(t) = 0$, all received signals are suppressed.

Sustained operation in the second mode leads to the self-clinching of the system, when all weighting coefficients tend to zero. But, if the modes rapidly alternate and the weights vary little during operation in each mode, the required direction of the main lobe is retained (at the operation in the first regime), and the power of interference is reduced to minimum (mostly, in the second mode). The useful signal in the second mode (switch is set to position II) arrives at the receiver input (R).

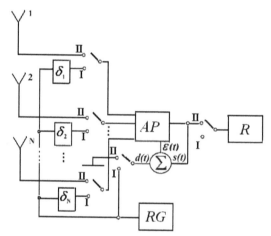

Figure 11.24 The structure circuit of dual-mode adaptation.

Digital simulation adaptive processing of signals confirms the procedure convergence and shows that this is an efficient method of the spatial filtration of interference with the useful signal retained. The experimental testing of adaptive system confirming its efficiency has simultaneously revealed the danger of suppressing the system operation by interference with a frequency close to that of the useful signal as well as necessity of protection, for example, by means of modulating the useful and reference signals with pseudo noise code [96].

Transparent Antennas

12.1 CURRENT DISTRIBUTION ALONG THE TRANSPARENT ANTENNA

Thin films of ITO (Indium-Tin-Oxide) placed on high-quality glass substrates are on the one hand electrically conductive and on the other hand optically transparent. They have high homogeneity of sheet resistance, which allows to use them as flat antennas for mobile communications and other applications. As an example in Figure 12.1, quoted in [97], the optical transparency of the ITO film for different resistivity R_{sq} is presented. It is specified in Ohm per square section, for light wavelength 550 nm. As it is seen from this figure the transmittance increases with increasing the film resistivity and becomes high enough (near 95%) if the film resistivity is greater than 5 Ohm/sq.

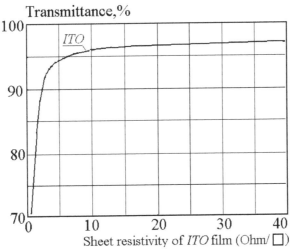

Figure 12.1 Example of the film transmittance dependence on its sheet resistivity.

In order to better understand the material constraints imposed by the low conductivity of ITO film (this conductivity is low in comparison with conductivity of printed cards and metal antennas), we examine the sheet resistivity of a film depending on its thickness d. This magnitude is designated as R_{sq1}. According to impedance boundary condition of Leontovich (see, for example, [19]), if the thickness d of a metal film is greater than its skin depth s of a current, the sheet resistivity is equal to

$$R_{sq1} = \frac{1}{s\sigma} \text{ Ohm}. \tag{12.1}$$

Here σ is its specific conductivity for constant current (in S/m), and s is given by

$$s = 1 / \sqrt{\pi f \mu \sigma}, \tag{12.2}$$

where f is frequency (in Hz), $\mu = \mu_0 = 4\pi \cdot 10^{-7}$ F/m is the absolute permeability, λ is wave length (in m). If the thickness d of a metal film is small in comparison with the skin depth s, the film's sheet resistivity is equal to

$$R_{sq1} = R_{sq} s / d = 1 / (d\sigma), \tag{12.3}$$

i.e. R_{sq1} does not depend on the frequency.

The resistivity of ITO films is substantially greater than the resistivity of printed cards and metal antennas, where copper or aluminum is used. For example, the sheet resistivity R_{sq1} of the transparent film $CEC005P$ is equal to 4.5 Ohm/sq. The specific conductivities of copper and aluminum are respectively $5.8 \cdot 10^7$ and $3.5 \cdot 10^7$ S/m, and hence in accordance with (12.1) and (12.2) the sheet resistivity of a copper plate with thickness greater than the skin depth at frequencies 1 and 5 GHz is equal to $6.9 \cdot 10^{-3}$ and $18.4 \cdot 10^{-3}$ respectively. Therefore, the resistance R_{sq1} of ITO transparent film is greater by several orders than the resistance of copper and aluminum.

In recent years, transparent films have been the subject of many works [98–100]. However these works were devoted to definition and improvement of characteristics of materials. Physical processes in transparent antennas, their electrical characteristics and their difference from characteristics of metal antennas with a high conductivity as a rule were not considered. Knowledge of law of a current distribution along the antenna axis is of great importance for understanding of physical processes in antennas. This knowledge allows defining the all main characteristics of the antennas. Therefore, the determination of this law is the basic problem of an analysis of any antenna. This postulate holds good in spite of elaboration of calculation programs such as program CST, since firstly these programs in the main allow calculation of input characteristics of antennas (and characteristics dependent on them). Calculation of a current distribution by means of these programs is difficult problem. Secondly such program does not permit to find out a reason of obtained current distribution, that is does not permit to take into account and to use features of antennas. Unfortunately, the character of the current distribution along transparent antennas even has not been considered in the papers published so far.

The flat transparent antenna is the linear antenna with nonzero (impedance) boundary conditions. In the case of the cylindrical metal antenna, if the surface impedance is large enough, it changes the propagation constant and the antenna current distribution already in the first approximation, i.e., significantly alters all electrical characteristics of the antenna [28]. Similarly, in the case of a flat transparent antenna it is necessary first of all take into account the surface impedance. The width of a flat transparent antenna can be taken in account afterwards, since the antenna width has smaller effect.

Let us write an equation for the current in a flat antenna in accordance with the integral equation for the current in a cylindrical antenna. Integral equation (2.53) for current $J(z)$ in a cylindrical impedance antenna is given in Chapter 2. Its solution is sought as a series (2.15) in powers of the small parameter χ. Substituting (2.15) into (2.53) and equating coefficients of the same powers of χ, we come to a set of integral equations and boundary conditions. If $Z/(2\pi a)$ is of the same order as $1/\chi$, so that the surface impedance affects the current distribution along the antenna in the first approximation, the set of equations takes the form of (2.54). In the case of a transparent antenna, the radiator is a thin rectangular plate (not a circular cylinder). The surface impedance is equal to $Z = R_{sq1}$, where R_{sq1} is the sheet resistivity of the transparent film, which for the film CEC005P of thickness 310 nm is equal to 4.5 Ohm/sq.

An impedance transmission line, which is equivalent to symmetrical radiator (dipole) is shown in Figure 5.2. An infinitesimal element dz of the line comprises inductance $d\Lambda = \Lambda_1 dz$ and capacitance $dC = C_1 dz$ (here Λ_1 and C_1 are the inductance and capacitance per unit length), and also the additional resistance $(Zdz/2\pi a)$. The wave propagation constant γ along such antenna is a complex magnitude, which in accordance with (5.17) is given by

$$\gamma^2 = k^2 - j2\omega\varepsilon\chi R_{sq1}/a = k^2 - j\Delta. \tag{12.4}$$

For transition from the cylindrical to a flat antenna it is necessary firstly to determine the parameter χ_1 and secondly to replace in (12.4) πa by the plate width b. As it is shown in Section 2.7, parameter χ is a cofactor in the expression for C_0. It equals to $\chi = C_0/(4\pi\varepsilon)$, where C_0 is capacitance between antenna wire (per unit length of this wire) and the surface of zero potential in the shape of a cylinder with radius $2L$, whose axis coincides with the antenna wire axis. The self-capacitance of a plate with length L and width b per unit length, is equal to (see [34])

$$C_{r1} = 8\varepsilon / \left[sh^{-1}(L/b) + (L/b) sh^{-1}(b/\langle L \rangle) \right],$$

if $1 \leq 2L/b \leq 10$, and to

$$C_{r2} = 2\pi\varepsilon/\ln(2.4L/b),$$

if $2L/b \geq 10$. From (12.4),

$$\gamma^2 = \gamma_0^2 \exp(-j\Delta/k^2). \tag{12.5}$$

Here $\gamma_0 = \sqrt[4]{k^4 + \Delta^2}$, $\Delta = 8\pi^2\varepsilon_0 f \chi_1 R/b$, i.e.

$$\gamma = \gamma_0 \exp(-j\varphi), \tag{12.6}$$

where

$$\gamma_0 = \sqrt[4]{k^4 + \Delta^2}, \varphi = 0.5 \tan^{-1}(\Delta/k^2). \tag{12.7}$$

We write the current distribution in the form of

$$J(z) = J(0) \frac{\sin\gamma_1(L-z)}{\sin\gamma_1 L}.$$

and represent the numerator as an imaginary component of exhibitor

$$\sin\left[\gamma_0 e^{-j\varphi}(L-z)\right] = \text{Im}\exp\left[-j\gamma_0 e^{-j\varphi}(z-L)\right] = \text{Im}\exp\left[-j\gamma_0(z-L)\cos\varphi - \gamma_0(z-L)\sin\varphi\right],$$

i.e.

$$\sin\gamma_1(L-z) = \exp\left[-\gamma_0(z-L)\sin\varphi\right]\sin\left[\gamma_0(L-z)\cos\varphi\right].$$

Similarly,

$$\sin\gamma_1 L = \exp(\gamma_0 L)\sin(\gamma_0 L\cos\varphi),$$

i.e.

$$J(z) = J(0)\exp(-\gamma_0 z\sin\varphi)\frac{\sin\left[\gamma_0(L-z)\cos\varphi\right]}{\sin(\gamma_0 L\cos\varphi)}$$

This expression means that the current along the antenna wire is distributed over sinusoidal law with the propagation constant $\gamma_1 = \gamma_0\cos\varphi$ and the exponential decay of the amplitude with the decrement (the rate of decrease) $\beta = \gamma_0\sin\varphi$:

$$J(z) = J(0)\exp(-\beta z)\frac{\sin\gamma_1(L-z)}{\sin\gamma_1 L}. \qquad (12.8)$$

Accordingly, the reactive component of the input impedance of the antenna, which is made in a transparent film, is equal to

$$X_A = -W_A\cot\gamma_1 L, \qquad (12.9)$$

where

$$W_A = \sqrt{\left(j\omega\Lambda_1 + R_{sq1}\right)/\left(j\omega C_{r1}\right)} = W_0\sqrt{1-2jb\Delta/k^2}. \qquad (12.10)$$

Here W_0 is the wave impedance of a metal antenna with the same dimensions. The radiation resistance is equal to

$$R_\Sigma = 40\gamma_1^2 h_e^2, \qquad (12.11)$$

where $\gamma_1 h_e \approx \tan\left(\dfrac{\gamma_1 L_e}{2}\right)$ is the effective length of antenna. It is easily seen that the

effective length and the radiation resistance of the transparent antenna close to analogous magnitudes of the metal antenna with the same dimensions.

The performed analysis leads to an important conclusion. From (12.10) and (12.11) it follows that the length of the radiating segment of the antenna is inversely proportional to δ and in a first approximation is independent on frequency. This means that increasing the antenna length for operation on lower frequencies is completely useless.

12.2   EXPERIMENTAL RESEARCH

In accordance with the theoretical results we compared two radiators which are shown in Figure 12.2 (dimensions are given in mm). One radiator is made in the metal with a perfect conductivity, and another radiator is made in film *CEC005P*.

Figure 12.3 shows the current distribution $J(z)$ along these radiators calculated using program CST. The current curve for a metal antenna is denoted by number 1, the current curve for a transparent antenna is denoted by number 2. The current decay in the metal antenna with perfect conductivity is absent. The current of the transparent antenna decays rapidly.

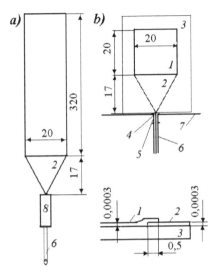

Figure 12.2 Antennas with a perfect conductivity (*a*) and from the film CEC005P with the flat metal triangle (*b*).
1—transparent film, 2—metal triangle, 3—glass substrate, 4—soldered joint, 5—connector, 6—cable, 7—disc, 8—balun.

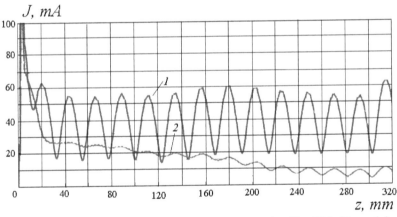

Figure 12.3 Current distributions along a metal (1) and a film ITO (2) model with a length 320 mm for a frequency 5 GHz.

In accordance with the theoretical results the manufactured model of the antenna from film *CEC005P* had a small height. Assuming that the frequency is equal to $f = 5 \cdot 10^9$ GHz, we obtain in accordance with given above formulas: $\chi_1 = 0.36$, $\Delta = 284$ $1/m^2$, $\gamma_1 \approx k$, $\beta = 1.4$ $1/m$, $|W_A| \approx |W_0|$. The current distribution along this model is presented in Figure 12.4 for a frequency 5 GHz. As the calculation showed, the decrement is greater than the calculated value. This circumstance is due to the fact that not only losses in the film caused the decrement, but also the proximity of the substrate and radiation resistivity.

The calculated curves for the active and reactive components of the antenna input impedance are given in Figure 12.5. Experimental values are given accordingly by circles and triangles. The model during measurement was mounted on a metal disk with a diameter of 0.5 m (Figure 12.6).

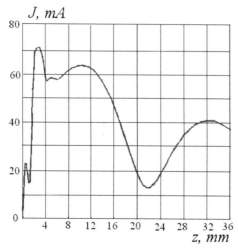

Figure 12.4 Current distributions along the model in Figure 12.2*b* for a frequency 5 GHz.

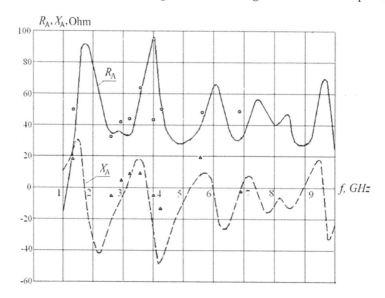

Figure 12.5 Active and reactive components of input impedance of antenna
model shown in Figure 12.2*b*.

The exponential decay of the current along an antenna means that the signal is created by an antenna segment of length $2/\beta$, which is adjacent to the feed point, and the current is virtually absent in the rest part of the antenna. Therefore, the input impedance of such antenna does not have a sharp resonance, and the effective length of the antenna is small. In order to improving matching of the transparent antenna with a cable and to raise the antenna efficiency, a metal triangle with the width equal to the width of the radiator is connected to the radiator's base, as shown in Figure 12.2*b*. This triangle permits to create a uniform current distribution across the whole width of the antenna. That increases the total current of the antenna and its radiated signal. The triangular segment of the described model has been constructed as a printed circuit. But the experiment has shown that it can be made from the same film *CEC005P*.

Experimental characteristics of this model are presented in Figure 12.7 (reflectivity), 12.8 (standing wave ratio). The magnitude of the vertical signal in the plane of antenna (1) and in the perpendicular direction (2) are shown in Figure 12.9. The antenna directional patterns in horizontal (*a*) and vertical plane (*b*) are given in Figure 12.10. The results of measurements show that this model has stable and rather high characteristics in the frequency range 2.5–4.5 GHz and higher.

Figure 12.6   Model of antenna on the disc.

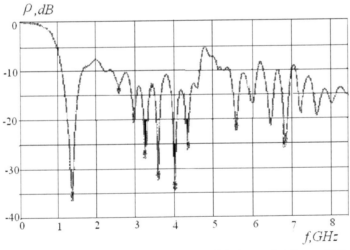

Figure 12.7   Reflectivity of antenna model shown in Figure 12.2*b*.

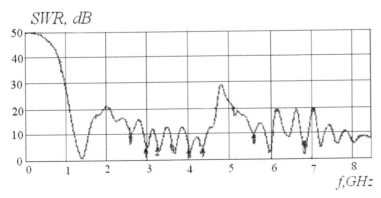

Figure 12.8   Standing wave ratio of antenna model shown in Figure 12.2*b*.

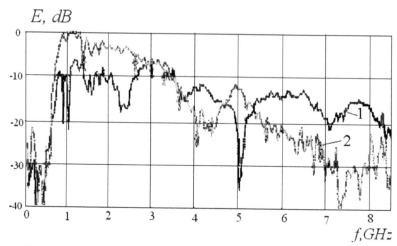

Figure 12.9 Vertical signal of antenna model shown in Figure 12.2b: 1—in the antenna plane, 2—in the perpendicular plane.

Additional experiments were made in order to be convinced that increasing the antenna length for operation on lower frequencies is ineffective. These experiments were carried out on the model of antenna (Figure 12.11a), created by its developers for operation at the frequency of 0.5 GHz. The model consists of a transparent plate 1 and a metal pad 2. The plate is flat rectangular glass substrate, coated with a thin film ITO. Model is placed on a metal disc 3. Compared to the wavelength (0.6 m in free space), the size of the plate is not too small: 0.3 m × 0.2 m. The metal pad is significantly smaller: it is made as a square metal plate with sides, 0.015 m in length. The measurement setup is presented in Figure 12.11b. Such models were used by different performers and gave unsatisfactory results of tests, which seemed inexplicable to their creators.

In the first experiment the fields of three models, shown in Figure 12.12a, were measured by a vertical receiving antenna located at a distance of 3 m. Model 1 is the complete antenna, model 2 consists of a vertical metal pad and a horizontal segment, similar to segment 1 in Figure 12.12a, and model 3 is only the vertical pad. Model 2 has on the upper end of the pad the horizontal load, whose impedance is close to the load of the model 1, but this location leads to a sharp decrease of the vertical component of the signal. The field of model 3 is even smaller. The results of measurements in decibels are given in Table 12.1. They show that the transparent film creates the major part of the radiated signal, since the signal of the pad is significantly weaker. However, the total signal is relatively small for an antenna of such height.

Table 12.1   Measured Fields of Three Models

Model	Structure	Field (dB)
1	Vertical film	−31
2	Horizontal film	−37.5
3	Pad only	−40

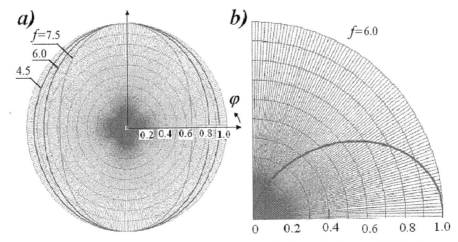

Figure 12.10   The directional patterns of antenna model shown in Figure 12.2b in horizontal (*a*) and vertical (*b*) plane.

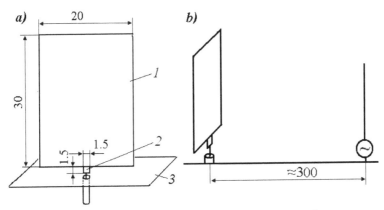

Figure 12.11   The antenna model for the frequency of 0.5 GHz (*a*) and measurements setup (*b*).

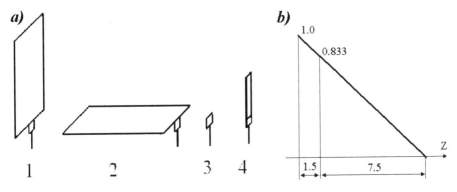

Figure 12.12   Experimental models (*a*) and effective length of the antenna (*b*).

The second experiment allowed determining what fraction of the signal is radiated by the film. Model 4 of Figure 12.12*a* was used for this purpose. In this case, the transparent film is replaced with a vertical copper strip connected in series with the pad. The width of the copper strip is equal to the width of the pad, and the length of the copper strip is chosen so that the signals radiated by models 1 and 4 have equal values. The measurements showed that in this case the total height of model 4 must be 0.09 m. Since model 4 is a uniform monopole, it allows determining the contribution of each part to the overall radiated signal. Assuming that the current along the monopole in the first approximation is linearly distributed (Figure 12.12*b*), one can calculate that the entire effective length of the antenna is 0.045 m and the effective lengths of the pad and the transparent film are 0.0137 and 0.0313 m, respectively. This means that the transparent film radiates 70% of the signal.

The experimental results confirmed the nature of the current distribution along the antenna and demonstrated the usefulness of a smooth transition from the wide transparent film to the central wire of the cable that to improve matching and to create a uniform current distribution across the whole width of the antenna. If these results are not used, then the transparent film radiates weakly in comparison with a conventional antenna of similar dimensions.

12.3 TRANSPARENT ANTENNA WITH METAL TRIANGLE

It is expedient to consider other methods of improving the level of matching. As is shown previously, the input impedance of the transparent antenna does not have a sharp resonance, and the effective length of the antenna is small. Low level of matching of the transparent antenna with a cable is an additional reason for its small efficiency. This disadvantage for example is inherent in the model of the antenna, shown in Figure 12.11*a*. In order to improve matching of the transparent antenna with a cable and to raise the antenna efficiency, a metal triangle was included in the base of the antenna, presented in Figure 12.2*b*.

In order for the current distribution along the rectangular and triangular segments to be uniform and the reflectivity be minimal on the segments boundaries and also at the point of cable connection, the wave impedances of these segments and the cable must be close to each other. The wave impedances of antenna segments will be the same, if these segments will be made from the same film and in the shape of a common triangle or a common flat cone (Figure 12.13*a*).

An expression for the wave impedance of the structure in the form of two back-to back flat cones (symmetric version) with a vertex angle 2α is given in [70]. For an asymmetric version (one flat cone and a plane) the wave impedance is half of this magnitude:

$$W_2 = 60\pi \, K(n)/K\left(\sqrt{1-n^2}\right),\qquad(12.12)$$

where $K(n)$ is the total elliptic integral of the first kind of argument $n = \tan^2(\pi/4 - \alpha/2)$.

The wave impedance of a standard cable is equal to 50 Ohm. It is practically impossible to ensure good matching of the flat cone with such a cable, since a vertex angle 2α of the cone must be equal approximately to 160° in order for the antenna wave impedance to be equal to the cable wave impedance (see Table 12.2). This will cause to a sharp increase of spurious currents between the antenna edges and the ground.

Table 12.2 Wave Impedances of a Metal Triangle

$\alpha°$	80	70	60	50	45	40	30	20	10
$K(k)/k(\sqrt{1-k^2})$	0.251	0.315	0.391	0.462	0.50	0.542	0.639	0.773	0.996
W_2, Ohm	47.3	59	73.7	87.1	94.2	102.1	120.5	145.8	187.8

As it is shown in Chapter 9, the wave impedance of a self-complementary antenna depends on the number of metal radiators therein and from the circuit of connection of these radiators to poles of the generator. If an antenna consists of two metal dipoles, the plates of which are fabricated in the form of metal radiators with an angular width 45° (Fig. 12.13b) and connected in pairs, its wave impedance is equal to 15π (47 Ohm). By adjusting the angular width of each radiator one can provide exact equality of wave impedances of the antenna and the cable. In Figure 12.14 the reflectivity of three antennas are compared with each other: the reflectivity of antenna with triangular segment (curve 1, this reflectivity is presented in Figure 12.7), the reflectivity of antenna with one transparent cone (curve 2, the length L of the arm is equal to 0.045 m), and the reflectivity of antenna with two transparent cones (curve 3, an arm length also is equal to 0.045 m). As it is seen from Figure 12.14, the antenna with two metal radiators provides a smooth change of the reflectivity in a wide frequency range.

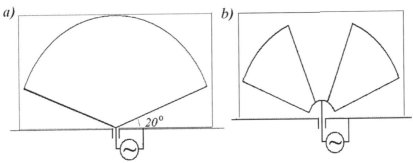

Figure 12.13 Asymmetrical self-complementary antennas of one (*a*) and two (*b*) metal flat cones.

This result shows the significant advantage of the antenna embodiment in the shape of self-complementary radiator with one or two flat cones.

The analytical and experimental study of a transparent flat antenna, which is made of ITO film, placed on high-quality glass substrate, shows that its characteristics differ significantly from the characteristics of conventional metal antennas—both thin and wide. This difference is due to the fact that low conductivity of the radiator leads to an exponential decay of current along the antenna axis and to a substantial shortening of the length of the radiating segment in comparison with the antenna length. As a result, the antenna effectiveness decreases drastically. If to use the existing films, it is impossible to create an effective antenna for operation at frequencies below 1 MHz. At frequencies above 1 MHz a rather efficient antenna can be created, if it is performed or as self-complementary antenna in the shape of one-two flat cones or in the shape of wide plate with triangular transition to the central wire of the cable.

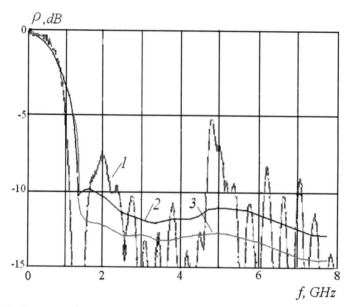

Figure 12.14 Reflectivity of antenna model with triangular segment (1) and with one (2) and two (3) metal cones.

Ship Antennas

13.1 WIRES ANTENNAS

This section is dedicated to separate issues of designing ship antennas. They are representatives of numerous groups of antennas deployed on mobile objects and as such have their own specific features, since they are placed in constrained conditions in close proximity to metal bodies of different shapes and dimensions.

An antenna of medium frequencies (of hectometer waves) is known as the main ship antenna. It must ensure the tuning of the main and emergency transmitters, and its efficiency during operation with the main transmitter must be sufficient in order to establish the electric field with strength 50 μV/m at a distance 150 miles. With allowance for details of medium-frequency waves' propagation, the antenna must create vertically polarized electromagnetic waves with the direction of maximum radiation along the ground surface. The directional pattern in the horizontal plane must be close to the circular one.

For this reason, the hectometer antenna is an asymmetrical vertical radiator. As a rule, its height is small in comparison with the wave length, i.e. the radiation resistance is low, and that leads to low efficiency. Accordingly, the underlying problem in the development of new antenna is the increase of its effective length. Therefore, much attention is paid to use of antennas with capacitance loads at the upper end, which permits to improve (to make more uniform) a current distribution along the antenna, in order to increase their effective length and radiation resistance.

Such antenna is excited as a rule in the base. It is called inverted-L antenna. Its circuit corresponds to Figure 13.1a, and the current distribution is given in Figure 13.1b. A vertical wire can be connected to the end or to the middle of the horizontal load.

As already mentioned, inverted-L antenna consists of a vertical segment and horizontal load (Figure 13.2). The vertical segment is performed in the form of a single wire or a fan of wires (i.e., several wires located in one plane and convergent to a

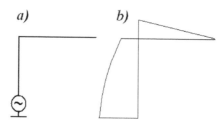

Figure 13.1 Inverted-L antenna (*a*) and a current distribution along it (*b*).

feed point) or a cylinder of wires (i.e., several wires located along the generatrices of a cylinder with cross-section in the shape of a circle with a radius 0.5–0.7 m). The horizontal load is stretched between ship masts and consists of one or more wires, which are located in a horizontal plane at a distance of 0.7–1.0 m from each other or along the generatrices of a round cylinder. The flexible antenna filaments—bronze basket (of type PAB) or copper basket (of type PAMG) are used as wires. The antenna design enables its rapid ascent and descent, as well as adjustment of a wires tension.

In principle one can use antennas with upper feed (Figure 13.3), when the transmitter connects to a wire located inside the mast, the upper end of which is connected with the horizontal load (see Figure 13.3*a*). The circuit is equivalent to placement of the exciting emf at the top of the antenna (at the vertex)—between the vertical radiator (the outer mast surface) and the load (a horizontal sheet or circular cylinder)—see Figure 13.3*b*. If the wire cannot be laid inside the mast, the variant with a shielded wire is feasible (see Figure 13.3*c*). The current distribution along a top-fed antenna is given in Figure 13.3*d*.

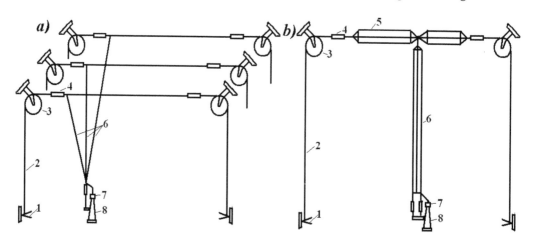

Figure 13.2 Ship wires antennas: inverted-L (*a*), T-antenna (*b*).
1—duck, 2—halyard, 3—block, 4—insulator chain, 5—horizontal load, 6—down-lead,
7—down-lead insulator, 8—antenna column.

In this case, the current antinode is in the radiator base. Since the current varies weakly near the antinode, then the effective length in the first approximation is equal to the geometric length of antenna, i.e. it is larger than in the case of antenna with feed in the base. But the radiation resistance decreases inside an antenna in the direction to base (to the point of connecting the exciting emf). In order to weaken this effect, it is necessary to increase the ratio of the internal mast diameter and the wire diameter.

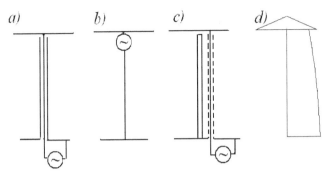

Figure 13.3 Antenna with the upper feeding: vertical wire inside a mast (*a*),
equivalent radiator (*b*), wire inside a shield (*c*), current distribution (*d*).

In the case of an antenna with feed in the base, a mast acts as a support and creates
an additional (parasitic) capacitance between the antenna and the ground, which causes
decrease of the radiation resistance. In the general case for analysis of the mast effect
one can rely on the program CST. In the particular case, when the antenna represents
a vertical wire without load, which is located in parallel to the mast and has the same
length (Figure 13.4*a*), one can use an explicit technique based on the theory of the
folded radiator. As seen from the figure, the radiating wire and the mast form the
folded radiator with shorting to ground, which is open at the upper end and consists
of wires with different diameters. By analogy with Section 3.2, one can connect in each
wire two voltage generators, whose total emf is equal to emf of the antenna exciter and
zero, and divide the radiator into two auxiliary circuits: a linear radiator of height L
with an equivalent radius and an open at the end long line with wave impedance W_l
(Figure 13.4*b*). The input admittance of the antenna near the mast is

$$Y_A = 1/Z_l + p^2/Z_e(a_e).\qquad(13.1)$$

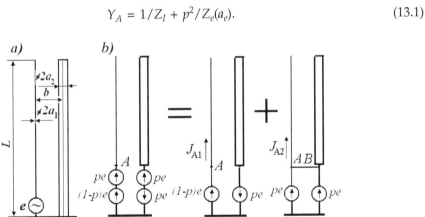

Figure 13.4 The relative disposition of a radiator and a mast (*a*) and dividing into two circuits (*b*).

Here, $Z_l = -jW_l \cot kL$ is the input impedance of the two-wire long line, Z_e is the
input impedance of the linear radiator, $p = C_{11}/(C_{11} + C_{22})$ is the fraction of in-phase
current in the excited wire, C_{11} and C_{22} are the self-capacitances of excited and passive
wires. Expression (13.1) is true, if distance b between the antenna and the mast is small
against the antenna length and the wavelength. If these values are commensurable,
it is expedient to use the induced emf method, in accordance with which the input

impedance of the excited radiator near the passive radiator is

$$Z_A = Z_{11} - Z_{12}^2/Z_{22}, \tag{13.2}$$

where Z_{11} and Z_{22} are the self-impedances of both radiators, and Z_{12} is their mutual impedance.

For the ship antennas as a rule inequalities are true:

$$\alpha = kL \ll 1, \quad \delta = kb \ll 1.$$

So in calculating active and reactive components of self- and mutual impedances one can be confined by first terms of expansion of integral function into a series. For example the components of a mutual impedance of monopoles are equal to

$$R_{12} = 10\alpha^2, \quad X_{12} = \frac{30}{\alpha^2}\left[4\sqrt{\delta^2+\alpha^2}-\sqrt{\delta^2+4^2}-3\delta+2\alpha\ln\frac{\delta\left(\sqrt{\delta^2+4\alpha^2}+2\alpha\right)}{\left(\sqrt{\delta^2+\alpha^2}+\alpha\right)^2}\right].$$

Figure 13.5 gives an example of radiation resistances R_Σ of wire antennas of the length 6.52 and 13 m (the wire radius is $a_1 = 3.7 \cdot 10^{-3}$ m) at the frequency 460 kHz as function of the distance b between the antenna and the mast for different radii a_2 of the mast. For comparison the magnitude R_{11} of the radiation resistance for the solitary antenna is plotted. Experimental values for the mast of radius $a_2 = 0.4$ m are shown with dots. The coincidence of experimental results with the calculations confirms the rightness of the obtained results.

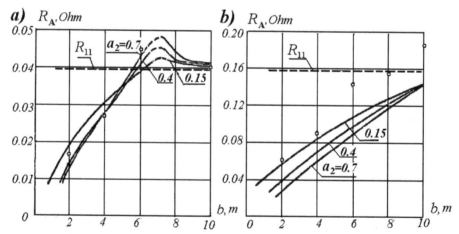

Figure 13.5 Radiation resistance of wire antenna with height 6.52 m (*a*) and 13 m (*b*) against the distance to the mast.

It is seen from the Figure that the active component R_A of the antenna input impedance drops sharply as the distance between the antenna and the mast decreases. Horizontal loads of the wires weaken the influence of the mast. But in this case also it is necessary to move the antenna away from the mast as far as possible (from 4 to 8 m, depending from the mast height).

Low radiation resistance and lower efficiency are not the sole drawbacks of wires antennas. One must add to them such drawbacks as a wide variation range of the input impedance, which hampers standardization of antennas types and complicates the onboard equipment. Besides, an antenna curtain (horizontal load) can break down as a result of an ice formation or a storm. The antenna can hinder cargo handling. The antenna can require mounting a second mast, which is not necessary for contemporary ship. For this reason antenna-masts have found their use as the main ships antennas. At first, three variants of such antennas appeared: 1) with guy ropes and guy wires, 2) free standing and 3) mounted on the mast. The first variant was throwed soon because of the great area occupied by the antenna. So in the first stage self-supporting (free standing) antenna-masts were manufactured only for great ships. Further, in order to reduce the cost of the antenna and use it on ships of small and medium tonnage, the antenna-masts with inductive-capacitive load was designed. It is mounted on the ship mast.

13.2 ANTENNA-MAST WITH INDUCTIVE-CAPACITIVE LOAD

The circuit of this antenna is presented in Figure 13.6a. The circuit corresponds to the variant with open vertical wire and can be used on board of an exploited ship. The antenna is excited in the base and differs from inverted-L antenna only by the type of load. Variants with top excitation are possible also. The load is created in the shape of a vertical structure, which is the mast extension. The mast supports the load, so for a given mast height the geometric height of the antenna increases. It means that the effective height of the antenna increases also and other characteristics are improved.

The antenna load (see Figure 13.6b) consists of four whip antennas connected in the base by a conducting ring. The system of four whip antennas is equivalent to a thick metal radiator with low wave impedance and high capacitance. Double-turn volumetric spiral is connected in series with the system of the whip antennas. The spiral increases the antenna electrical length. The system of whip antennas creates the capacitive component of the load, and the double-turn volumetric spiral creates the inductive component of it. Both elements decrease the input reactance of the antenna and increase its effective length. Use of tilted whip antennas allows decreasing, if necessary, the total height of the structure. The access of man to the elements of load is foreseen at the time of parking ship in port and of the calm weather. The lightning arrestor (spark gap) is installed at the point of antenna wire leading to the radio deck house.

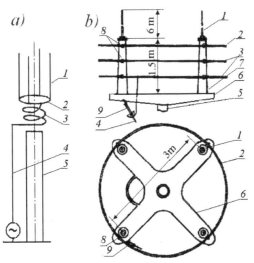

Figure 13.6 Antenna with inductive-capacitive load: circuit (a), device of the load (b). 1—tilted whip antenna, 2—conducting ring, 3—double-turn volumetric spiral, 4—open vertical wire, 5—mast, 6—work platform, 7—dielectric column, 8—base insulator, 9—rod insulator.

The antenna-mast with inductive-capacitive load was proposed in 1966 and was improved in 1970 [101]. The specimen of antenna was mounted on the board of cargo ship *Konstantin Shestakov* with displacement 3500 ton (Figure 13.7). The antenna is placed on the ship's upper bridge, on the mast of height 9.5 m and of diameter 0.3 m. The static capacitance of an antenna is 442 pF, the natural wave length is 240 m, and the resistance at frequency 400 kHz is 4.3 Ohm.

Figure 13.7 Antenna-mast with inductive-capacitive load on the upper bridge of cargo ship *Konstantin Shestakov*.

In later years, the antennas, in accordance with a similar circuit, were constructed in other countries (Figure 13.8). They include Norwegian antenna AS9 (in the version with possible inclination of the antennas, AS9ST) and antenna 938G-1 of the firm Collins, USA. The capacitive load in them, as in the described antenna, is made in the form of whip antennas installed on the top of the mast. The inductive load is made by means of a coil or spiral connected in series with a system of whip antennas and vertical wire.

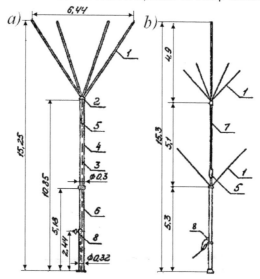

Figure 13.8 Antenna-masts with inductive-capacitance load: 938G-1 (*a*), AS9 (*b*).
1—whip radiator of fiberglass with wicker copper mesh, 2—bronze cap, 3—central copper wire, 4—fiberglass mast, 5—inductor, 6—aluminum tube, 7—fiberglass tube, 8—antenna lead-in.

They are built as free-standing structures of fiberglass and have the total height about 15.3 m. Combining their antennas with a conventional ship mast is not provided.

Consider the methodology of calculating the electrical characteristics of the antenna mast on the example of the antenna with inductive-capacitive load. The equivalent circuit of the variant with excitation in the base is shown in Figure 13.9. The antenna consists of three segments: (1) the systems of whip antennas of a height L_1 with the wave impedance W_1, (2) the volumetric spiral of a height s with the inductance Λ and (3) the vertical wire of a length h with the wave impedance W_2. The wave impedances of each section can be determined by the method of Howe.

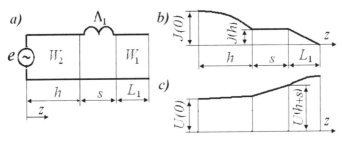

Figure 13.9 Equivalent circuit of the antenna with excitation in the base (*a*), current (*b*) and voltage (*c*) distribution along it.

The input impedance of an equivalent transmission line is

$$Z_l = -jW_2 \cot k(h + l_e), \tag{13.3}$$

where

$$l_e = \frac{1}{k} \tan^{-1}\left(\frac{W_2}{W_1 \cot k_0 L_1 - \omega\Lambda}\right).$$

Expression (13.3) allows approximately determine the reactive component of the antenna input impedance, as well as its natural wavelength $\lambda_0 = 2\pi/k_0$, where k_0 is the solution of the transcendental equation

$$\tan k_0 h = \frac{1}{W_2}(W_1 \cot k_0 k_1 - k_0 c\Lambda)$$

Here c is velocity of light. If the electrical length of the mast and the whip antennas is low ($k_0 h$, $k_0 L_1 \ll 1$),

$$\lambda_0 = 2\pi\sqrt{\frac{L_1}{W_1}(hW_2 + c\Lambda)}, \tag{13.4}$$

The current distribution along each section of antenna shown in Figure 13.9 is described by the expressions

$$J(z) = \begin{cases} j(h) \sin k(L-z)/\sin kL_1, & h+s \le z \le L, \\ j(h), & h \le z \le h+s, \\ j(0) \sin k(h+l_e-z)/\sin k(h+l_e), & 0 \le z \le h. \end{cases} \tag{13.5}$$

Here $J(h)$ is the current of spiral, $L = L_1 + s + h$ is the total height of the radiator, $J(0)$ is the current in antenna base:

$$J(h) = J(0) \sin kl_e/\sin k(h + l_e).$$

Electrical field strength E_z is calculated in accordance with (1.68) and input impedance—in accordance with (1.50). An effective height of antenna is

$$h_e = \frac{J(h)}{J(0)\sin kL_1}\int_0^{L_1}\sin k(L_1-z)dz + s\frac{J(h)}{J(0)} + \frac{1}{\sin k(h+l_e)}\int_0^h \sin k(h+l_e-z)dz,$$

or taking into account that $kh \ll 1$

$$h_e = \frac{\sin kl_e}{\sin k(h+l_e)}\left(\frac{L_1}{2}+s\right) + \frac{1}{k\sin k(h+l_e)}[\cos kl_e - \cos k(h+l_e)]. \qquad (13.6)$$

As is seen from Figure 13.9, the current is maximal in the antenna base

$$J(0) = \sqrt{P_A/R_A}, \qquad (13.7)$$

where P_A is the power delivered to the antenna. The voltage is maximal between the base of a whip antenna and a grounded mast

$$u(h+s) = u(0)\frac{\cos kl_e}{\cos k(h+l_e)} + J(h)\omega\Lambda, \qquad (13.8)$$

where $u(0) = J(0)|Z_A|$ is the voltage in the antenna base.

The methodology of calculating characteristics of a top-fed antenna is given in [102].

13.3 INFLUENCE OF METAL SUPERSTRUCTURES ON THE ANTENNA PERFORMANCE

An important feature of any mobile object, in particular, a ship, is a confined area for antenna placement, in this connection the antennas are installed near diverse metal structures of complex shape, such as masts, superstructures, pipes, etc. Analysis of their impact on the antenna characteristics is difficult even in the simplest cases. The superstructure can be considered as an additional passive radiator. Solving the set of Kirchhoff equations for the totality of radiators, we can find the current in each one. The difficulty lies in large transverse dimensions of the superstructure, i.e. the calculation of its self-impedance and mutual impedances between it and other radiators, which is based on the thin antenna theory, yields too rough approximation. The analysis method based on replacing the metal body with a system of thin wires [33, 83] is more efficient.

We shall begin with a single superstructure of a regular shape. Figure 13.10 gives the circuit of the wire structure, equivalent to a thick superstructure shaped as a round cylinder of finite length, next to which a whip antenna is placed. It is assumed that the ground surface is perfectly conducting, and the structure is symmetrical with respect to this surface. The round cylinder is replaced with the wire structure of eight equidistant wires located along the cylinder generatrices and the radii of its covers. Since the antenna excites mainly the longitudinal current component in the superstructure, then, for simplicity of calculation the horizontal circular wires are disregarded. Diameters of wires and the whip antenna are assumed the same. The coordinate origin coincides with the superstructure center. The dimensions of the figure are given in meters.

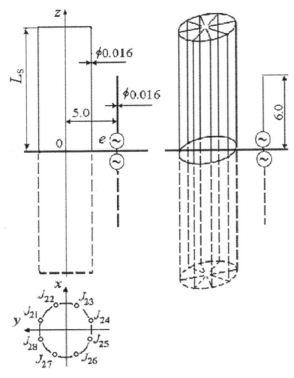

Figure 13.10 Whip antenna near a cylindrical superstructure.

Table 13.1 presents input impedance and maximal directivity (with respect to maximal directivity of the quarter-wave monopole) for a whip antenna of height 6 m near a superstructure of diameter 5 m and of heights 6 and 20 m on the three high frequencies. These characteristics are also given for the case when the superstructure is absent ($L_S = 0$). Calculating the amplitudes and phases of the currents in a base of each wire of the superstructure (at the ground surface) shows that their values are symmetrical with respect to axis y, i.e. $J_{21} = J_{28}$, etc. At the same time currents in the wires on the side of the whip and on the opposite side differ substantially both in amplitude and in phase.

Figures 13.11 and 13.12 show calculated directional patterns in horizontal plane and vertical plane xOz respectively on the three frequencies of HF region. Together with the calculated curves the experimental values are presented. The coincidence of the calculated and experimental data is good.

Table 13.1 Characteristics of a Whip Antenna Near a Superstructure

f, MHz	Z_A, Ohm			D_{max}		
	$L_S = 0$	6	20	0	6	20
6.0	$6.0 - j353$	$3.7 - j351$	$1.8 - j350$	1.86	2.46	4.32
12.5	$39.7 + j20.9$	$24.7 + j38$	$17.0 + j28.2$	2.00	5.39	5.32
19.0	$289 + j424$	$356 + j505$	$185 + i448$	2.35	4.76	6.80

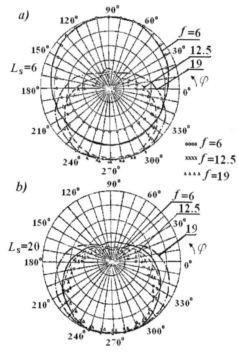

Figure 13.11 Horizontal directional patterns of the whip near the metal
cylinder of height 6 (*a*) and 20 m (*b*).

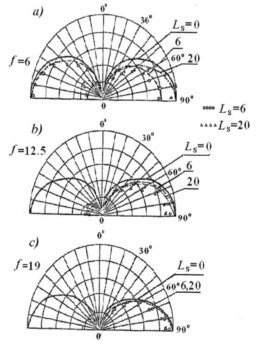

Figure 13.12 Vertical directional patterns of the whip antenna near the metal cylinder
on the frequency 6 (*a*), 12.5 (*b*) and 19 MHz (*c*).

The calculations show that the antenna characteristics to a high degree depend on the superstructure presence and on its height. The radiation resistance decreases (excepting the region of parallel resonance). The radiation in the direction to the superstructure decreases sharply (excepting lower frequencies and the structure, which is shorter than the whip antenna). Superstructure height significantly affects the properties of the radiator in the entire frequency range.

The good agreement of calculation with experiment confirms indirectly the rightness of the wire structure choice. As is seen from Figure 13.10, in this structure the circular wires are absent, i.e. calculation does not take into account cross currents induced on the cylinder surface. Calculating electrical characteristics of the whip antenna located near the superstructure with height 6 m, whose equivalent structure was supplemented by horizontal wires, was performed in order to verify the rightness of model. The mentioned wires were located in parallel planes at a distance 2 m from each other and each of them had the shape of a regular polygon inscribed in a circumference with a radius equal to the radius of the cylinder.

The calculation results show that the directional pattern of the antenna located near the structure with additional horizontal wires, is almost identical the directional pattern without these wires: directivity differs by 1–2%, the input impedance varies in the range of 5–10%. The currents in the horizontal wires are substantially smaller than currents in the vertical conductors, for example, at a frequency of 6 MHz they smaller by a factor of 10^5.

The calculation allows to find the minimal height of a superstructure, at which the antenna characteristics coincide with the characteristics of an antenna, located near an infinitely high superstructure. This height exceeds the antenna height approximately by a quarter of the wavelength. In actual practice, the superstructure shape is different from the cylindrical. Comparison of influence of superstructures differently shaped, e.g., with circular or polygonal cross-section, but of close dimensions, which are located on equal distance from the antenna axis, shows that their influence on the antennas properties remains basically the same.

The variety of variants of antennas' placement on ships requires typifying these variants in order to allow forecasting their characteristics, taking into account the effect of closely spaced metal structures. It is expedient to list these structures that distort the characteristics of adjacent antennas. This is primarily antennas with large transverse dimensions, such as antenna-masts and radar antennas. Usually, the problem is reduced to determining the effectiveness of antennas placed close to the two superstructures, such as a chimney and a ship mast or a ship mast and the free-standing antenna-mast. An additional superstructure causes, as the analysis shows, a further decrease of a radiation resistance and an increase of directivity.

The described technique, which allows estimating an effect of the metal superstructure situated close to the antenna on characteristics of this antenna, can also be used for the analysis of such effect on the characteristics of the linear array of whip antennas (see Section 5.6).

It should be emphasized that the wire structure, which is used in the calculations as the electrodynamics equivalent of a metal object, should correspond to the physical meaning of the problem. For example, the wires of a structure must be located along the supposed lines of a current. This allows upon the given accuracy of calculation to reduce significantly the number of wires and, correspondingly, the amount of computation (and vice verse, to increase the accuracy of results upon the same volume of calculations).

Defining the necessary number of wires, which provides an equivalence of electrodynamics properties of the model and the original, is an important question upon using a considered procedure. Each wire is divided on the segments (short dipoles). The lengths of these segments must not exceed 0.2λ. A further decrease in the segments' length and increase in their number has almost no effect on the accuracy of calculations. By decreasing the segments' length up to 0.01λ the calculation accuracy decreases again.

In this connection, the characteristics of the whip antenna located near the round metal cylinder were considered. The round metal cylinder was replaced by the wire structure of different numbers of wires. The results of calculating directional patterns and experimental verification showed that in this concrete problem the number of wires must be chosen so that the distance between the wires was less than 0.08λ. If the number of wires are more than the specified number, the shape of the directional pattern does not change practically, and the input impedance changes slightly—in the range of 5–10%.

Various authors called different values for the minimal number of wires of the structure, which provides an equivalence of electrodynamics properties of the model and the original object [83]. The number of segments, and hence the number of basis functions is accordingly changed. The distance between the wires, which was selected in the above described problem, is 0.08λ.

Similar results were obtained during calculating the characteristics of antennas installed at the edge of a ship's deck or on a sail yard, and also for symmetric dipole on the axis of a trough with finite length. The latter problem occurs if the antenna is located near a metal body and mounted flush with the body, i.e. does not rise above its surface. Another variant of the same problem occurs if the radiator is placed in the dielectric capsule floating along the sea surface.

Sometimes the authors' conclusions are surprising. For example, in [103] the results of measuring field of a quarter-wave monopole mounted on square screens of the same size are presented. One screen is fabricated in the form of metal sheet, and others—in the form of wire grid with square meshes, dimensions of which in the different screens are different. The given results show that the dimensions of the meshes should not exceed 0.06–0.08λ. But the authors argue contrary to the presented graphs that the size of the mesh can be increased up to 0.1λ.

13.4   ANTENNA FOR COAST RADIO CENTER

An antennas for coast radio centers provides a communication with the ships in the ranges of high and medium frequencies. A project of antenna was developed on the basis of the theory of self-complementary antennas in two versions—planar and volumetric.

Both variants were intended for creating an antenna of height 50 m with distance 100 m between supports. In the high-frequency region the antenna characteristics are similar to the characteristics of a self-complementary structure. In the medium-frequency region this antenna is a variant of the folded radiator. The first embodiment (see Figure 13.13a) provides bi-directional radiation and the second (see Figure 13.13b)—unidirectional radiation with increased directivity.

Figure 13.13a shows a wire structure, used in the calculation of a flat vertical antenna. Dimensions are given in meters. It is considered that a ground is perfectly conducting and a structure is symmetric with respect to its surface. The central vertical wire is located along an axis of the antenna symmetry. The antenna consists of two sectors. Each wire is regarded as an insulated conductor (circuit), which adjoins in the end point to the central wire or to the previous side wire.

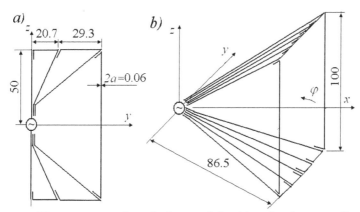

Figure 13.13 Wire structures for calculation of flat (*a*) and volumetric (*b*) antennas.

Input characteristics of the flat vertical antenna are presented in Figure 13.14*a*, the directional pattern in the horizontal plane in Figure 13.14*b*, in the vertical plane—in Figure 14.14*c*. The TWR was calculated for a cable with wave impedance 75 Ohm. Together with the calculated curves in the figure the experimental values are shown by points (in the form of squares, circles and triangles). The experimental values are obtained for a model executed on the scale of 1:50. The coincidence of the calculated and experimental data in the first part of the range is quite good, in the second—rather qualitative.

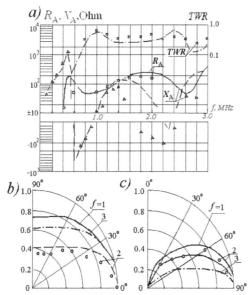

Figure 13.14 The input impedance (*a*), the directional patterns in horizontal (*b*) and vertical (*c*) plane of the flat vertical antenna.

The results of calculation and experiment show that increasing the number of wires in the antenna allows to reduce the reactive component of the input impedance and to raise the level of TWR. It is expedient to connect the wires of triangular radiator with each other by horizontal connecting wires. Increasing diameters of side shunts (supports) also helps to improve matching.

In Figure 13.15 an experimental TWR is given for an antenna with a triangular radiator of 9 wires. Curve 1 corresponds to the antenna with one additional horizontal connecting wire, curve 2—with four connecting wires. Curve 3 is given for the antenna with four connecting wires and with a diameter of each support increased to 0.7 m. The last option allows to obtain TWR more 0.4 at frequencies from 0.9 to 5.3 MHz, i.e., bandwidth ratio is equal to 5.9. The project was recommended to realization.

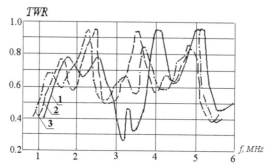

Figure 13.15   Experimental TWR of different antenna variants.

The directional pattern in the horizontal plane has a shape of an oval elongated in a direction perpendicular to the antenna plane (see Figure 13.14*b*). The width of the main lobe on the level of 0.7 at frequencies up to 3 MHz is more 80°. In the vertical plane the directional pattern is flattened against the ground and has a width from 20 to 40° (see Figure 13.14*c*).

Using a volumetric antenna with an inclined triangular radiator (see Figure 8.12*b*) allows to create unidirectional radiation and also to expand the range to the side of lower frequencies. The wire structure used for the calculation of this antenna is shown in Figure 13.13*b*, and electrical characteristics—in Figure 13.16. From the drawings it is seen

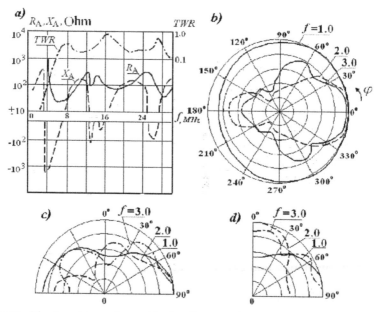

Figure 13.16   The input impedance (*a*), the directional patterns of the volumetric vertical antenna in horizontal (*b*) and vertical planes *xOz* (*c*) and *yOz* (*d*).

that increasing the dimensions of a triangular radiator shifts to the left of the resonances. Radiation is increased in the direction of the triangle's inclination. Directional patterns in mutually perpendicular vertical planes are substantially different from each other.

Table 13.2 shows the directivity of a flat and volumetric antenna (relative to isotropic radiator).

Table 13.2 Maximal Directivities of Antennas

Type of antenna	f = MHz	3 MHz
Flat	5.9	6.4
Volumetric	7.1	6.9

13.5 INFLUENCE OF CABLES ON A RECIPROCAL COUPLING BETWEEN COAXIALLY DISPOSED RADIATORS

Section 4.4 was entirely devoted to the creation of antennas, which provides radiation in a plane perpendicular to an antenna axis in wider frequency range. In other words, serious attention was paid to the question of creating the vertical antenna with required directional pattern in the vertical plane. An issue of obtaining the required directional pattern in the horizontal plane is very often not less important. A single symmetric about axis vertical antenna has in the horizontal plane a circular directional pattern. But if two antennas are located nearby, then their mutual effect leads to a deterioration of their electrical characteristics, in particular to the distortion of their horizontal directional pattern.

So at high frequencies, for example, in VHF-UHF ranges, coaxial installation of antennas is used widely in order to decrease the mutual influence of vertical antennas located close to each other. This radiators arrangement greatly reduces their mutual influence. In addition, mast or other construction, on which antennas are mounted, is expensive structure and should be used as much as possible. If the mast is made of dielectric material (plastic) it does not affect the radiators properties. But the mast is not the only cause of distorting characteristics of antennas. It should also reduce the influence of the cables of antennas located above, as well as other metallic elements. A similar problem occurs when one must build a phased antenna array, if the flat vertical metal reflector, behind of which it is possible to hide the cables, is absent in this array.

Considering the different ways of counteracting this unpleasant effect, it should be clearly understood that the placement of cables around the radiator is better than placing radiators around the cables, and that only one radiator should be located in each storey. This method of solving the problem allows to approximate the characteristics of an antenna mounted on a mast to the characteristics of an antenna in free space, i.e. allows to reduce to the utmost the influence of cables on its properties.

A typical placement of vertical radiators on the mast is presented in Figure 13.17. Different options of placement of cables and other metal elements around radiators are considered in [104]. It assumed that dimensions of cable cross section do not exceed 0.1 λ. Results of analysis are given further for the following variants:

(1) metal elements in the form of several metal rods evenly located around the radiator (Figure 13.18),
(2) metal elements in the shape of a meander, a horizontal segment of which lies in a plane passing through the radiator center; the midpoint of a segment coincides with this center (Figure 13.19),

(3)   metal elements in the form of several metal rods evenly located around each
      radiator; the upper and lower segments of the rods are displaced relative to one
      another by a half of interval between the rods (Figure 13.20),
(4)   metal element in the shape of a vertical cylindrical spiral (Figure 13.21).

     It is well known that the directional pattern of a vertical radiator 1, placed next to
the mast 2 (see Figure 13.17a) is distorted the stronger, the more cross section of the
mast. Figure 13.17b shows the experimental directional patterns in the H-plane for the
radiator located at a distance $R = 0.19\lambda_{max}$ from a vertical metal rod of height λ_{max} with
a cross-section in the shape of a corner with a side $0.025\lambda_{max}$. As can be seen from the
figure, the directional patterns have areas where the signal level is sharply reduced (up
to 0.3 of a maximum). It is unacceptable for antennas of mobile objects.

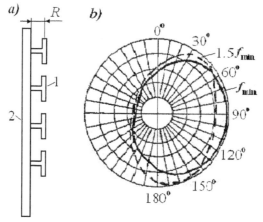

Figure 13.17   Vertical radiators next to the mast (a) and their directional
patterns in horizontal plane (b).

     Directional pattern can be improved if radiator 1 is placed between two metal rods
2 (see Figure 13.18a). Measurements show that the directional patterns of the radiator,
located between the two rods of mentioned type at a distance $R = 0.19\lambda_{max}$ from each
rod, remain uniform ($E \geq 0.7E_{max}$) in the range from f_{min} to $1.5\,f_{min}$ (see Figure 13.18b).

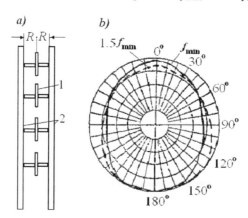

Figure 13.18   Vertical radiators between two metal rods (a) and their directional
patterns in horizontal plane (b).

Let the metal element have the shape of a meander, and the midpoint of its horizontal segment coincides with the center of the radiator (see Figure 13.19a). It allows to improve the directional pattern in the H-plane in comparison with the pattern of radiator located next to the mast, and at the same time permits to reduce the impact of the metal element on the input impedance in comparison with the impact on the input impedance of the radiator placed between the rods. As can be seen from Figure 13.19b, in using this metal element, fabricated of rods of mentioned type, with a length of horizontal segment $2R = 0.38\lambda$ the directional pattern remains uniform in the frequency range from f_{min} to $1.5\,f_{min}$.

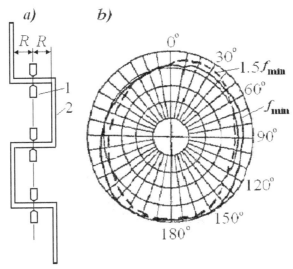

Figure 13.19 Metal element in the shape of a meander (a) and directional patterns of the radiator in horizontal plane (b).

The variant shown in Figure 13.20a is a modification of the variant shown in Figure 13.18a. It allows to increase the number of metal elements. As the measurements show, the higher a number M of metal rods, placed around the radiator, the wider the frequency range, in which the directional pattern is close to the circular. On the other hand increasing M and decreasing the radius R of free space around the radiator causes a growth of the input impedances of the radiator and makes antenna matching with a cable more difficult. Figure 13.20b contains experimental data about the growth of the active component of the input impedance as a function of R and M in the range from f_{min} to $4f_{min}$. As the curves show, in order to avoid large transverse dimensions of the structure and great growth of input impedances (and therefore the high SWR), it is needed that the radius of the circumference be equal to $R = 0.33\sqrt{\lambda_{max}\lambda_{min}}$ and the number of M not less than three.

In order to expand the frequency range of the radiator surrounded by vertical rods, the rods must be divided by a plane passing through the middle of each radiator (see Figure 13.20a) onto the upper 2 and lower 3 segments. Further they should be shifted relative to each other by half of interval between the rods (by an angle $\alpha = \psi/2 = \pi/M$). It is equivalent to doubling the number of vertical rods in accordance with their effect on the directional pattern ($M_e = 2M$) and does not increase SWR. Connecting links 4 connect the rods with each other and with the center of the radiator.

Experiments show that this form of metal elements, fabricated of metal rods with a cross-section in the shape of a corner with a side $0.01\lambda_{max}$, allows obtaining the uniform directional pattern in the H-plane ($E \geq 0.7E_{max}$). These elements almost do not effect on the directional pattern in E-plane. They as a rule increase SWR in comparison with the radiator located in the free space not more than 12% in a wide frequency range from f_{min} to $4f_{min}$, if $M_e = 2M = 6$ and $R = 0.33(\lambda_{max}\lambda_{min})^{1/2}$.

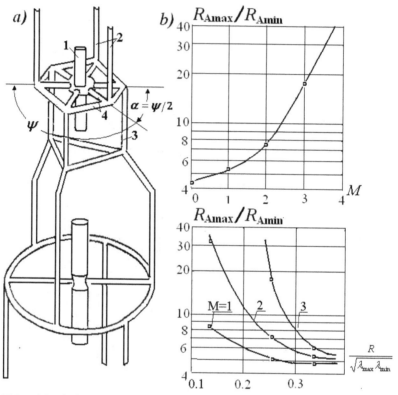

Figure 13.20 Metal elements in the form of several metal rods (*a*) and the radiator resistance as a function of R and M (*b*).

Metal element in the shape of a vertical cylindrical spiral (see Figure 13.21*a*) deserves special attention, since the spiral has a minimal effect on the electrical characteristics of vertical radiators. It is expedient to choose close values of a spiral pitch S and a spiral radius R, since with a growth of S/R the spiral begins to damage radiator characteristics, and with decreasing S/R a length of a cable, placed into a spiral tube, becomes excessively large, and this means large losses in the cable and a great weight of the device.

Measured characteristics of radiators with $SWR \leq 2$ showed that the spiral with $S \approx R$ has a very weak effect on the directional pattern in H-plane (less than 2 dB) (SWR increases to a maximum of 15%). The influence on radiator resistance took place when $R/\lambda = 0.04 - 0.6(\lambda_{max}/\lambda_{min} = 15)$, but it allows to use an antenna in a wide frequency range, if the spiral parameters do not change.

Figure 13.21*b* compares vertical directional patterns of the radiator in free space (solid lines) and of the radiator placed along the axis of spiral made of a cable with an outer diameter $0.0033\lambda_{max}$ (dotted lines). Figure 13.22 presents the results of measuring mutual coupling between two identical radiators located along the axis of spiral with $R = S = 0.04\lambda_{max}$ at a distance, which is equal to $0.1\lambda_{max}$ and $0.2\lambda_{max}$. In a wide frequency range the magnitude of mutual coupling between the radiators is close to the same value between the radiators in the free space. It is very substantial from the point of view of electromagnetic compatibility.

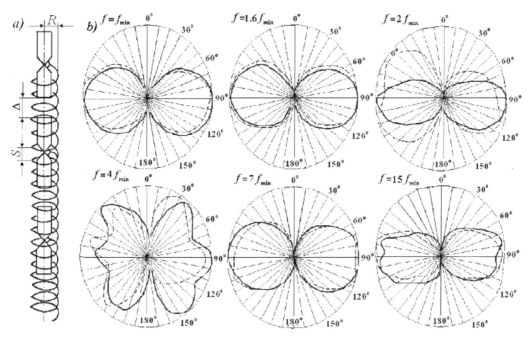

Figure 13.21 Metal element in the shape of a vertical cylindrical spiral (*a*) and experimental directional patterns of a radiator in a vertical plane (*b*).

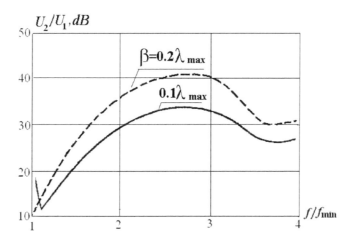

Figure 13.22 Mutual coupling between radiators located along a spiral axis.

The results of calculating input impedances and directional patterns in a system of two radiators are presented for following three variants of the cable placement (Figure 13.23):

(a) the cable is absent,

(b) the cable is placed vertically (excluding the short horizontal segment),

(c) the cable is placed along the cylindrical spiral and the mentioned radial segment.

The cylinder radius and the distance between the structure axis and the vertical cable are taken the same and equal to $0.042\lambda_{max}$. These results demonstrate the weak effect of a spiral cable upon the electrical characteristics of radiators.

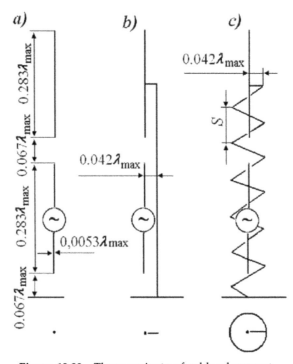

Figure 13.23   Three variants of cable placement.

The equivalent circuits of the radiating structures for each variant are shown in Figure 13.23. It is considered that the upper radiator is passive and short-circuited in the middle. Radii of all wires for the sake of simplicity are taken equal to the radius of the passive radiator ($0.0027\lambda_{max}$).

Figure 13.24a shows curves for TWR of the lower radiator in a range $\lambda_{max}/\lambda_{min}$ = 4 ($S = R$). As is seen from the figure, placement of the cable along the cylindrical spiral (variant c) allows to obtain a higher level of matching. The calculated directional patterns in the horizontal plane for different variants of the cable placement and different diameters of the spiral are given in Figure 13.24b. The minimal values $f(\varphi) = E/E_{max}$ of the directional patterns are given in Table 13.3. Figures and table clearly confirm that the placement of the cable along the cylindrical spiral allows to weaken significantly the distortions of the directional patterns.

The results of analyzing characteristics of the antennas located in the form of several floors show that the arrangement of cables along the cylindrical spiral, axis of which coincides with axis of radiators, facilitates the solution of problem of their electromagnetic compatibility.

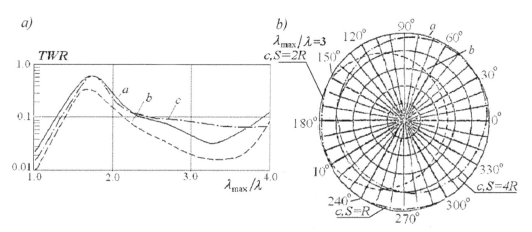

Figure 13.24 Level of matching (*a*) and directional patterns (*b*) of lower radiator.

Table 13.3 Minimal Values of Directional Patterns

λ_{max}/λ	Variant a	Variant b	Variant c		
			$S = R$	$S = 2R$	$S = 4R$
1.0	1.0	0.786	1.0		
1.5	1.0	0.705	0.997		
2.0	1.0	0.600	0.999	0.994	0.905
2.5	1.0	0.554	0.997		
3.0	1.0	0.514	0.985	0.986	0.749
3.5	1.0	0.462	0.991		
4.0	1.0	0.424	0.982		

14

Log-Periodic Antennas

14.1 SELF-COMPLEMENTARY STRUCTURE AND THE ABILITY TO "CUT-OFF" CURRENTS AS THE BASIS OF LOG-PERIODIC ANTENNA

Log-periodic dipole antennas (LPDA) have been used widely in the past decades. Their creation represents a significant step in the development of wide-range directional antennas [105–108]. These antennas belong to the class of frequency-independent antennas. They are based on the principle of complementarities and the ability automatically to "cut-off" the current. LPDA provides directional radiation along its longitudinal axis and retains shape of the directional pattern over a wide frequency range. They have also constant input impedance.

In accordance with the principle of electrodynamics similarity any radiator has the same electrical characteristics at different frequencies, if its geometric dimensions vary with frequency in proportion to the wavelength (in the first approximation the requirement about corresponding change of the material conductivity may be neglected). Not only the tunable antennas, but also the antennas whose shape is completely determined by the angular dimensions, conform to the principle of electrodynamics similarity. In this case changing of scale does not change the antenna, i.e. a radiator shape and dimensions in wavelengths are the same at different frequencies.

Antennas having the property of the automatic "cut-off" currents arouse among frequency-independent antennas a particular interest. This property means that the field at each frequency is radiated by a current along a small antenna segment, which is called by the active area, and that the electric current outside the boundaries of this area is quickly attenuated. Here, coordinates and dimensions of radiated segment are rigidly related with the magnitude of a wavelength. If the frequency was changed, the antenna segment, radiating the field, shifts along the antenna. The electrical dimensions of the area, both longitudinal and cross, remain constant and ensure the invariability of the characteristics. Thus, the antenna has the constant input impedance and invariable directivity characteristics in an infinitely wide band.

If the antenna has finite dimensions, its frequency range is finite, but in this finite range the antenna has the properties of an infinite antenna. The maximal wavelength depends on the maximal cross dimension of the antenna (on its width), and the minimal wavelength depends mostly on the accuracy of the structure manufacturing near the excitation point.

LPDA (Figure 14.1) is a collection of elements (of wires), dimensions of which form a geometric progression with denominator $1/\tau$.

$$R_{n+1}/R_n = l_{n+1}/l_n = 1/\tau. \tag{14.1}$$

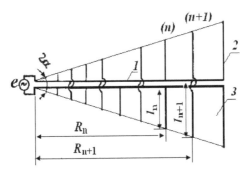

Figure 14.1 Log-periodic dipole antenna.
1—longitudinal wire, 2—transverse wire, 3—interval between the transverse wires.

Here R_n is the distance from the vertex of the angle α to dipole n, l_n is the arm length of dipole n, α is the angle between the antenna axis and the line passing through the dipoles ends (see Figure 14.1). Accordingly, the antenna's electrical characteristics are repeated at frequencies forming the geometric progression with the same denominator. It means that directivity characteristics and input impedance of the antenna are periodic functions of logarithm of frequency f, i.e. if the electrical characteristics are drawn as a function of $ln\ f$, their values are repeated with period equal to $ln\ \tau$. From here the antenna name is selected.

Weak variation of antenna's characteristics within the period is an indispensable condition of a weak frequency dependence of these characteristics. In order to meet this condition, this period must be small. But this is insufficient.

LPDA shown in Figure 14.1 consists of two structures situated in one plane. Each structure is shaped as a straight wire, with the linear conductors attached to it at right angles alternately from the left and from the right. Their lengths increase with the growing distance from the excitation point in accordance with the law of geometric progression. Such an antenna is a simplified and modified variant of a flat log-periodic structure shown in Figure 14.2, which is the self-complementary structure, i.e. it consists of metal plates and slots coinciding with each other in shape and dimensions. The input impedance of a flat infinite self-complementary structure is purely active, independent of the frequency, and is equal to 60π Ohm (see Chapter 8). Designing log-periodic antenna in the form of a self-complementary or similar structure ensures a small variation of electrical characteristics of the antenna within one period of oscillation.

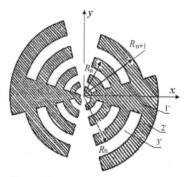

Figure 14.2 Flat self-complementary log-periodic antenna.
1'—metal sector, 2'—metal strip, 3'—slot strip.

Each of the two structures, forming a LPDA (see Figure 14.1), differs from the structure, which forms an arm of a flat log-periodic antenna (see Figure 14.2). The metal sector 1' is replaced with the longitudinal wire 1, the metal strip 2' situated along the arc of a circumference is replaced with transverse wire 2, tangent to the arc, and the slot 3' is replaced with the interval 3 between the transverse wires. Such construction is essentially simpler for implementation and, at the same time, its electrical characteristics are close to the electrical characteristics of original construction.

Rotation of one metal structure (of one arm of the antenna) around the y-axis (see Figure 14.2) through angle π and placing both structures in one plane allows providing unidirectional radiation. The unidirectional log-periodic antenna shown in Figure 14.1 may be interpreted as a linear array of symmetrical radiators. These radiators have monotonically changing lengths and are excited by a two-wire long line. A generator is connected in the line from the side of the shorter radiators.

A reasonable implementation of an antenna design, which requires no special balun, is shown in Figure 14.3. The cable is placed inside one of two tubes forming a two-wire distribution line. The cable sheath and the tube of distribution line form a single unit, and an inner conductor of the cable is connected to the second tube at the antenna vertex. This design provides a shortcut circuit of the distribution line. It is implemented at a distance $\lambda_{max}/8$ from the base of the first dipole. Here λ_{max} is the maximal wavelength.

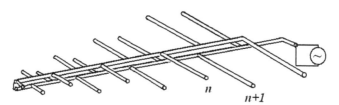

Figure 14.3 Design of LPDA.

Unfortunately, the opinion that the log-periodic structure itself provides constant input impedance is widespread. In [75] it is said that "there are serious misunderstandings. It seems, that such ignorance can be attributed to the term 'log-periodic antenna' for the self-complementary log-periodic antenna, without reference to the most important fact that it is a derivative of the self-complementary structure...In order to correct such misunderstandings, experimental tests have been done by taking a conically-bent modified antenna, which is arranged in the log-periodic manner as shown in Figure 14.4a (the figures numbers are replaced—B.L.). As the most straightforward arrangement of non-self-complementary log-periodic structure, the antenna shown in Figure 14.4*b* was constructed, where one wing of the two half-structures of the antenna is upside-down... The measured values of input resistance for these two antennas are compared in Figure 14.5, and a significant difference is apparent between them, in spite of the fact that the two wings of both antennas are identical.

The input resistance of the incorrectly arranged log-periodic structure, which is shown in Figure 14.5 by the dotted curve and crosses, varies distinctly in a log-periodic manner for varying frequency, though the constant-resistance property is satisfactory for the self-complementary antenna (its input resistance is shown by the solid curve and circles). From the results described above, it can be concluded that the origin of the broad-band property of the "log-periodic antenna" is not in its log-periodic shape, but rather in the aspect of the shape that is derived from the self-complementary structure.

It is hardly necessary to add anything to the above words.

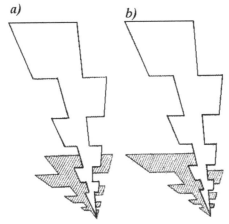

Figure 14.4   Two kinds of log-periodic antenna: self-complementary antenna (*a*), antenna, which is built by anti-complementary method (*b*).

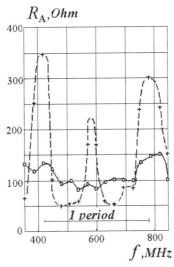

Figure 14.5   Input resistances of a self-complementary antenna (ooo) and of an antenna, which is built by anti-complementary method (+++).

Further we consider the active area of LPDA with the view of explaining the principle of its operation. The area consists of dipoles with the arm length close to $\lambda/4$. In their input impedance an active component is predominant, and the reactive component is small. In actual practice the number of dipoles forming the active area is usually equal to five. For the sake of simplification we assume only three dipoles, with the arm length of central dipole being $\lambda/4$.

As is seen from Figure 14.1, the upper arms of the dipoles connected alternately to one or another conductor of the distribution line. That is equivalent to crossing conductors of the long line on the segments between the dipoles. With allowance for this crossing the electrical current in the larger dipole outstrips in phase the current in the resonance radiator, and the current in the shorter dipole lags behind the current in the resonance radiator, i.e. the larger dipole acts as a reflector, and the shorter dipole acts as a director. As a result, the fields of individual radiators are summed in the direction toward the excitation point (to the side of shorter dipoles) and cancel each other in the opposite direction.

The waves in the distribution line, reflected from the dipoles of the active area, cancel each other to a large degree, since the reactive components of the input impedances of short and large dipoles are opposite in sign. This explains a high level of matching of the active area with the distribution line. In addition the electrical length of the line from the feed point to the active area remains unchanged during the frequency change. Therefore, an impedance of active area transformed to the antenna input is the same at different frequencies as well.

The dipoles located outside the active area are excited weakly due to the great reactive impedance. The short dipoles at the beginning of the structure practically fails to radiate, since the fields created by them summed almost in anti-phase because of crossing wires and the proximity of dipoles to each other (as compared with the wavelength). As a result, the *EM* wave along this segment of line does not weaken, i.e. the distribution of currents and voltages at the line segment between the excitation point and the active area is close to that of the traveling wave mode. The short dipoles act as capacitances shunting the distribution line and thereby decreasing slightly its wave impedance. The long dipoles situated behind the active area radiate weakly too, since, first, their input impedances are great and, second, the power of the *EM* wave at that segment of line drops substantially as a result of attenuation in the active area.

14.2 THE METHOD OF LPDA CALCULATION

The method of LPDA calculation [109] is based on antenna presentation in the form of a parallel connection of two multipoles (Figure 14.6), one of which describes a system of dipoles and is defined by matrix $[Z_A]$ of mutual impedances, and the other describes the distribution line with matrix $[Y_l]$ of admittances. For each cross-section n of the structure, where the dipole is connected in parallel with the distribution line, the following equations are true:

$$J_{nA}Z_{nA} = J_{nl}/Y_{nl}, \quad J = J_{nA} + J_{nl}, \tag{14.2}$$

i.e. $J_{nl} = J_{nA}Z_{nA}Y_{nL}$. Here J_{nA} is the current at the dipole input, Z_{nA} is the input impedance of the dipole (with allowance for coupling with neighboring dipoles), J_{nl} is the current of the distribution line, Y_{nl} is the admittance of the line in the cross-section n, and J is the extraneous current at given point. It should be noted that in calculating J_{nA}, the mutual coupling with neighboring dipoles is accounted, and in calculating Y_{nl} it is considered that the distribution line is shorted at the terminals of neighboring dipoles (according to Kirchhoff's law other sources of emf are replaced by short circuit).

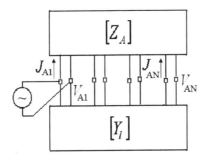

Figure 14.6 Equivalent circuit of LPDA.

The first equation of (14.2) is written for the total voltage along closed circuit, the second equation is written for the total current in this point. From here,

$$J = (1 + Z_{nA}Y_{nl})J_{nA}. \tag{14.3}$$

Accordingly, a matrix equation for the column-vector $[J_A]$ of the dipole's input current is written in the form

$$[J] = ([E] + [Z_A][Y_l])[J_A], \tag{14.4}$$

where $[E]$ is the identity matrix, $[J]$ is the column-vector of currents feeding the lines, which connects multipoles with each other. Since the extraneous current is only at the distribution line input, in the first cross-section (it is equal to J_0), then $[J] = \begin{vmatrix} J_0 \\ 0 \\ \dots \\ 0 \end{vmatrix}$. Solving

equation (14.4), we find the column-vector $[J_A]$, and then matrix $[V_A] = [Z_A][J_A]$ of the voltages at the dipoles inputs. The first element of the matrix at the input of the shortest dipole is the voltage. If the exciting current J_0 is equal to 1, this first element is equal to the input impedance of the antenna.

In [109], the elements of the matrix $[Z_A]$ are calculated, in fact, by means of the induced emf method. Later on, to obtain more exact results, the matrix elements were calculated by means of the integral equation's solution with the help of the Moment Method [110]. The difference between the approximate and the exact method is particularly noticeable, if the LPDA consists of thin radiators or has a wide angle at the antenna's vertex. The energy in such an antenna propagates along the distribution line beyond the boundaries of the active area and excites the long dipoles.

When designing LPDA, it is important to choose the geometric dimensions so that the electrical characteristics changed weakly in a range from f to τf. The magnitude τ and all antenna characteristics depend essentially on the parameter σ, which is equal to the distance between the half-wave dipole and the neighbor shorter dipole (in wavelengths):

$$\sigma = 0.25(1 - \tau) \cot \alpha. \tag{14.5}$$

In fact, it is dependent on the angle α in view of τ. As is shown in [109, 111], the characteristics change weakly, if $\tau > 0.8$ and $0.05 \leq \sigma \leq 0.22$. Under these conditions, the currents of the dipoles located near the resonant (half-wave) radiator reach a maximum and the wave along the distribution line is so attenuated in the active area that the follow dipoles practically do not radiate.

In [112] on the basis of generalization of data available in the literature, the optimum relationship of the above mentioned basic parameters is defined in the form:

$$\sigma/\tau = 0.191. \tag{14.6}$$

This ratio does not depend on the values of α, l_n/a_n and Z_0. Here a_n is the radius of dipole n, $Z_0 = 60ch^{-1}[(D^2 - 2a^2)/(2a^2)]$ is the wave impedance of the distribution line, a is the radius of the distribution line's wires, and D is the distance between axes of these wires. Substituting (14.6) into (14.5), authors of [112] obtain the simple expressions connecting the optimal parameters τ and σ with the antenna dimensions:

$$\tau = 1/(1+0.765\tan\alpha) = L/[L+0.765(l_1 - l_N)], \quad \sigma = 1/(4\tan\alpha + 5.23). \tag{14.7}$$

The value L in these expressions is the distance between the first and the last (N) dipole.

The antenna with $\sigma/\tau = 0.191$ has a narrow directional pattern and high front-to-back ratio. Figures 14.7 and 14.8 corroborate these statements. They show the given in [112] calculated beam width for LPDA with $Z_0 = 100$ Ohm and $l_n/a_n = 177$ in the planes E and H and also front-to-rear ratio depending on the parameters τ and σ. SWR of the same antenna with the optimal parameters τ and σ depending on the values l_n/a_n and Z_0 is presented in Figures 14.9 and 14.10. Magnitude of SWR in a properly designed LPDA is typically smaller than 1.5.

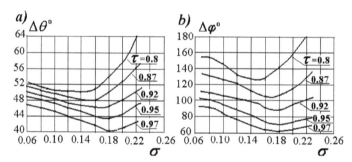

Figure 14.7 Dependence of half-power beam width of LPDA with $Z_0 = 100$ Ohm and $l_n/a_n = 177$ in the planes E (*a*) and H (*b*) on the parameters τ and σ.

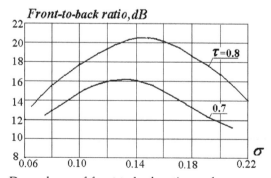

Figure 14.8 Dependence of front-to-back ratio on the parameters τ and σ.

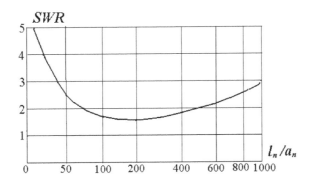

Figure 14.9   Dependence of *SWR* of the antenna with optimal τ and σ on value l_n/a_n.

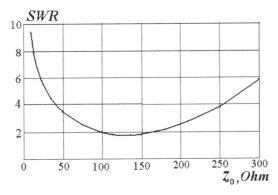

Figure 14.10   Dependence of *SWR* of the antenna with optimal τ and σ on Z_0.

Under antenna development it is necessary to take into account that the arms of each dipole are connected to different conductors of the distribution line, and so they are not coaxial. To decrease the influence of misalignment on the antenna pattern, one must reduce the distance between the conductors' axes: it should not exceed $0.02\,\lambda_{min}$. Here λ_{min} is the minimum wavelength.

14.3   DECREASING TRANSVERSE DIMENSIONS OF LPA

Log-periodic antennas have rather large overall dimensions. In order to decrease transverse dimensions, it is expedient to shorten the longest dipoles using loads of different kind or structures with the slowing-down, i.e. the same manners, which are used for reducing the monopole's and dipole's length. Different variants of shortened monopoles are presented in Figure 14.11. Among them, inverted-*L* and *T*-radiators (*a, b*) and antennas with concentrated inductive loads (*c*) are. Slowing-down is employed in a helical (*d*) and meandered (*e*) antennas and in monopoles of fractal shape of Koch (*f*). It should be noted that the slowing factor is always less than the increase of the wire length.

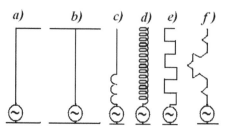

Figure 14.11 Variants of shortened monopoles: inverted-L antenna (*a*), T-antenna (*b*), antenna with concentrated load (*c*), helical antenna (*d*), meandered antenna (*e*), antenna of Koch fractal shape (*f*).

Slowing-down allows shortening the monopole, i.e. to reduce the length of the monopole by a factor m for the given frequency of the first resonance or to decrease the resonance frequency by a factor m for the given length of the monopole. But the radiation resistance at the resonance frequency in consequence of the length reduction decreases by a factor m^2, and the antenna wave impedance is increased by a factor m. And both impair matching with the cable of each element of LPDA and the antenna on the whole.

Figure 14.12 demonstrates the results of a rigorous calculation of SWR and gain for log-periodic antennas with linear and helical dipoles. Parameters of the antenna are the following: $N = 15$, $\tau = 0.92$, $\alpha = 10°$, $l_n/a_n = 100$, $l_n/\rho_n = 20$ (here ρ_n is the radius of helical dipole n), $Z_0 = 150$ Ohm. The helical dipole arm consists of five turns; the wire length is twice as large than the straight dipole's length. The value of SWR is calculated in a cable with wave impedance 100 Ohm. The relative length l_N/λ of the largest dipole's arm is used as the argument.

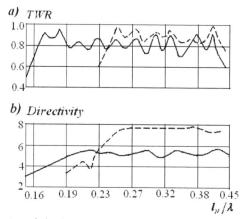

Figure 14.12 Characteristics of the log-periodic antennas with helical dipoles (solid curve) and straight dipoles (dotted curve): traveling-wave ratio (*a*), directivity (*b*).

As can be seen from Figure 14.12, the level of $TWR \geq 0.7$ for the antenna with the helical dipoles is maintained in the range $0.163 \leq l_N/\lambda \leq 0.425$. The dotted curves in the Figure correspond to the log-periodic antenna with straight dipoles. The Figure shows that, if both antennas have the same dimensions, LPDA with the helical dipoles and a double wire length has an operation range, expanded by half in the direction of low frequencies in comparison with the range of ordinary antennas. The useful effect is accompanied by decreasing match level and some deterioration of directivity, caused by a higher Q of helical dipoles.

To increase the parameter τ up to 0.95 and the number of dipoles up to 24, in the considered example we shall obtain the antenna, the characteristics of which are almost the same as the characteristics of an antenna with straight dipoles and the transverse dimensions reduced by half. Thus, it is theoretically possible to reduce its transverse dimensions at the cost of increasing the dipoles' number and at the same time to maintain characteristics of the log-periodic antenna. But practically acceptable designs are obtained, if the transverse dimensions are reduced no more than two or three times.

Attempts to decrease longitudinal dimensions of an antenna by using slowing-down in the distribution line or at the expense of additional dipoles connection, failed, since violation of geometric progression's relationships and increase of the dipoles number causes, as a rule, sharp deterioration of electrical characteristics and gives insignificant decrease of overall dimensions.

The variant of log-periodic antenna, which operates in two adjacent frequency bands and allows making the antenna shorter than the antenna designed for operation in the total range, is described in [112]. Basically the authors' proposal reduces to the use of linear-helical dipoles, i.e. radiators, each of which consists of straight and helical dipoles arranged coaxially and having a common feed point (Figure 14.13).

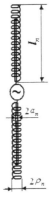

Figure 14.13 The linear-helical dipole.

The dipoles length is the same, but the helical wire length is twice as much as the straight rod's length. Linear-helical dipole in contrast to straight and helical dipole has two serial resonances, and the ratio of the resonant frequencies for the same dipole's length is equal to the slowing factor of the helical dipole.

As is well known, the resonant dipole and its nearest neighbors create an active area, passing through which the electromagnetic wave, whose frequency is close to the resonant frequency, actively radiates energy. LPDA with linear-helical dipoles has two active areas, and they provide a signal radiation in two bands of the frequency range. The experimental check of log-periodic antenna with linear-helical dipoles, described in [112], confirm that this proposal is promising. The antenna is designed for operation in the frequency range from 250 to 1250 MHz. The length of mock-up is equal to 0.44 m; the dipole maximum length is 0.42 m. The test results are given in Figures 14.14–14.17.

From Figure 14.14 it is seen that the TWR in the cable with wave impedance 75 Ohm is greater than 0.3 in the ranges 252–610 and 645–1250 MHz, at the frequencies 613 and 625 MHz its value decreases to 0.17 and 0.18, respectively. The front-to-back ratio is greater 8 dB (see Figure 14.15). The half-power beam width (both in the plane E and in the plane H) in the lower part of range is wider than in the top (see Figure 14.16).

Accordingly, here the antenna directivity is smaller. Directional radiation exists from 260 to 1250 MHz (see Figure 14.17). Only at 550 MHz this ratio falls sharply to 2 dB.

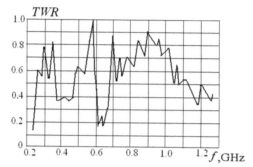

Figure 14.14   *TWR of antenna with linear-helical dipoles.*

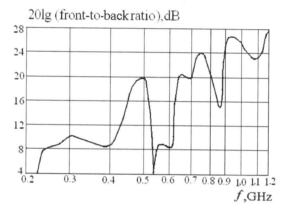

Figure 14.15   Front-to-back ratio of antenna with linear-helical dipoles.

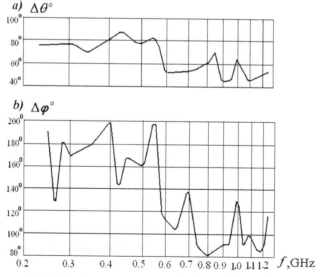

Figure 14.16   Pattern of antenna with linear-helical dipoles in the plane *E* (*a*) and in the plane *H* (*b*).

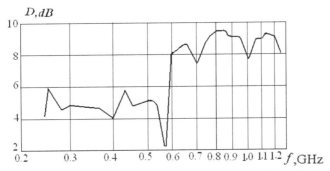

Figure 14.17   Directivity of antenna with linear-helical dipoles.

If electrical characteristics of log-periodic antenna with straight dipoles and with linear-helical dipoles are similar, then the length of the antenna with straight dipoles is greater in 1.8 times. If only helical dipoles are used, the length of antenna is greater than the length of the antenna with linear-helical dipoles 1.3 times. In addition *TWR* of the proposed antenna in the upper part of the range is smaller on the average by 4 dB. Decrease of the antenna dimensions is obtained at the cost of *TWR* and directivity reduction in the narrow band in the middle of the operation range. This reduction is caused by the transfer of the active region from the helical elements of LPDA to straight elements.

14.4   DECREASING THE LENGTH OF LOG-PERIODIC ANTENNA

The length of the log-periodic antenna can be reduced by increasing the angle α between the antenna axis and the line passing through the dipoles ends. This option seems the most simple and natural. But, as it is seen from (14.5), increase of α, if τ is constant, leads to decrease of the distance between the dipoles and to the growth of their mutual influence, and as a result to decrease of directivity and active component of input impedance and to the deterioration of the frequency-independent characteristics.

One can increase the angle α by another manner. LPDA consists (see Figure 14.1) of two asymmetric structures located in the same plane and excited in opposite phases. If these structures are located at an angle $\psi > \alpha$ to each other, as is shown in Figure 14.18, the resulting three-dimensional structure will incorporate two distant from each other planar structures. The monopoles are connected alternately from left and from right to the conductor of the distribution line. The distance between the monopoles,

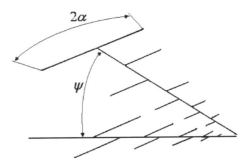

Figure 14.18   Volumetric antenna of two structures.

situated on one side of the conductor, is almost twice as large as in a planar LPDA. This reduces their mutual influence and allows to increase the angle α. However, this antenna occupies a great volume, and that makes difficult its installation and changes its characteristics. This, for example, increases inadmissibly input resistance, creates additional trouble for antenna's utilization.

An asymmetrical coaxial log-periodic antenna, described in [112], does not have these disadvantages. Two-wire distribution line in this antenna is replaced by a coaxial line, and dipoles are replaced by monopoles. Antenna as an assembly is shown in Figure 14.19. The antenna consists of two structures, circuits of which are given in Figure 14.20. The first of them (Figure 14.20a) is a straight conductor. The wires' segments of required length located in one plane connected to defined points of this conductor at the right angle alternately from left and from right. This conductor is the central wire of the coaxial distribution line and the wires' segments are monopoles, which are excited by means of this conductor.

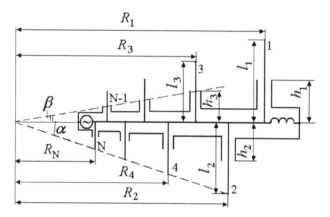

Figure 14.19   The circuit of asymmetrical coaxial log-periodic antenna.

The second structure (Figure 14.20b) is designed as a long cylindrical tube with short tubes embedded in it, which are opened inside and outside. The long tube is the outer shell of the coaxial distribution line, the short tubes are the outer coaxial shells, which partially cover the monopoles connected to the inner conductor of the distribution cable. As a result, monopoles' are the radiators with a feed point displaced from the base. As one can be seen from Figure 14.19, the first structure is inserted into the second one, so that their axes coincide.

In accordance with the usual practice of designing log-periodic antenna, its dimensions must correspond to the geometric progression with ratio $1/\tau$.

$$R_{n+1}/R_n = l_{n+1}/l_n = h_{n+1}/h_n = 1/\tau. \tag{14.8}$$

Here h_n is the distance from the axis of the distribution line to a feed point of radiator n. Other values are defined previously. In addition, it is necessary that the ratio of the shell diameter to the central conductor diameter ensured coincidence of this segment wave impedance with the radiator resistance on the frequency of the first serial resonance.

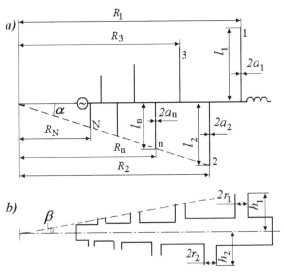

Figure 14.20   Internal (*a*) and external (*b*) structures, from which asymmetrical coaxial log-periodic antenna consists.

From the above it follows that the two-wire distribution line is replaced in the proposed antenna by a coaxial cable, and the dipoles are replaced by monopoles connected to the inner conductor of this cable. The outer shell of the cable is used as a ground. This shell in turn serves as a ground for the monopoles excited in the anti-phase, and that substantially distinguishes this ground from a large metal sheet. This means that the proposed structure realizes an asymmetrical version of the usual log-periodic antenna (symmetrical version of such antenna is implemented as the antenna LPDA). Consequently, it is possible to increase significantly the angle α and to shorten the antenna without fear of directivity decrease and deterioration of frequency-independent characteristics. Since the radiating elements of the antenna are the monopoles, then by analogy with the LPDA, where the dipoles play a similar role, it is expedient to name this antenna by LPMA.

The principle of a symmetrical antenna's operation was reviewed earlier by means of the analysis of processes in its active area. The processes in the active area of an asymmetrical antenna practically do not differ, since the waves in a coaxial distribution line are similar to the waves in a two-wire line and are depended on the monopole influence, which is similar to the influence of dipoles on the waves in a symmetrical structure. In the surrounding space the equally excited dipoles and monopoles produce in the same fields.

Mock-up of an asymmetric log-periodic antenna designed in order to operate in the range of 200–800 MHz, has been manufactured and tested. Antenna characteristics were measured by authors for the two variants of its mounting on the metal mast (Figure 14.21): the cantilevered variant, when the radiators are arranged vertically (*a*), installation on the mast top, where the radiators are mounted horizontally, and the gravity center coincides with the mast axis (*b*). Distribution line was formed in the shape of a truncated pyramid with a square cross-section and the inner conductor made in the shape of a horizontal plate of variable width.

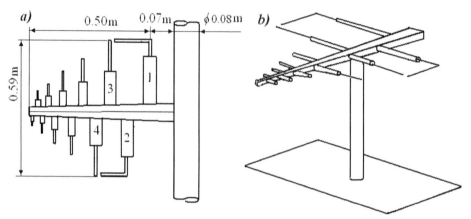

Figure 14.21 Asymmetrical antenna on the metal mast: cantilevered variant (*a*), installation on the mast top (*b*).

Experimental check confirmed that LPMA regardless of the variant of installation has frequency-independent electrical characteristics. The cantilevered variant gave the following results. The average magnitude of half-power beam width in the operation range 200–800 MHz is equal to 70° in the plane *E* (vertical) and 124° in the plane *H* (horizontal). Typical patterns in both planes are shown in Figure 14.22. Back-to-front ratio does not exceed 0.15. The directivity value is 6.8 dB.

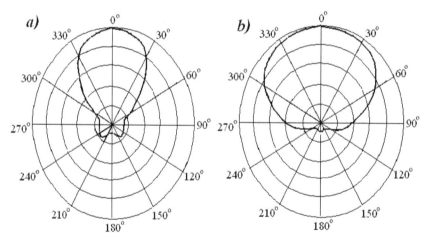

Figure 14.22 Typical directional pattern of asymmetrical antenna in the plane *E* (*a*) and in the plane *H* (*b*).

The experiment shows that the antenna does not lose the directional characteristics up to frequency 2.75 GHz. In Figure 14.23, TWR of antenna is given in the cable with the wave impedance 50 Ohm in the range from 0.8 to 2.8 GHz. One can be seen that in the range from 1.5 to 2.4 GHz, TWR is equal to 0.3–0.7. The beam width in the plane *E* in this range is 35–40° (i.e. by half less than that in the main operation range), and in the plane *H* it is equal to 115–140°. Accordingly, in additional range directivity is higher by a half than in the main range.

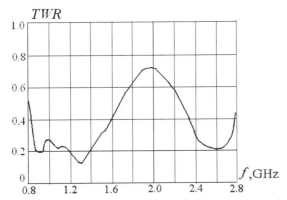

Figure 14.23 *TWR of antenna in the cable with the wave impedance 50 Ohm.*

The authors' point of view on the causes of additional operating range's emergence is absent in [112]. From our point of view, the reason is obvious enough, if we to take into account the calculations and measurements results for LPDA with linear-helical dipoles presented by the authors. Here each radiator consists of two connected in parallel elements with different resonant frequencies. Actually, in this LPDA along the distribution line two dipoles' structures are set. The dipoles' dimensions are defined by two different angles—α and β (see Figure 14.20)—between the axis of the antenna and the line passing through the ends of the radiators. Each structure provides the required electrical characteristics within its operating range. The sharp deterioration of characteristics occurs at the boundary of the ranges.

Similar result during measurements of LPMA is caused by the fact that each radiator of the antenna consists of two connected in parallel elements: the monopole and the short tube, i.e. the segment of coaxial cable, surrounding the monopole. Structure dimensions are chosen so that the distance from the central conductor axis to the end of the tube is equal to half of the monopole length. But that does not mean that the length of a single element is half of the other element's length because the length of one element is equal to the short tube's length (it is necessary to subtract the radius of the distribution tube). Therefore, the average frequency of the additional range (1.95 GHz) is greater approximately four times (not two times) than the average frequency of the main range (0.5 GHz).

When the mock-up of LPMA is placed on the mast top, TWR in the cable with a wave impedance 75 Ohm does not fall below 0.3 in the range from 115 to 800 MHz (Figure 14.24), i.e. the lower frequency decreased by a factor 1.6.

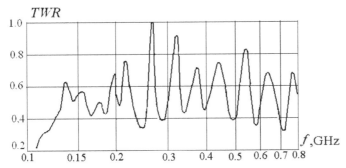

Figure 14.24 *TWR of antenna, placed on the mast top.*

Since for the sake of decreasing the transverse dimensions of the antenna, the first monopole was bent at right angle at the height of the third monopole and the second monopole was bent at the height of the fourth monopole, the third and the fourth monopole have the largest lengths. At a frequency 115 MHz, the length of the third and fourth monopole was 0.227λ.

Typical directional pattern of antenna placed on the mast top, in the E plane at frequencies above 300 MHz, is similar to the directional pattern shown in Figure 14.22. When the frequency decreases, the back lobe grows. At frequencies below 160 MHz, the directional pattern with two lobes turns and becomes perpendicular to the distribution line axis. This rotation is caused by the fact that at low frequencies the outer tube of the distribution line becomes the main radiator.

Different Issues

15.1 ANTENNAS FOR UNDERGROUND RADIO COMMUNICATION AND OBSERVATION

Reliable underground radio communications is required for many applications, e.g. for communication between people, working below ground and on the ground surface [113, 114]. The need for such communication is especially obvious in the case of underground accident, for example, in the mines.

In addition to the need in communication between an underground zone and a ground surface, a channel between two subterranean points is also often required. This channel is more complicated, and its difficulties grow due to the fact that the signal, radiated by an underground source, is subjected to strong attenuation. Besides that it propagates in two ways [115, 116]. The first pathway goes in a vertical direction to the earth surface, located above the transmitting antenna. Further the signal propagates in air along the earth surface, and finally, passing vertically downwards through earth layers, reaches the receiving antenna. The second pathway is horizontal, i.e. the signal propagates through the earth in a horizontal direction.

The ratio between the two signals at the receiving point depends on a distance between the correspondents, on a placement depth of transmitter and receiver, and on electromagnetic characteristics of the ground layers along the channel. As a rule, the signal, propagating by the first way, is greater. On the other hand, the signal, propagating by the second way, may be used in order to secure observation of underground channel and detecting its change. In that case, the second signal must be substantially greater than the first signal. Or it is necessary be able to extract the second signal from the mixture of signals.

Losses in the earth dramatically increase at high frequencies, and the signal, propagating through the earth, is quickly attenuated. Using low frequencies requires radiating structures of large dimensions. Proceeding from these contradictory requirements, one must find a compromise solution. When developing an underground communication system it is necessary also to take into consideration that a horizontal component of the signal propagating in the earth is attenuated slower than a vertical component. Since the efficiency of an antenna, located close to the earth is small, it is expedient to create a big air cavity around the transmitting antenna. It should be also noted that the presence of a metal sheet underneath of the transmitting antenna increases the first signal, propagating in air along the earth, and hence placement of a metal sheet under the antenna is harmful for the system of observation. Published works show that today it is possible to create an underground communication at a distance of 10 km by means of a transmitter of power 100–200 W, operating at frequencies 150–200 kHz [117].

Based on the foregoing, the placement of the transmitting and receiving stations in such a system may look as shown in Figure 15.1. In order to increase the signal magnitude, the transmitting antenna must be directional. For this purpose, it is expedient to use an antenna array, consisting of two or three horizontal dipoles with arm length, close to a quarter of wavelength. These dipoles may be active or passive. In the first case all dipoles are excited by a generator, and changing the emf phases permits to alter the radiation direction. If only one radiator is excited, then for changing the direction of radiation one must re-tune the dipoles operating in passive mode. The dipoles are placed in the air cavities, located in an earth on a distance of about a quarter of wavelength from each other.

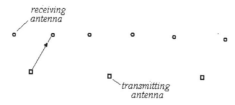

Figure 15.1 Placement of transmitting and receiving antennas.

As it is rightly pointed out in [116], due to the high attenuation of electromagnetic waves for calculating fields in media with relatively high conductivity it is necessary to use rigorous expressions. Therefore, in accordance with Maxwell's equations for a perfectly conducting filament, used as a model of a symmetrical radiator, z-component of its electrical field in the system of cylindrical coordinates (ρ, φ, ζ) is calculated in accordance with (1.29). The electrical current along the radiator is

$$J(\zeta) = J(0)\frac{\sin[k(L - |\zeta|) + \varphi]}{\sin(kL + \varphi)}. \tag{15.1}$$

If in the expression (15.1) $\varphi = 0$, the current is distributed in accordance with sinusoidal law (Figure 15.2a). At $\varphi = \pi/2$, it is distributed in accordance with cosine law (Figure 15.2b). Using relations between derivatives of R and R_+ with respect to ζ and z, and integrating the expression by parts twice, we come to (1.30). The field of the radiator in an equatorial plane ($z = 0$) is

$$E_0 = j\frac{60J(0)}{\varepsilon_r \sin(kL + \varphi)}\left[\sin\varphi(1 + jkR_{01})L\frac{\exp(-JkR_{01})}{R_{01}^3} - k\cos\varphi\frac{\exp(-jkR_{01})}{R_{01}} + \right.$$

$$\left. k\cos(kL + \varphi)\frac{\exp(-jk\rho)}{\rho}, \right. \tag{15.2}$$

where ε_r is a relative dielectric permittivity of the medium, $R_{01} = \sqrt{\zeta^2 + \rho^2}$, and ρ is the distance from an observation point to a middle of the radiator.

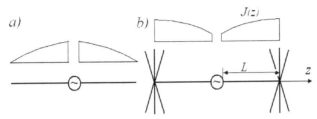

Figure 15.2 Sinusoidal (*a*) and cosine (*b*) current distributions along the radiator arm.

Let's consider a particular task of propagating a direct and reflected signal in a region of a radius 1 km. Fields in this region are the far fields, since the main component of the field is close to $1/\rho$. The losses in the earth lead to increasing dielectric permittivity and to decreasing the signals. The signal magnitude depends on the multiplier $\exp(-jk\rho)$, where k is equal to

$$K = \omega\sqrt{\varepsilon\mu(1-j\sigma/\omega\varepsilon)}. \tag{15.3}$$

Here ε, μ and $\sqrt{\varepsilon\mu}$ are permittivity, permeability and propagation constant in a lossless medium, σ is a medium conductivity. Since the wavelength in a lossless medium is $\lambda = \lambda_0/\sqrt{\varepsilon_r}$, where λ_0 is the wavelength in a free space, then

$$k = \omega\sqrt{\varepsilon\mu}\sqrt{1-j60\lambda_0\sigma/\sqrt{\varepsilon_r}}. \tag{15.4}$$

It is easy to see that for the real soil conductivity the second term under the square root is much greater than the first term, i.e.

$$k = \omega\sqrt{\varepsilon\mu}(1-j)\sqrt{30\lambda_0\sigma/\varepsilon_r}. \tag{15.5}$$

Hence it follows that the signal, which propagates belowground, decays exponentially with a decrement, equal to

$$\alpha = 2\pi\left(\sqrt{\varepsilon_r}/\lambda_0\right)\sqrt{30\lambda_0\sigma/\varepsilon_r} = 2\pi\sqrt{30\sigma/\lambda_0}. \tag{15.6}$$

If $\sigma = 0.5$ S/m, $\lambda_0 = 200$ m, then $\alpha = 1.72$ 1/m. If a distance is $\rho = 10$ m, signal attenuation is equal to $\exp(\alpha\rho) = 3\cdot10^7$ times.

Let's compare field magnitudes created in the far region by radiators with sinusoidal and cosine current distribution. Assuming that the input currents of the radiators in both cases have the same amplitudes and introducing the notation $A = -j60J(0)\exp(-jk\rho)/\varepsilon_r$, we obtain at $\varphi = 0$ (sinusoidal distribution) and $\varphi = \pi/2$ (cosine distribution) accordingly

$$E_{0I} = \frac{A}{\sin kL}\frac{k}{\rho}(1-\cos kL), \quad E_{0II} = \frac{A}{\cos kL}\left(\frac{L}{\rho^3}+j\frac{kL}{\rho^2}-\frac{k\sin kL}{\rho}\right). \tag{15.7}$$

If $\lambda_0 = 200$, $L = 50$, $\rho = 2\cdot10^3$, the fields are equal to $E_{01} \approx 0.4A$, $E_{0II} \approx A$.

In order to provide a cosine current distribution or distribution close to it, one must decide the complicated task of creating wires' structure on the antenna ends or realize a good contact with the ground. Temporarily leaving this task to the side, we shall consider electrical characteristics of a radiator with such current distribution depending on arm length of the radiator. An antenna input impedance is $Z_A = R_\Sigma + jX_A$, where for the linear radiator in the first approximation one can write

$$R_\Sigma = 20(kh_e)^2, \quad X_A = jW \tan kL. \tag{15.8}$$

Here $W = 120[\ln(2L/a) - 1]$ is the wave impedance, L is the arm length, a is the arm radius. Since $J(\zeta) = J(0) \cos k(L - |z|)/\cos kL$, the effective length h_e of the radiator is equal to

$$h_e = \frac{2}{J_A(0)} \int_0^L J_A(z)dz = \frac{2}{k} \tan kL. \tag{15.9}$$

Let, for example, $\lambda = 200$, $a = 0.1$ (dimensions in meters). If $L = 0.15\lambda = 30$, then $X_A = j892$, $h_e = \dfrac{2.75}{k}$, $R_\Sigma = 20(kh_e)^2 = 151$. When wave impedance W_c of the cable is equal to 100, the reflectivity $\rho = \sqrt{\dfrac{(R_\Sigma - W_c)^2 + X_A^2}{(R_\Sigma + W_c)^2 + X_A^2}}$ is 96, and standing wave ratio is equal to $SWR = \dfrac{1 + \rho}{1 - \rho} = 49$. If $L = 0.2\lambda = 40$, then $X_A = j2102$ Ohm, $h_e = 6.16/k$, $R_\Sigma = 759$ Ohm, $\rho = 0.97$, $SWR = 66$. It is obvious that both variants with such matching level are no acceptable, and the situation is still worse, when the arm length is close to a wavelength quarter.

In addition to the problem of matching, it is necessary to take into account that the difference between the fields created by antennas with sinusoidal and cosine current distribution at the same radiation power, is relatively small, and that the field decreases rapidly with a distance. Therefore the use of cosine distribution is inexpedient.

In order to decrease the dimensions of the antenna, one can employ a flat self-complementary antenna with rotational symmetry (see Section 9.2), for example with three metal dipoles (see Table 9.1 and Figure 15.3). Each arm of such antenna consists of a three plates. As a result, firstly the wave impedance of such antenna is essentially smaller unlike that of a flat self-complementary antenna with an arm of one plate. This antenna allows to provide a high level of matching with a standard cable. And secondly the antenna arm at the frequency of the first series resonance is shorter than the quarter of the wave length, i.e. one may decrease the antenna dimensions. Since the signal must have horizontal polarization, the antenna must be located in a vertical plane along the horizontal axis and be perpendicular to the direction to the placement area of the receiving antennas.

Another possible type of radiator is the curvilinear V-dipole with capacitive loads (see Section 6.5).

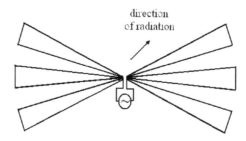

Figure 15.3 The circuit of the flat self-complementary antenna with rotational symmetry.

Analysis of the structure, which can be used for underground communication, shows that realization of this communication system is possible. For this embodiment it is expedient to use the medium frequencies. For the signal transmitting one must use an antenna array, consisting of two or three horizontal radiators. As the horizontal radiator, one may use a directional antenna of new type: flat self-complementary antennas with rotational symmetry or curvilinear V-dipole with capacitive loads.

The research results confirm the opportunity of creating a communication channel between two subterranean points and securing observation of underground channel and detecting its change.

15.2   MEASURING AN ANTENNA GAIN IN A FRESNEL ZONE

It is known that when measuring directional pattern and gain of antenna, an observation point should be placed in a far zone (Fraunhofer zone). If the antenna dimensions are big in comparison with a wavelength, the far zone boundary lies at the distance

$$R_0 = 2a^2 / \lambda, \tag{15.10}$$

where a is the maximum dimension of the antenna and λ is the wavelength.

At $R < R_0$ the observation point falls within a Fresnel zone. In this zone, the form of the directional pattern depends on a distance between an antenna center and a spherical surface, which passes through the observation points. (This form is different from the form of the directional pattern within the Fraunhofer zone.) Also in this zone, the radial component of field is rather great.

When measuring an antenna's characteristics, the limited dimensions of a range and also insufficient sensitivity of the measuring equipment frequently requires to place receiving and transmitting antennas in the Fresnel zone (for example, during the gain measurement). The directional pattern' dependence on a distance entails the corresponding gain change. It is necessary to solve the task: How big is the error in a gain measurement for such an antenna's placement?

The given problem in relation to linear radiators with different laws of current distribution was examined in [118]. The results were presented as the curves, which can be used as nomograms. In [119] is presented the method for calculating the gain of a rectangular aperture in the Fresnel zone, and the results for a uniformly exited antenna are given. This solution has analytical restrictions imposed on the calculation accuracy of fields created by the separated parts of aperture:

(1) Radii R_1 from these parts to an observation point are parallel to the radius R, connecting the aperture's center with an observation point. It means (Figure 15.4) that the signals are coming to the observation point from the aperture, which is a straight segment with the length $a = 2R \tan (\alpha/2)$. Here α is the angle, at which the segment a is seen from the observation point. Actually the signals come to the observation point from the curve line, i.e. from the segment in the form of an arc with the length $b = \alpha R$. Since $a > b$, the signal in the observation point is higher than veritable;

(2) Magnitude $1/R_1 = 1/(R + \Delta R)$ is replaced by $1/R$, i.e. the influence of a difference $\Delta R = R_1 - R$ on the signal amplitude is neglected.

Note. The author did not change the text of this section, written in accordance with the theorem of complex power, since this section considers a lossless medium.

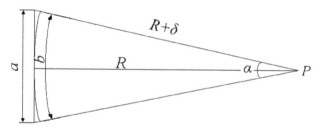

Figure 15.4 The field of a broadside array.

Here in the analysis method (it is different from the method described in [119]), a flat antenna is replaced by an equivalent no flat (convex) structure. The distance of any element of this structure from the plane, which passes through the observation point P parallel to the flat antenna, is equal to the distance from the corresponding element of the flat antenna to the observation point. Then one can consider that rays from separate points of antenna to the observation point P are parallel. The suggested method is equivalent to the method of Polk, but it is distinguished from one by greater obviousness, since geometric interpretation of approximate solution is used herein. This permits to select a shape and dimensions of convex structure in order to remove the first analytical restriction and change the integrand denominator in order to remove the second analytical restriction and to refine results obtained by Polk.

In this section, simple expressions are obtained for calculating measurement error of an antenna's gain. The error depends on the dimensions of the antenna, on a wavelength and a distance from measuring the antenna, as well as on the distribution law of a field along the antenna's aperture. Measurement errors induced by repeated indirect reflections many times, reflections from ground and antenna's nonidentity are not considered here.

Let us assume that the antenna is represented by a broadside array that consists of a number of radiating elements with the identical phase (Figure 15.5). In the observation point P, located along a normal to the array center, the fields from antenna edges will lag in phase from a central element field by $\varphi = 2\pi\delta/\lambda$, and, as it is clear from figure,

$$(R_0 + \delta)^2 = R_0^2 + a^2/4. \tag{15.11}$$

For $\delta \ll a, R_0$

$$R^2 + 2R\delta \cong R^2 + a^2/4, \tag{15.12}$$

that is

$$\delta = a^2/(8R). \tag{15.13}$$

If the allowable difference of phase path of the individual antenna elements is equal to $\pi/8$, i.e. $\delta = \lambda/16$, then from expression (15.13) follows equality (15.10).

If the radiuses from the antenna points to the observation point are parallel, than in order to take into account delay of fields of the different antenna elements, the antenna equivalent circuit should correspond to Figure 15.5, and any radius R_1 is equal

$$R_1 = R + \Delta R, \tag{15.14}$$

where likewise (15.12)

$$\Delta R = x^2/(2R). \tag{15.15}$$

So, in order that the delay of fields from different antenna elements are identical in both circuits, the line of intersection of the convex antenna area depicted in Figure 15.5 with the plane xOz must be a piece of parabola (with a length a_1). The end point of a parabola is removed away axis x by a distance $\delta_1 = a_1^2/(8R)$. Let us suppose that in a first approximation, the parabola length is equal to

$$a_1 = b_1\left[1 + 2/3(2\delta_1/b_1)^2\right],\tag{15.16}$$

where b_1 is the projection of this segment to axis x. Then

$$b_1 = a_1[1 - \delta_1/(3R)].\tag{15.17}$$

In case of a rectangular array with the sides a_1 and a_2

$$R_1 = R + (x^2 + y^2)/(2R).\tag{15.18}$$

The area projection length to axis y is

$$b_2 = a_2[1 - \delta_2/(3R)],\tag{15.19}$$

where $\delta_2 = a_2^2/(8R)$.

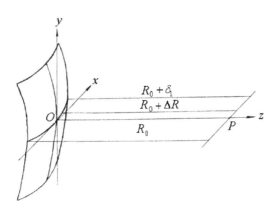

Figure 15.5   The equivalent circuit of a broadside array.

The replacement of the plane antenna field (with the rays, converging to the point P), by the field of the convex area (with the parallel rays) allows significantly simplify the field calculations. On the other hand the field of the uniformly excited area of a convex shape, represented on Figure 15.5, differs from a field of a plane area with parallel rays that corresponds to an arrangement of the observation point in the far zone. The bigger are ratios a_1/R and a_2/R the bigger is this difference. The difference between those fields allows to estimate the magnitude of the error at the measurement in the Fresnel zone. In order to demonstrate this, we calculate the field of the elementary (infinitesimal) dipole of length dx, whose electrical current I_0 is constant along the dipole length. The field is equal to

$$E_\theta = jZ_0 \frac{kL_0e^{-jkR_1}}{4\pi R}\sin\theta\,dx,\tag{15.20}$$

where $Z_0 = 120\pi$ is the characteristic impedance of a free space and $k = 2\pi/\lambda$ is the propagation constant of a wave in an air. Accordingly, the field of the direct dipole with length a_1 and current I_0 is equal along a normal to a radiator (for $\theta = \pi/2$) to

$$E_{\theta0} = jZ_0 \frac{kl_0e^{-jkR}}{4\pi R}b_1.\tag{15.21}$$

The field of the dipole, bent along the parabola with an arm $b_1/2$ is

$$E_{\theta 1} = jZ_0 \frac{kI_0 e^{-jkR}}{4\pi R} J,$$

(15.22)

where

$$J = 2 \int_0^{b_1/2} \frac{e^{-jk\Delta R}}{1 + \Delta R/R} dx = 2 \int_0^{b_1/2} \frac{e^{-jkx^2/(2R)}}{1 + x^2/(2R^2)} dx.$$

(15.23)

Note that in order to correct the first analytical restriction, the top limit $a_1/2$ of integral J is replaced by $b_1/2$. In order to correct the second analytical restriction, the value of $\Delta R/R$ is added to the integrand denominator 1.

Within the scope of a used physical model, which considers that values $Kx^2/(2R)$ and $x^2/(2R^2)$ are small and limiting by magnitudes of the same order of smallness, we shall find:

$$J = 2 \int_0^{b_1/2} [1 - x^2(1+jkR)/2R^2 - k^2x^4/8R^2] dx = b_1[1 - a_1^2/(24R^2) - k^2 a_1^4/(640R^2) - jka_1^2/(24R)],$$

(15.24)

that is in view of (15.17)

$$E_{\theta 1} = E_{\theta 0}[1 - a_1^2/(12R^2) - k^2 a_1^4/(640R^2) - jka_1^2/24R)].$$

(15.25)

For the direct dipole with length a_1 the Pointing's vector P_{10} is equal to

$$P_{10} = \mathrm{Re}(E_{\theta 0} H_{\varphi 0}^*) = \frac{1}{\eta} |E_{\theta 0}|^2.$$

(15.26)

For the curved dipole

$$P_{11} = \mathrm{Re}(E_\theta H_\varphi^*) = \frac{1}{\eta} |E_{\theta 0}|^2 \mathrm{Re} \frac{JJ^*}{a_1^2},$$

(15.27)

where

$$\mathrm{Re}(JJ^*/a_1^2) = 1 - a_1^2/(6R^2) - k^2 a_1^4/(320R^2) + k^2 a_1^4/(576R^2) = 1 - a_1^2/(6R^2) - k^2 a_1^4/(720R^2),$$

(15.28)

i.e.

$$P_{11} = P_{10}[1 - a_1^2/(6R^2) - a_1^4/(18R^2\lambda^2)].$$

(15.29)

For the convex uniformly excited area with the sides a_1 and a_2

$$P_1 = P_0 \prod_{i=1}^{2} [1 - a_i^2/(6R^2) - a_i^4/(18R^2\lambda^2)],$$

(15.30)

where P_0 is the power density created by the plane area along the normal to it. The expression (15.30) allows determining a ratio between maximum directivity factors of convex and plane antennas (with uniformly excited areas)

$$D_1/D_0 = \prod_{i=1}^{2} [1 - a_i^2/(6R^2) - a_i^4/(18R^2\lambda^2)].$$

(15.31)

If, for example, the observation point is located at the boundary of the Fresnel and Fraunhofer zones, then for $a_1 = a_2$

$$a_i^2 = R_0 \lambda / 2, \quad a_i^2 / (6R^2) = \lambda / (12R), \quad a_i^4 / (18R^2 \lambda^2) = 1/72, \quad (15.32)$$

and

$$D_1 / D_0 = [0.9861 - \lambda / (12R)]^2 \approx 0.9724 - \lambda / (6R). \quad (15.33)$$

The formula above takes into account, that $\lambda / (6R) \ll 1$. Thus, the error at the directivity measurement in the far zone boundary in a comparison with measurements in depth of a far zone, is about 2.8% (0.12 dB). The result, close to this (0.06 dB), is obtained in work [120] for the circular aperture. On a twice-smaller distance the error will significantly increase:

$$a_i^2 = R\lambda, \quad a_i^2 / (6R^2) = \lambda / (6R), \quad a_i^4 / (18R^2 \lambda^2) = 1/18, \quad (15.34)$$

that is

$$D_1 / D_0 = [0.944 - \lambda / (6R)]^2 \approx 0.892 - \lambda (3R) \quad (15.35)$$

(it is more than 0.5 dB).

The error ΔD in decibels in the general case is equal to

$$10 \log_{10} (D_1 / D_0) = 4.3 \ln (D_1 / D_0) = 4.3 \sum_{i=1}^{2} [1 - a_i^2 / (6R^2) - a_i^4 / (18R^2 \lambda^2)]. \quad (15.36)$$

Since at $\alpha \ll 1$ $\ln(1 + \alpha) = \alpha$, then

$$\Delta D_1, dB = -5.7[(\delta_1 + \delta_2) / R + 2.7(\delta_1^2 + \delta_2^2) / \lambda^2]. \quad (15.37)$$

In accordance with the presented analysis the second term in parentheses of expression (15.36) is caused by increase of distance from a point of convex area to an observation point, and the third term is caused by a phase's difference of the signals. Both reasons result in total signal attenuation and in a diminution of the measured directivity. The calculations demonstrate that the third term is greater than the second, but the second term is not small and affects the magnitude of the gain.

If we neglect the first term in (15.37), for a linear radiator of length a we obtain

$$\Delta D_1, dB = -15.4 \delta_1^2 / \lambda^2 = -15.4(a^2 / 8R\lambda)^2. \quad (15.38)$$

Let $R = Na^2 / \lambda$. Then $\Delta D_1 = -0.24/N^2$, that is at $N = 0.5; 1; 2; 4$ the magnitude ΔD_1 is equal to 0.96; 0.24; 0.06; 0.01. This is entirely in agreement with results obtained in [118] by the direct calculation. At that

$$\log_{10} |\Delta D_1| = -(0.624 + 2\log_{10} N), \quad (15.39)$$

that is $\log_{10} |\Delta D_1|$ and $\log_{10} N$ are related by a linear correlation. This circumstance is also noted in [118].

Let us compare the obtained results with the results presented in [119]. The proposed method for the most part is equivalent to the method of Polk, but is more understandable which saves from making occasional mistakes. Besides that, ΔR in [119] is taken into account only in order to calculate a signal phase. In the proposed method, an influence of ΔR on the signal amplitude is also taken into account. That allows at the calculation of fields to limit oneself consistently by the meaning of the same order of smallness. An accuracy of proposed method of calculation corresponds to determination accuracy of a distance between a convex antenna element and an observation point. Furthermore, here the difference between the length b_1 of the straight segment connecting the parabola ends and the length a_1 of a parabola herself is taken into account. This difference allows to consider the structures, where radiuses are not parallel. Integration in [119] that goes in our notation from $-a_1/2$ to $a_1/2$ causes increase of the antenna area dimensions and decrease of the calculation accuracy, i.e. resulting ratio D_1/D_0 is higher than the veritable.

The values of a ratio D_1/D_0 for the in-phase square aperture, calculated by the proposed method and described in [119] are presented in Table 15.1. They confirm the previous remarks. It is seen from the table that at the boundary of far zone (more precisely, at the distance, which is considered usually as a boundary of a Fresnel zone and a Fraunhofer zone) the magnitude of gain error, calculated by the proposed method and by the method of Polk, is practically the same and is close to 3%. Analogous result is obtained in [121].

On the other hand, as easy to prove, at $R \ll 2a^2/\lambda$ the proposed method does not give a reasonable result, since for functions under the integral sign, a small number of a series terms is used in calculation. But this case corresponds to a gain measurement in a depth of Fresnel zone (far from the boundary with the Fraunhofer zone), where the measurement accuracy is small, since a radial component of field and an oscillating power are great and increase the error magnitude.

Table 15.1 Ratio D_1/D_0 for a Square In-phase Aperture

$R/(2a^2/\lambda)$	The proposed method	[119]
1	$0.9724 - \lambda/(6R_0)$	0.9726
0.5	$0.884 - \lambda/(3R)$	0.895
0.25	$0.605 - 2\lambda/(3R)$	0.641
0.167	$0.250 - \lambda/R$	0.383

Graphs for the gain error as functions of the magnitude $\alpha = a/\sqrt{2\lambda R}$ at different values of a_2/a_1 are given in [122]. They are made in accordance with the article of Polk. It is necessary to note that the maximal value of α in these figures is 1.5 that corresponds to the magnitude $R/(2a^2/\lambda) = 0.111$. At that point the relative error is close to 100%.

The expression (15.30) allows determining the error of the measuring gain in case of uniformly exited radiators. Examples of such radiators are antenna arrays with uniform excitation of rows and uniform excitation of elements in each row.

In some cases, the field along an antenna aperture is distributed in accordance with the cosine law (for example, field along a horn aperture). Let us consider a field of the curved dipole with an arm $a_1/2$ and current $I = I_0 \cos(\pi x/a_1)$. The field's magnitude is calculated as earlier by the formula (15.25). However

$$J_1 = 2 \int_0^{b_1/2} \frac{\exp[-jkx^2/(2R)]}{1+x^2/(2R^2)} \cos(\pi x/a_1)dx. \tag{15.40}$$

If to limit oneself, as was done earlier, by magnitudes of the first order of smallness, we shall find:

$$J_1 = \int_0^{b_1/2} [1-x^2(1+jkR)/(2R^2)-k^2x^4/(8R^2)][\exp(j\pi x/a_1)+\exp(-j\pi x/a_1)]dx. \tag{15.41}$$

Integration of the expression for J_1 gives:

$$J_1 = 2a_1/\pi \left[1 - a_1^2/(40R^2) - k^2 a_1^4 (160\pi^2 R^2) - jka_1^2/(40R) \right]. \tag{15.42}$$

The field of a direct dipole with length a_1 and with cosine distribution of the current along the dipole is

$$E_{\theta 2} = 2E_{\theta 0}/\pi. \tag{15.43}$$

For the curved dipole at $\theta = \pi/2$

$$E_{\theta 3}/E_{\theta 2} = J_1\pi/(2a_1) = 1 - a_1^2/(40R^2) - k^2 a_1^4/(160\pi^2 R^2) - jka_1^2/(40R). \tag{15.44}$$

Accordingly, the power density, created by a planar area, if the field along one of its sides (with length a_1) is distributed in accordance with the cosine law, and along other side (with length a_2) is fixed, is equal to

$$P_2 = 2P_0/\pi. \tag{15.45}$$

The power density of no planar area with the same distribution of fields

$$P_3 = P_2[1 - a_1^2/(20R^2) - a_1^4/(40R^2\lambda^2)][1 - a_2^2/(6R^2) - a_2^4/(18R^2\lambda^2)], \tag{15.46}$$

i.e. error in the gain's measurement, in decibels is

$$\Delta D_3 = -5.7[(0.3\delta_1 + \delta_2)/R + (1.2\delta_1^2 + 2.7\delta_2^2)/\lambda^2]. \tag{15.47}$$

As it is seen from (15.37) and (15.47), here an error in gain's measurement is smaller than in case of a uniformly exited antenna.

If the excitation falls down to both edges of an array under the linear law, the magnitude ΔD decreases more strongly. It is known, that the field E_θ of a direct dipole of length a_1 with linear current distribution along an arm $I = I_0(1 - 2x/a_1)$ is equal to half of the field of dipole with a current I_0, i.e.

$$E_{\theta 4} = E_{\theta 0}/2. \tag{15.48}$$

For the curved dipole

$$J_2 = 2 \int_0^{b_1/2} \frac{e^{-jk\Delta R}}{1 + \Delta R/R} (1 - 2x/a_1)dx = J - \Delta J, \tag{15.49}$$

where J was calculated earlier, and ΔJ is equal to

$$\Delta J = 4/a_1 \int_0^{b_1/2} x \frac{e^{-jkx^2/(2R)} \exp[-jkx^2/(2R)]}{1 + x^2/(2R^2)} dx = 4/a_1 \int_0^{b_1/2} x[1 - x^2(1 + jkR)/(2R^2)]dx, \tag{15.50}$$

i.e.

$$\Delta J = a_1/2[1 - 7a_1^2/(48R^2) - k^2 a_1^4/(384R^2) - jka_1^2/(16R)], \tag{15.51}$$

whence

$$J_2 = a_1/2[1 - a_1^2/(48R^2) - a_1^4/(48R^2\lambda^2) - j\pi a_1^2/(24R\lambda)], \tag{15.52}$$

and

$$E_{\theta 5}/E_{\theta 4} = 2J_2/a_1 = 1 - a_1^2/(48R^2) - a_1^4/(48R^2\lambda^2) - j\pi a_1^2/(24R\lambda). \tag{15.53}$$

If P_4 is the power density created by a planar area with currents, falling down to its edges under the linear law, then

$$P_4 = P_0/4. \tag{15.54}$$

The power density of convex antenna with the same distribution is

$$P_5 = P_4 \prod_{i=1}^{2} [1 - a_i^2/(24R^2) - a_i^4/(41R^2\lambda^2)]. \tag{15.55}$$

Accordingly, the error of gain's measurement, in decibels is

$$\Delta D_5 = -4.3[(\delta_1 + \delta_2)/3R + 1.6(\delta_1^2 + \delta_2^2)/\lambda^2] \tag{15.56}$$

Thus, if the antenna aperture is exited not uniformly and the field falls down to array edges, the error of measurements decreases.

Figure 15.6 presents the calculation results of the magnitude ΔD, in decibels, for the distance $R = 5$ m between the antenna and the observation point. The magnitude ΔD is given as a function of frequency for different antenna dimensions (in meters), including 0.3 m, 0.4 m and 0.5 m. For square antennas with side a_i, the magnitude ΔD is given as a continuous curve, for linear ones with the lengths a_1 magnitude, ΔD is given as dashed line. Figure 15.6a presents the curves for antennas with uniform excitation and Figure 15.6b shows the curves for antennas with not uniform excitation, falling down to edges under the linear law.

The results of the measurement of the error magnitude for the square antenna with a side 0.5 m at three frequencies of a range are denoted by circles. Antennas under study are planar in-phase broadside arrays of vertical half-wave micro-strip radiators with operating frequencies 6, 8 and 10.4 GHz.

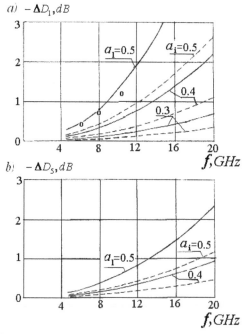

Figure 15.6 The calculation results for the magnitude ΔD: for antennas with uniform excitation (a), for antennas with no uniform excitation, falling down to edges under the linear law (b).

The gain was measured at the distance 5 m and in the far zone. Measurements at the distance 5 m were performed in anechoic chamber of length 7.5 m and at an open test bench of the length about 30 m. The measurements in the far zone are accomplished also at this open test bench. The distance between an antenna under study and a measuring antenna was 25 m. The error of measurement is assumed to be the difference of the mentioned magnitudes. They are pointed in Figure 15.6a by circles. Measurements were made by the standard procedure using the device Vector Network Analyzer.

The proposed method of magnitude error calculation at a measurement of directivity of two-dimensional antenna placed in a Fresnel zone is different from the well-known method by a greater clarity and allows defining more exactly an error magnitude by means of more rigorous calculating dimensions of radiating surface and by means of taking into consideration an effect of distances between antenna points and observation point on the amplitude of received signal.

If the considered antenna is a broadside antenna with the constant phase and the measuring antenna located in its Fresnel zone, a measured gain is less always than the true, since fields' phases from antenna edges do not coincide with a field phase from its middle. The measurement error is determined by sizes of both antennas. Therefore the measuring antenna must have small dimensions. With a frequency growth, if the antenna's dimensions (geometric dimensions and dimensions in wavelengths) are constant and a distance between them is the same, the error is increasing. In the case of not uniform excitation, which is falling down to the edges, the error is smaller than in case of uniform excitation.

15.3 MULTI-CONDUCTOR CABLES

The theory of electrically coupled lines (see Section 3.1) permits to show, that the mutual coupling between lines in multi-conductor cables results in the emergence of the electromagnetic interference (cross talk) in communication channels and that the asymmetry of excitation and loads causes the emergence of the common mode currents in the lines.

In order to determine the signal magnitude at the end of a multi-conductor cable located inside a metal shield, it is necessary to calculate the electrical characteristics of the lines. The values of voltage across loads placed at the ends of an adjacent line can be used as a measure of such distortions [123]. The rigorous method of calculating the mutual coupling between lines enables the development of a simple and effective procedure of preventing interference.

Electromagnetic interference in communication channels (imbalance of a cable) is caused not only by cable asymmetry, but also by asymmetry of excitation and load, which provokes the emergence of the in-phase currents in cables (common mode currents). The rigorous calculation method of the electrical characteristics of multi-conductor cables enables to determine these currents. Compensation of the in-phase currents allows to decrease the EM radiation and susceptibility to the external fields.

We employ a rigorous method—first for calculating characteristics of a two-wire line located inside a metal screen and then for mutual coupling between lines. The lines are considered uniform. The electromagnetic waves are considered transverse (TEM), and the cable diameter is considered small in comparison with the wavelength.

A single pair of wires (twisted pair) inside a metal cylinder can be modeled as two wires of radius a, situated at a distance b from each other inside a metal cylinder of radius

R and length L (Figure 15.7). Wire radius a and distance b in multi-conductor cables are small in comparison with cylinder radius R, i.e. $a, b \ll R$, so the wave impedance of the line is constant along its length (when the axial lines of the twisted pair and the cylinder do not coincide, and the inequality is not true, the wave impedance varies along the line). We assume that the wires are straight and take into account the twisting by increasing length L of the equivalent line.

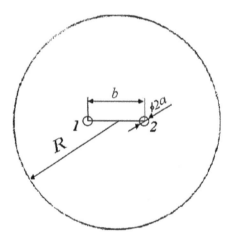

Figure 15.7 Two wires inside a cylinder.

Since the helical pitch along which each wire is located is greater than the diameter b of the helix, inductance Λ per unit length undergoes a slight change at the replacement of a helical wire by a straight wire. The wire capacitance per unit length also varies only slightly, i.e., the twisting wires have no effect on the wave impedances of a structure. The line asymmetry in a real cable can cause in the two-wire line a change of the wave impedance and a change of the input impedance.

The implementation of each two-wire line in the form of a twisted pair (helix) is another cause of cable asymmetry. The twisted pair is the design, which leads to a difference of the average distances between different wires and to the mutual coupling (cross talk) between two two-wire lines surrounded by a single screen, even if the exciting emf of each line and the line load are symmetric.

The equivalent circuit of a single line inside the screen is shown in Figure 15.8. The two-wire line is located above the ground (inside a metal cylinder). The current and the potential along the nth wire of an asymmetrical line of N parallel wires situated above ground, in the general case, are determined by (3.3). The boundary conditions for the currents and potentials in this circuit are

$$i_1(0) + i_2(0) = 0, \ u_1(0) = u_2(0) + i_1(0)Z, \ i_1(L) + i_2(L) = 0, \ u_1(L) = e + u_2(L). \quad (15.57)$$

Here, Z is the impedance of the line load. Substituting expressions (3.3) in the first and second equalities of set (15.57), we find:

$$I_2 = -I_1, \ U_2 = U_1 - I_1 Z.$$

Taking into account formulas (3.6), we find from the third equation of set (15.57) that

$$U_1 = I_1 Z \left(\frac{1}{W_{22}} - \frac{1}{W_{12}} \right) \Big/ \left(\frac{1}{W_{11}} + \frac{1}{W_{22}} - \frac{2}{W_{12}} \right) = I_1 Z \frac{\rho_{11} - \rho_{12}}{\rho_{11} + \rho_{22} - 2\rho_{12}}.$$

And, from the fourth equation, we obtain

$$I_1 = e / [Z \cos kL + j(\rho_{11} + \rho_{22} - 2\rho_{12}) \sin kL].$$

The input impedance of a two-wire line inside a metal screen (the load impedance of generator e) is $Z_l = e/i_1(L)$. Substituting magnitude $i_1(L)$ from expression (3.3) and using the relationships between e, I_1, I_2, U_1, U_2, we find that

$$Z_l = W \frac{Z + jW \tan kL}{W + jZ \tan kL}, \tag{15.58}$$

where $W = \rho_{11} + \rho_{22} - 2\rho_{12}$.

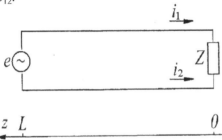

Figure 15.8 The equivalent circuit of a single line inside a screen.

It is readily seen that expression (15.58) coincides with the expression for the input impedance of a two-wire long line that has no losses and is loaded at the end by impedance Z. It is located in free space and is characterized by wave impedance W. The asymmetry of line leads to a difference of the electrodynamics and electrostatic wave impedances of the wires ($\rho_{11} \neq \rho_{22}$, $W_{11} \neq W_{22}$)

The calculation of currents $i_1(z)$ and $i_2(z)$ shows that the currents in a two-wire line are identical in magnitude and opposite in sign:

$$i_1(z) = -i_2(z) = I_1 \cos kz + j I_1 (Z/W) \sin kz.$$

In a wire pair, there are only anti-phase currents. There are no in-phase currents in the wires, because the emf and load impedance are placed between the line wires. The appearance of the in-phase currents can be caused by the connection of an additional emf or an additional load between one wire of a line and the screen.

To find the potential coefficients α_{ns}, one should take into account the following fact. If the system consists of two identical conductors (a wire and its image) and the structure is electrically neutral, the mutual partial capacitance coincides with the capacitance between the conductors [34] and is equal to

$$C = 1 / [2(\alpha_{11} - \alpha_{12})],$$

where α_{11} is the self-potential coefficient, and α_{12} is the potential coefficient of the image. The conductor-to-ground capacitance is twice as much as the capacitance between the two conductors: $C_l = 2C$. For two wires of radius a, located inside the metal cylinder of a radius R at a distance b from each other, symmetrically with respect to the cylinder axes (see Figure 15.7), we can write, using (4.20) from [34],

$$\alpha_{11} = \alpha_{22} = \frac{1}{2\pi\varepsilon} ch^{-1} \frac{R^2 + a^2 - b^2/4}{2Ra}.$$

Here ε is the permittivity of the medium inside the cable. If wire radius a and distance b are small in comparison with the cylinder radius R, then, in the air,

$$\rho_{11} = \rho_{22} \approx 60 \ln(R/a). \tag{15.59}$$

Similarly, using (4.22) from [34], we find:

$$\rho_{12} \approx 60 \ln(R/\sqrt{ab}), \tag{15.60}$$

i.e., the wave impedance of a lossless two-wire line, symmetrically situated inside a metal cylinder, is a half of the wave impedance of the same line in free space:

$$W_0 = \rho_{11} + \rho_{22} - 2\rho_{12} \approx 60 \ln(b/a). \tag{15.61}$$

If the wires inside a metal cylinder of a radius R are located asymmetrically, e.g., they are displaced to the right by distance Δ (Figure 15.9), then

$$\alpha_{11} = \frac{1}{2\pi\varepsilon} ch^{-1} \frac{R^2 + a^2 - (b/2 - \Delta)^2}{2Ra},$$

so at $a, b \ll R$

$$\alpha_{11} = \frac{1}{2\pi\varepsilon} \ln\left\{\frac{R}{a}\left[1 + \frac{\Delta(b-\Delta)}{R^2}\right]\right\} = \frac{1}{2\pi\varepsilon}\left[\ln\frac{R}{a} + \frac{\Delta(b-\Delta)}{R^2}\right], \quad \alpha_{22} = \frac{1}{2\pi\varepsilon}\left[\ln\frac{R}{a} - \frac{\Delta(b+\Delta)}{R^2}\right].$$

In this case, the wave impedance of the line is

$$W = W_0 - \frac{120\Delta^2}{R^2}. \tag{15.62}$$

That is one of the possible causes of changing lines wave impedance inside the screen.

If the distance between the wires is increased by value Δ, then at $\Delta \ll b$

$$\rho_{12} \approx 60 \ln\frac{R}{\sqrt{a(b+\Delta)}} \approx 60\left(\ln\frac{R}{\sqrt{ab}} - \frac{\Delta}{2b}\right). \tag{15.63}$$

This is the second cause. As can be seen from (15.62) and (15.63), a change in the distance between wires has a greater effect on the wave impedance of the line than the displacement of wire relative to the cylinder axis.

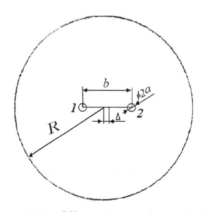

Figure 15.9   Offset wires inside a cylinder.

Figure 15.10 shows the equivalent circuit of two coupled two-wire lines inside a screen. One of the lines is excited by generator e and has on the opposite end the complex load Z_1. The loads Z_2 and Z_3 are connected in the wires at both ends of the other line. It is necessary to emphasize that such circuit has the most general nature. If, for example, generator e_1 is located at the end of the second line (at point $z = L$), the currents and voltages created by generator e are calculated considering that Z_3 is equal to the input impedance of generator e_1.

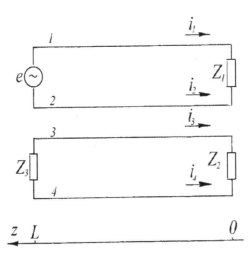

Figure 15.10 The equivalent circuit of two coupled lines placed inside a screen.

We take into account inequalities $a \ll b \ll d$, R (here d is the distance between the axes of twisted pairs). The wires inside the screen form the bunch, whose diameter in many cases is small in comparison with the diameter of the metal screen. But when the bunch consists of many wires, its diameter is close to the screen diameter. However, it is necessary to take into account that the maximal mutual coupling exists between adjacent lines. Therefore, by analyzing the mutual coupling between them it is possible to consider in the first approximation that $d \ll R$.

As was stated at the beginning of this section, cable asymmetry leads to mutual coupling (cross talk) between two two-wire lines. The reason for such asymmetry is the structure of each two-wire line in the form of a twisted pair (helix). The placement of the line conductors in different variants of winding is shown in Figure 15.11. If, in the initial cross-section of the cable, the ends of helices 1 and 3 are located at the same point of their section (we shall call it the initial point) and the ends of helices 2 and 4 are displaced along the perimeter of cross-section by π from this point, it means that the distance between wires 1 and 3 (and also between wires 2 and 4) is $D_{13} = D_{24} = d$ along all length of the cable, whereas the distance between wires 1 and 4 (and also between wires 2 and 3) varies along wires from $d + b$ to $d - b$. For example, the distance between wires 1 and 4 (see Figure 15.11a) is

$$D_{14} = \sqrt{(d + b\cos\alpha)^2 + b^2 \sin^2 \alpha} \approx d + b\cos\alpha + \frac{b^2 \sin^2 \alpha}{2d}$$

(here α is angular displacement of points 1 and 4 along the perimeter of cross-section), i.e., the average distance between these wires

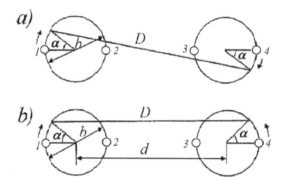

$$(D_{14})_0 = \frac{1}{\pi}\int_0^{\pi} D\,d\alpha = d + \frac{b^2}{4d}, \tag{15.64}$$

differs from distance d. The potential coefficients as well as the electrodynamics and electrostatic wave impedances vary accordingly.

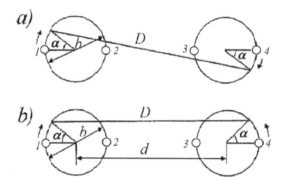

Figure 15.11   Distance between wires 1 and 4 at the same (a) and the opposite direction winding of wire 4 (b).

If, at the initial cross-section of the cable, the ends of helixes 3 and 4 are displaced along the perimeter of the cross-section by $\pi/2$ and $3\pi/2$ from the initial point, respectively, the distance between wire 1 and wire 3 (or wire 4) is

$$D_{13(4)} \approx d + \frac{b}{2}(\cos\alpha \pm \sin\alpha) + \frac{b^2}{8d}(\sin\alpha \mp \cos\alpha)^2. \tag{15.65}$$

Here, the upper sign applies to wire 3, and the lower sign – to wire 4. The average distance between the wires from this equation is

$$(D_{13(4)})_0 \approx d \pm \frac{b}{\pi} + \frac{b^2}{8d}, \tag{15.66}$$

i.e., the displacement of the helix ends of the two-wire line by $\pi/2$ changes substantially the average distance between wires. Difference between $(D_{13})_0$ and $(D_{14})_0$ increases from value $b^2/4d$ to $2b/\pi$, where $b \ll d$.

In order for the average distance D_0 between wires 1 and 4 does not differ from d, it is necessary to wind wire 4 in the opposite direction to the direction winding of other wires. In this case (see Figure 15.11b)

$$D = d + b \cos\alpha, \; D_0 = d. \tag{15.67}$$

The electrodynamics wave impedances of the structure at the same direction winding are

$$\rho_{11} = \rho_{22} = \rho_{33} = \rho_{44} = \rho_1 = 60\ln(R/a), \; \rho_{12} = \rho_{34} = \rho_2 = 60\ln(R/\sqrt{ab}),$$
$$\rho_{13} = \rho_{24} = \rho_3 = 60\ln(R/\sqrt{ad}), \; \rho_{14} = \rho_{23} = \rho_4 = 60\ln\left|R/\sqrt{a(d+b^2\pi/4d)}\right|. \tag{15.68}$$

In the case of lines located at finite distance H from the cable axis, we find

$$\rho_1 = 60\ln\frac{R(1-H^2/R^2)}{a}.$$

These expressions for other quantity ρ_n remain valid. This means that the wave impedance of a two-wire line, which has no losses and is situated inside a metal cylinder at a distance H from its axis, is in accordance with Equation (15.62)

$$W = 60 \ln\left[b(1 - H^2/R^2)^2/a \right],$$

i.e., the wave impedance of this line decreases as the result of its displacement from the cable axis. When H is small and equal to Δ, we arrive at the expression (15.63).

According to (3.4), the electrostatic wave impedances are

$$W_{ns} = \begin{cases} \Delta_N/\Delta_{ns}, & n = s, \\ -\Delta_N/\Delta_{ns}, & n \neq s, \end{cases} \tag{15.69}$$

where $\Delta_N = |\rho_{ns}|$ is the $N \times N$ determinant, and Δ_{ns} is the cofactor of the determinant Δ_N. For a structure made of four wires, in accord with (15.68) and (15.69),

$$W_{11} = W_{22} = W_{33} = W_{44} = W_1 = \Delta_4/\Delta_{11}, \ W_{12} = W_{34} = W_2 = -\Delta_4/\Delta_{12},$$

$$W_{13} = W_{24} = W_3 = -\Delta_4/\Delta_{13}, \ W_{14} = W_{23} = W_4 = -\Delta_4/\Delta_{14}. \tag{15.70}$$

The current and potential of wire n of an asymmetric line, consisting of N parallel wires located above ground, are found from expression (3.3). The boundary conditions for the currents and voltages in the circuit shown in Figure 5.10 are

$$i_1(0) + i_2(0) = 0, \ i_3(0) + i_4(0) = 0, \ u_1(0) = u_2(0) + i_1(0)Z_1, \ u_3(0) = u_4(0) + i_3(0)Z_2,$$

$$i_1(L) + i_2(L) = 0, \ i_3(L) + i_4(L) = 0, \ u_1(L) = e + u_2(L), \ u_3(l) = u_4(L) + i_3(L)Z_3. \tag{15.71}$$

Substituting expressions (3.18) in the equations of system (5.15), we find

$$I_1 = \frac{e}{Z_1 \cos kL + 2j[\rho_1 - \rho_2 + (\rho_3 - \rho_4)A] \sin kL}, \ I_3 = AI_1, \tag{15.72}$$

where

$$A = \frac{4(\rho_3 - \rho_4) + Z_1 Z_3(1/W_3 - 1/W_4)}{-4(\rho_1 - \rho_2) + Z_2 Z_3(1/W_1 + 1/W_2) + j2(Z_2 - Z_3) \cot kL}.$$

If $\rho_3 = \rho_4$ (and, accordingly, $W_3 = W_4$), then $A = 0$, the current at the beginning of the second line is zero. In this case, the presence of the second two-wire line has no effect on the first line. This result obviously corroborates the fact that exactly the asymmetry of cable leads in mutual coupling (cross talk) between two two-wire lines.

Knowing all parameters in expressions (3.3), one can calculate the load impedance of the generator e:

$$Z_l = e/i_1(L) = 2\frac{Z_1 + 2j[\rho_1 - \rho_2 + A(\rho_3 - \rho_4)] \tan kL}{2 + j[Z_1(1/W_1 + 1/W_2) - AZ_2(1/W_3 - 1/W_4)] \tan kL} \tag{15.73}$$

and the currents in the wires of the second (unexcited) line:

$$i_3(z) = I_1 A \cos kz + j\frac{I_1}{2}[AZ_2(1/W_1 + 1/W_2) - Z_1(1/W_3 - 1/W_4)] \sin kz, \ i_4(z) = -i_3(z). \tag{15.74}$$

The sum of the currents is zero, i.e., as in the case of one line placed into the screen, there is no in-phase current since the emf and the loads connected only between wires of each line.

The voltages across passive loads are

$$V_1 = i_1(0)Z_1 = I_1Z_1 V_2 = i_3(0)Z_2 = I_1AZ_2,$$

$$V_3 = i_3(L)Z_3 = I_1Z_3 \left\{ A \cos kL + j\frac{1}{2}[AZ_2(1/W_1 + 1/W_2) - Z_1(1/W_3 - 1/W_4)]\sin kL \right\}. \quad (15.75)$$

As an example, consider a structure from two pairs of wires inside the screen with sizes (in millimeters): $a = 0.2$, $b = 0.5$, $d = 2$, $R = 2$. For the identical loads $Z_1 = Z_2 = Z_3 = 100$ Ohm, ratio A of the currents at the beginning of the second (unexcited) line and the first line amounts to 0.13. If the values of the loads are equal to the wave impedance of the single two-wire line inside the metal screen, i.e., in accordance with equation (5.6), $Z_1 = Z_2 = Z_3 = 55$ Ohm, the ratio of the currents is substantially increased ($A = -0.76$).

The absolute values of the currents as functions of kz are plotted in Figure 5.6. Here, k is the propagation constant of a wave in a medium, z is the coordinate along the line (see Figure 5.4).

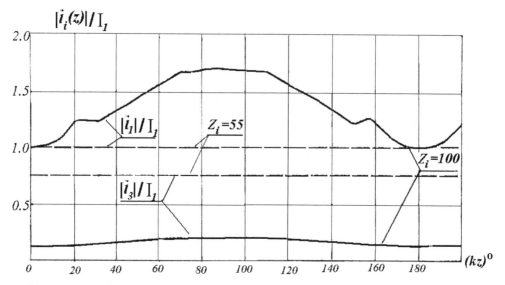

Figure 15.12 The absolute values of the currents in the excited and unexcited wires.

Consider the effect of loads placed between the wires and the screen, using a two-wire line as an example (Figure 15.13). It differs from the circuit shown in Figure 5.8 by connection of its wires near the generator to a screen through complex impedances Z_1 and Z_2, whose values depend on the circuit of a line excitation. In a realistic circuit the secondary winding of the transformer can act as emf e, exciting a two-wire line. In this case, stray capacitances of the winding to ground (to the cable screen) act as impedances Z_1 and Z_2.

The boundary conditions for the currents and potentials in the circuit shown in Figure 15.13 are

$$i_1(0) + i_2(0) = 0, \quad u_1(0) = u_2(0) + i_1(0)Z,$$

$$i_1(L) + i_2(L) + \frac{u_1(L)}{Z_1} + \frac{u_2(L)}{Z_2} = 0, \quad u_1(L) = e + u_2(L). \quad (15.76)$$

Substituting expressions (3.3) in the equations of set (15.76), we find the input impedance of a two-wire line

$$Z_l = \frac{e}{i_1(L)+u_1(L)/Z_1} =$$

$$\frac{Z\cos kL + j(\rho_{11} + \rho_{22} - 2\rho_{12})}{1+U_1/I_1[1/Z_1 + j(1/W_{11} - 1/W_{12})\tan kL]+ j[Z/W_{12}+(\rho_{11} - \rho_{12})/Z_1]\tan kL} \quad (15.77)$$

and the sum of the currents in the line wires is

$$i_s(z) = i_1(z) + i_2(z) = jI_1[Z(1/W_{12} - 1/W_{22})+U_1/I_1(1/W_{11}+1/W_{22} - 2/W_{12})]\sin kz. \quad (15.78)$$

Therefore, connection of the loads results in the emergence of the in-phase current in the wires and in the emergence of the current along the inner surface of the cable screen, equal in magnitude but opposite in direction.

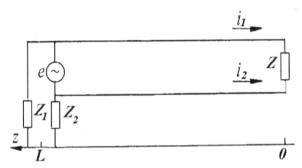

Figure 15.13   The equivalent circuit of a single line with loads connected between the wires and the screen.

For two wires of the same radius situated symmetrically to the cylinder axis

$$Z_l = \frac{Z+2j(\rho_1 - \rho_2)\tan kL}{1+j\dfrac{Z\rho_2}{\rho_1^2 - \rho_2^2}\tan kL + j\dfrac{1}{2(\rho_1+\rho_2)}[Z+(\rho_1+\rho_2)C]\tan kL+\dfrac{1}{2Z_1}[Z+(\rho_1+\rho_2)C+2j(\rho_1 - \rho_2)\tan kL]}, \quad (15.79)$$

and

$$i_s(z) = j\frac{eC\sin kz}{Z\cos kL + j2(\rho_1 - \rho_2)\sin kL}. \quad (15.80)$$

where

$$\rho_{12} = \rho_2, \quad \frac{1}{W_{11}} = \frac{1}{W_{22}} = \frac{\rho_1}{\rho_1^2 - \rho_2^2}, \quad \frac{1}{W_{12}} = \frac{\rho_2}{\rho_1^2 - \rho_2^2}, \quad \frac{1}{W_{11}} - \frac{1}{W_{12}} = \frac{1}{\rho_1+\rho_2},$$

$$C = \frac{2(\rho_1 - \rho_2)-jZctgkL}{\dfrac{2Z_1Z_2}{Z_1-Z_2}-j(\rho_1+\rho_2)\dfrac{Z_1+Z_2}{Z_1-Z_2}ctgkL}$$

It is not difficult to verify that, if $1/Z_1 = 1/Z_2 = 0$, quantity C is zero and the expressions for U_1 and Z_l coincide with the similar expressions for the circuit without loads between wires and screen. From the presented results it is easy to obtain also the expressions for the cases when there is only one load, for example, $1/Z_1 = 0$.

The above analysis confirms that the cause of the emergence of the in-phase currents in wires of line is the asymmetry of its excitation due to the connection of complex impedances, e.g., stray capacitances of the secondary transformer winding, between the wire of line and ground (the cable screen). Asymmetry of loads at $z = 0$, i.e., at the far end of line, produces similar results. The in-phase currents in the excited line induce the in-phase currents in wires of the adjacent unexcited line, even if it is totally symmetric (with respect to ground and the excited line). Removal of the excitation and load asymmetry in the excited line results in the disappearance of these currents in wires of excited and unexcited lines.

In order to reduce or eliminate the in-phase currents, it is necessary to violate the asymmetry, e.g., to neutralize the effect of stray capacitances to ground (to the cable screen). To this end, in [124] it was proposed to compensate the current through stray capacitance with the current equal in magnitude and opposite in direction, which is created by an additional transformer winding.

As is shown in Section 3.1, the theory of electrically coupled lines is based on the telegraph equations and on the relationship between the potential coefficients and the coefficients of electrostatic induction. The z-axis is selected in parallel to the wires, and the dependence of the current on coordinate z is adopted as $\exp(\gamma z)$, where γ is the complex propagation constant of the wave along the wires.

In the case, when the wires and the medium have no losses, electrostatic W_{ns} and electrodynamics ρ_{ns} wave impedances between wires n and s are real-valued quantities determined by equations (3.4), and $\gamma = jk$ is a purely imaginary quantities (k is the propagation constant of the wave in the medium).

As follows from the above, electrodynamics wave impedance ρ_{ns} is proportional to the self- or mutual inductance of wires segment, i.e. is proportional to the reactance, which is connected in series with wires. Electrostatic wave impedance W_{ns} is proportional to the mutual capacitance between wires, i.e. to the susceptance between them. Therefore, it is expedient to connect the resistance of losses in a wire (e.g., the skin-effect loss) in series with the inductance and the leakage conductance—in parallel with the mutual capacitance.

In order to take into account the loss in the medium and in the wires, one must consider that wave impedances W_{ns} and ρ_{ns} and the propagation constant k are complex values. If the inductance of the wire n per unit of its length is Λ_0 and its active resistance is R_0, its impedance per unit length is $j\rho_{nn} = j\omega\Lambda_0 + R_0$, i.e. the self-electrodynamics wave impedance of a loss wire is equal to

$$\rho_{nn} = \rho_0(1 - jR_0/\rho_0), \tag{15.81}$$

where $\rho_0 = \omega\Lambda_0$ is the electrodynamics wave impedance in the absence of losses and R_0 is the total loss resistance in the wire n and in the metal screen per unit length.

For the mutual electrodynamics wave impedance between wires n and s, we obtain

$$\rho_{ns} = \rho_{ns0}(1 - jR_{ns0}/\rho_{ns0}), \tag{15.82}$$

where $\rho_{ns0} = \omega M_{ns0}$, M_{ns0} is the mutual inductance between wires n and s per unit length, and R_{ns0} is the loss resistance in both wires per unit length.

Similarly, for the admittance between wires n and s per unit length we find: $jW_{ns} = j\omega c_{ns0} + G_{ns0}$, i.e. the electrostatic wave impedance in a medium with losses is

$$W_{ns} = W_{ns0}(1 - jG_{ns0}/W_{ns0}), \tag{15.83}$$

where $W_{ns0} = \omega C_{ns0}$, C_{ns0} is the mutual partial capacitance between wires n and s per unit length, and G_{ns0} is the leakage conductance per unit length.

Thus, the evaluation of the electrical performances of the coupled lines with losses can use the results obtained for the lossless lines by substitution of the complex wave impedances into the earlier expressions in accordance with equations (15.81)–(15.83). Here, the losses in wires and losses in an imperfectly conducting metallic tube (screen) is taken into account.

A rigorous method for the calculation of the characteristics of two-wire lines inside a metal screen allows revising the mechanism of mutual coupling between lines in multi-conductors cables. It permits to determine the values of voltage (interferences) across impedances located at the beginning and the end of the adjacent line at the given power in the main line. The reason of cross talks is the asymmetry of the cable structure (the different average distant between wires) and, accordingly, the asymmetric wave impedances. The avoidance of the asymmetry must reduce cross talks in multi-conductor cables, i.e. will allow to increase the carrying capacity of a channel. This is true also for multi-conductor connectors.

The reason for the emergence of the in-phase currents in the lines of a multi-conductor cable is the asymmetry of excitation and loads. As it was noted, the compensation of this asymmetry will allow to decrease the *EM* radiation and to reduce its susceptibility to external fields.

15.4 MAGNETIC IMPEDANCE ANTENNAS

Electric linear antennas with surface impedance, which are excited by concentrated emf, are considered in Sections 1.5 and 2.5. On their surfaces nonzero (impedance) boundary conditions in the form of expression (3.41) are performed. It is shown that the surface impedance Z, affected the distribution of electric current along the antenna already in the first approximation, significantly changes all characteristics of the antenna.

In addition to electrical impedance antennas there are magnetic impedance antennas. Boundary conditions on their surfaces have a form

$$\frac{E_\varphi(a,z)}{H_z(a,z)}\bigg|_{-L\leq z\leq L} = -Z(z), \tag{15.84}$$

where E_φ and H_z are components of an electric and magnetic field, z and φ are cylindrical coordinates of the point on the antenna surface, $Z(z)$ is the surface impedance, which is complex in general case. Examples of such antennas are in particular a magnetic core, excited by a loop, and a longitudinal slot in the conductive cylinder. A magnetic current, excited by the magneto motive force, flows along an axis of such antenna.

Magnetic impedance antennas of finite length were considered for the first time by A.A. Pistolkors using the method of eigen functions. A simpler and clearly evident method, based on the theory of impedance antennas, was applied in [125]. This method allows not only to define the structure of the electromagnetic field, and to establish the types of excited waves, but permits also to calculate the input impedance of the antenna.

Integral equation for the magnetic current J_m flowing along z-axis of an impedance magnetic radiator is written in [28]:

$$\frac{d^2 J_M}{dz^2} + \left(k^2 - 2jk\frac{Z_0\chi}{aZ}\right)J_M = -4\pi j\omega\varepsilon_0\chi\left[H_z^{ex}(z) + G(J_M,z)\right]. \tag{15.85}$$

Here k is the propagation constant of wave in the air, μ_0 is permeability of the core, Z_0 is the wave impedance of free space, $\chi = 1/[2 \ln(2L/a)]$ is a small parameter of the theory of thin antennas, L and a are the length and the radius of the antenna arm, $H_z^{ex}(z)$ is the extraneous magneto motive force, $G(J_M, z)$ is the functional, which takes account of radiation.

It is obviously, the magnitude

$$k_1 = \sqrt{k^2 - 2jk\frac{Z_0\chi}{aZ}} \tag{15.86}$$

has the meaning of the propagation constant of the magnetic current along the magnetic antenna. From (3) it follows that when the condition $Z/Z_0 \sim \chi/(ka)$ is valid, the surface impedance affects the current distribution along the antenna already in the first approximation. If the impedance Z is purely reactive, and

$$k^2 - 2jk\frac{Z_0\chi}{aZ} < 0, \tag{15.87}$$

the propagation constant is imaginary, i.e., the magnetic current will not propagate along the antenna (the current will attenuate quickly with increasing distance from a source).

The results of an analysis of electrical impedance antennas allow to write, using analogy, expressions for the magnetic current distribution and for the input admittance of magnetic antenna, excited by a concentrated magneto motive force e_M. If the feed point is located in the antenna middle, the distribution of the magnetic current in the first approximation has the form:

$$J_M = j\,\mathrm{Im}\,J_M = j\chi\frac{240\pi^2 k}{k_1\cos k_1 L}e_M\sin k_1\left(L - |z|\right). \tag{15.88}$$

The input admittance in this approximation is purely reactive

$$Y = jW_1\cot k_1 L, \tag{15.89}$$

where $W_1 = \dfrac{60}{\chi}\dfrac{k_1}{(120\pi)^2 k}$ makes sense of the wave impedance.

In the second approximation the magnetic current at the feed point is equal to

$$J_M(0) = j\chi\frac{240\pi^2\alpha}{\beta\cos\beta}e_M\sin(\beta - |t|) + \chi^2\frac{120\pi^2\alpha^2}{\beta^2\cos^2\beta}e_M\Theta(\beta,\alpha), \tag{15.90}$$

where

$$\Theta(\beta,\alpha) = \frac{1}{2}\left(\frac{\beta}{\alpha} + \frac{\alpha}{\beta} + j\frac{2ml}{\alpha}\right)e^{2j\beta}Ei(-2jm) - \frac{1}{2}\left(\frac{\beta}{\alpha} + \frac{\alpha}{\beta} - j\frac{2ml}{\alpha}\right)e^{-2j\beta}Ei(2jl) -$$
$$\left(\frac{\beta}{\alpha} + \frac{\alpha}{\beta} + j\frac{ml}{\alpha}\right)(1 + e^{2j\beta})Ei(-jm) + \left(\frac{\beta}{\alpha} + \frac{\alpha}{\beta} - j\frac{ml}{\alpha}\right)(1 + e^{-2j\beta})Ei(jl) +$$
$$\left[\left(1 + \frac{1}{2}e^{2j\beta}\right)\left(\frac{\beta}{\alpha} + \frac{\alpha}{\beta}\right) + j\frac{ml}{\alpha}\right]\ln m - \left[\left(1 + \frac{1}{2}e^{-2j\beta}\right)\left(\frac{\beta}{\alpha} + \frac{\alpha}{\beta}\right) - j\frac{ml}{\alpha}\right]\ln l +$$
$$j\left[\left(\frac{\beta}{\alpha} + \frac{\alpha}{\beta}\right)\sin 2\beta + \frac{2ml}{\alpha}\right]\ln\frac{\gamma}{2\alpha} - (1 + e^{-2j\alpha} + 2\cos^2\beta) + e^{-j\alpha}\cos\beta + j\frac{\beta}{\alpha}\sin 2\beta$$

Here we use the notation: $\alpha = kL$, $\beta = k_1L$, $t = k_1z$, $m = \beta + \alpha$, $l = \beta - \alpha$, $Ei(jx) = Cix + jSix$, and $\ln\gamma = C = 0.5772...$ is Euler's constant. Accordingly, the input admittance is

$$Y = e_M/J_m(0) = Z_E/(120\pi)^2, \tag{15.91}$$

where Z_e is the input impedance of the electrical antenna having the same dimensions and the same value k_1 of the propagation constant, as magnetic antenna has.

If the magnetic antenna is made (Figure 15.14) in the form of a ferrite core, excited by the loop, then the impedance, caused by the magnetic radiator in the loop, is equal to the ratio of the induced in the loop emf (of the magnetic current J_M) to the current J_l flowing along the loop (to the magneto motive force e_M)

$$Z_{ind} = J_M(0)/e_M = 1/Y = (120\pi)^2/Z_E. \tag{15.92}$$

In order to calculate the input impedance of short symmetric radiators ($2L \leq \lambda/2$), one may use the expression

$$Z_e = 80(k/k_1)^2 \tan^2 \frac{k_1L}{2} - j\frac{60}{\chi}\frac{k_1}{k}\cot k_1L. \tag{15.93}$$

Directional pattern of the magnetic radiator in analogy with electrical radiator has the form

$$F(\theta,\varphi) = \frac{\sin\theta}{k_1^2/k^2 - \cos^2\theta}[\cos(kL\cos\theta) - \cos k_1L] \tag{15.94}$$

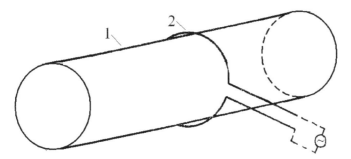

Figure 15.14 The magnetic antenna in the form of a ferrite core excited by a loop.

The surface impedance of the ferrite core can be found by solving the diffraction problem for a circular infinitely long homogeneous core. If to consider excitation of core by convergent cylindrical wave, we obtain the system of Maxwell equations in cylindrical coordinates:

$$\frac{\partial H_z}{\partial \rho} = -j\omega\varepsilon E_\varphi, \quad \frac{1}{\rho}\frac{\partial}{\partial \rho}(\rho E_\varphi) = -j\omega\mu H_z,$$

where ε and μ are absolute dielectric permittivity and magnetic permeability of the ferrite. Taking into account the magnitude of the field E_φ at $\rho = 0$, we find

$$Z = -\frac{E_\varphi}{H_z}\bigg|_{\rho=a} = j120\pi\sqrt{\mu_r/\varepsilon_r}\frac{J_1(ma)}{J_0(ma)}. \tag{15.95}$$

Here $m = k\sqrt{\mu_r \varepsilon_r}$, ε_r and μ_r are the relative permittivity and permeability (with allowance for demagnetization factor), $J_1(ma)$ and $J_0(ma)$ are the Bessel function. If a core is thin ($ma \ll 1$), the surface impedance and the propagation constant are equal to

$$Z = j60\pi\mu_r ka, \quad k_1 = \sqrt{k^2 - \frac{4\chi}{\mu_r a^2}}. \tag{15.96}$$

As it follows from (15.96), the magnetic current is propagating along the core, if k_1 is a real magnitude, i.e. the restriction

$$(2\pi/\lambda)^2 > 4\chi/(a^2\mu_r) \tag{15.97}$$

is satisfied. At the same time, as A.A. Pistolkors showed, the directional pattern of a radiator is changed: it ceases to be similar to the directional pattern of the loop. The equality of the left and right parts of (15.97) corresponds to the critical wavelength. If the wavelength is greater than the critical wavelength, the magnetic current decays rapidly along the antenna, that is, antenna has a low efficiency, similar efficiency of the loop antenna. To improve the efficiency it is necessary to provide the propagation of the magnetic current along the antenna, i.e. one must to turn it into a linear radiator. If the core has an arbitrary diameter, singular points are determined by the expression (15.95). In particular if $J_0(ma) = 0$, then $k_1 = k$, i.e. the ferrite core becomes an ideal magnetic radiator. If $k_1 = k$, then $E_\varphi = Z = 0$. This corresponds to the ideal metal surface.

Another embodiment of impedance magnetic antenna is shown in Figure 15.15. It is the narrow longitudinal slot into the conductive cylinder. Distributed capacitance is created between the edges of a slot. The capacitance can be increased, if capacitors are located along the slot. This antenna like a ferrite core is excited by a loop. The antenna design provides a capacitive character of surface impedance and thus the real magnitude of the propagation constant for the magnetic current along the antenna.

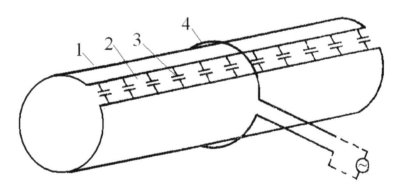

Figure 15.15 Magnetic antenna in the form of a narrow slot cut in the conductive cylinder.

An analysis of the antenna characteristics is made in [128]. For the sake of generality, it is assumed that the cylinder is filled by the ferrite core having a relative permeability μ_r. We consider the practical interest case, when the radius a of a cylinder is small compared with the wavelength in the ferrite. If the antenna is divided into segments of unit length, then the surface impedance of each ring is equal to

$$Z_1 = \frac{1}{j\omega C_1} + j\omega\Lambda_1 = \frac{1}{j\omega C_1}\left(1 - \frac{\omega^2}{\omega_0^2}\right). \tag{15.98}$$

Here $C_1 = NC_0/(2L)$ is the capaciance of the segment, $\Lambda_1 = 60\pi\mu_r$, $ka \cdot 2\pi a$ is the inductance of the segment, $\omega_0 = 1/\sqrt{\Lambda_1 C_1}$ is the natural frequency, N is the number of capacitors included along the antenna length $2L$, C_0 is the capacitance of one capacitor taking into account capacitance between the edges of the slot segment. The expression for the square of the propagation constant along the antenna

$$k_1^2 = k^2 + 4j\pi\omega\mu_0\chi / Z_1 = k^2 + 4\pi\mu_0\chi C_1\omega^2/(\omega^2/\omega_0^2 - 1) \tag{15.99}$$

shows that there is a critical frequency for a wave propagating along the slot in this case too. When $\omega = \omega_0$, the antenna characteristics are similar to characteristics of an ideal magnetic radiator. However, as the frequency increases, the magnitude of the propagation constant increases also. As a result, a great number of current half-waves are placed along the antenna. Their fields mutually cancel each other, and the real length of the radiating antenna quickly decreases. Therefore, this antenna is effective in a narrow frequency band. However, this antenna has been applied as a receiving VHF antenna [129]. Unfortunately, in this article, these antennas are considered as loop antennas. This approach does not allow explain the dependence of the first resonant frequency on the antenna length and also the existence of additional resonances.

The results of an experimental research of the antenna in the form of a narrow longitudinal slot in the metal cylinder with additional capacitors, located along the slot, are given in the figures. Figure 15.16 shows the input impedance of the antenna of length 2 m. The radius of the metal cylinder is 11 cm, the width of the slot is 5 cm. Capacitors with capacitance 51 pF are soldered at the distance 2.5 cm from each other. Numeral 1 denotes the results of the calculation, numeral 2 denotes the experimental curve, and numeral 3 is the impedance of a loop. Figure 15.17 represented the far field for the same variants. Figure 15.18 shows the experimentally obtained dependence of the first resonant frequency of the antenna on its length. The measurements were performed on models of antennas with a cylinder radius 2.1 cm and the width of the slot 1 cm. Capacitors with capacitance 25 pF are soldered along the slot at a distance 2 cm from each other.

The experiments clearly confirm the theoretical results.

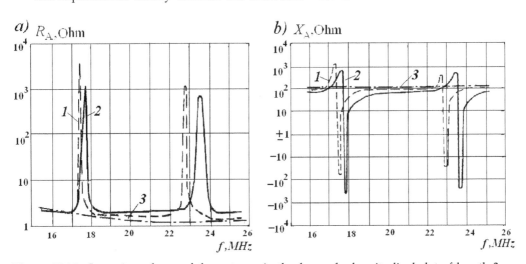

Figure 15.16 Input impedance of the antenna in the form of a longitudinal slot of length 2 m. 1—calculated curve, 2—experimental curve, 3—loop.

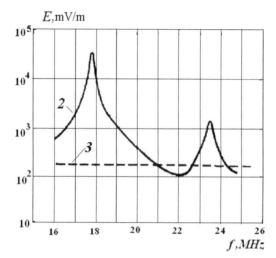

Figure 15.17 Field of the antenna in the form of a longitudinal slot of length 2 m. 1—calculated curve, 2—experimental curve, 3—loop.

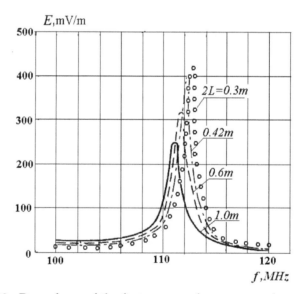

Figure 15.18 Dependence of the first resonant frequency on the antenna length.

References

1. King, R.W.P. 1967. The linear antenna—eighty years of progress. *Proc. IEEE* 1: 2–16.
2. Maxwell, J.C. 1864. A dynamical theory of the electromagnetic field. *Proc. Royal Soc. of London*: 531.
3. Hertz, H. 1887. Uber sehr schnelle electrische Schwingungen. *Wiedemanns Ann. Phys.*: 421 (*in German*).
4. Hertz, H. 1888. Die krafte electrischer Schwingungen behandelt nach der Maxwellshen theorie. *Wiedemanns Ann. Phys.*: 1 (*in German*).
5. Stratton, J.A. 1941. *Electromagnetic Theory*. McGraw-Hill, New York. USA.
6. Poynting, J.H. 1884. On the transfer of energy in the electromagnetic field. *Phil. Trans. R. Soc. Lond.*: 343.
7. Vainshtein, L.A. 1988. *Electromagnetic Waves*. Sovetskoye Radio, Moscow. USSR (*in Russian*).
8. Lavrov, G.A. 1975. *Mutual Effect of Linear Radiators*. Sviyaz, Moscow. USSR (*in Russian*).
9. Levin, M.L. 1947. About one new method of finding the thin antenna characteristic reactance. *Izvestiya AN USSR, ser. phys.*, 2: 117–133 (*in Russian*).
10. Aharoni, J. 1946. *Antennae: An Introduction to their Theory*. The Clarendon Press. Oxford. USA.
11. Aizenberg, G.Z. 1962. *Short-Wave Antennas*. Sviyazizdat, Moscow. USSR (*in Russian*).
12. Kontorovich, M.I. 1951. Some remarks in connection with the induced emf method. *Radiotechnika* 2: 3–9 (*in Russian*).
13. Balanis, C.A. 2005. *Antenna Theory: Analysis and Design*. Wiley & Sons, New York. USA.
14. Kraus, J.D. 1988. *Antennas*. McGraw-Hill, Boston. USA.
15. Levin, B.M. 1992. Once again about induced emf method. *Radio electronics and communications* 2–3: 17–23 (*in Russian*).

16. Elliott, R.S. 2003. *Antenna Theory and Design*. Wiley-IEEE Press, New York. USA.
17. Hallen, E. 1938. Theoretical investigations into the transmitting and receiving qualities of antennae. *Nova Acta Regiae Soc. Sci. Upsaliensis, ser.* IV 4: 1–44.
18. Leontovich, M.A. and M.L. Levin. 1944. On the theory of oscillations excitation in the linear radiators. *Journal of Technical Physics* 9: 481–506 (*in Russian*).
19. Levin, B.M. 2013. *The Theory of Thin Antennas and Its Use in Antenna Engineering*. Bentham Science Publishers. USA.
20. Brown, G.H., R.F. Lewis, and J. Epstein. 1937. Ground systems as a factor of antenna efficiency. *Proceedings JRE*: 753–787.
21. Nadenenko, S.I. 1946. Selecting dimensions of the antennas grounding system. *Radiotechnics* 2: 38–47 (*in Russian*).
22. Wait, J.R. and W.A. Pope. 1954. The characteristics of a vertical antenna with a radial conductor ground system. *Applied Scientific Research* 3: 177–195.
23. Maley, S.W. and R.J. King. 1964. Impedance of a monopole antenna with a radial-wire ground system on an imperfectly conducting halfspace. Part II. *Journal of Research NBS/USNC-URSI* 2: 157–163.
24. Levin, B.M. and V.P. Razumov. 1979. Ground resistance. *Antennas* 27: 125–133 (*in Russian*).
25. Braude, B.V. and E.G. Alexandrova. 1966. Questions of projecting and methods of calculating parameters of long-wave and superlong-wave antennas. *Antennas* 1 (*in Russian*).
26. Pocklington, H.C. 1897. Electrical oscillations in wire. *Cambridge Philosophical Society Proc.*, London: 324–332.
27. King, R.W.P. 1956. *Theory of Linear Antennas*. Harvard University Press, Cambridge. USA.
28. Miller, M.A. 1954. Application of uniform boundary conditions to the theory of thin antennas. *Journal of Technical Physics* 8: 1483–1495 (*in Russian*).
29. Markov, G.T. and D.M. Sazonov. 1975. *Energy*, Moscow. USSR (*in Russian*).
30. Levin, B.M. 1982. Integro-differential equation for two radiators. *Proc. CNII MF* 269: 69–81 (*in Russian*).
31. Popovic, B.D. 1973. Theory of cylindrical antennas with lumped impedance loadings. *The Radio and Electronic Engineer* 3: 243–248.
32. Djordjevic, A.R., B.D. Popovic and M.B. Dragovic. 1979. A method for rapid analysis of wire antenna structures. *Archiv für Electrotechnic (W. Berlin)* 1: 17–23.
33. Richmond, J.H. 1966. A wire-grid model for scattering by conducting bodies. *IEEE Trans. Antennas Propagat.* 6: 782–786.
34. Iossel, Yu.Ya., E.S. Kochanov and M.G. Strunsky. 1981. *Calculation of Electrical Capacitance. Energoisdat*, Leningrad. USSR (*in Russian*).
35. Pistolkors, A.A. 1947. *Antennas*. Sviyazizdat, Moscow. USSR (*in Russian*).
36. Kusnezov, V.D. 1955. Shunt radiators. *Radiotechnics* 10: 57–65 (*in Russian*).
37. Levin, B.M. 1976. Impedance folded radiator. *Antennas* 23: 80–90 (*in Russian*).
38. Leontovich, M.A. 1945. Theory of forced electromagnetic oscillations in thin conductors of arbitrary cross section and its applications to calculation of some antennas. *Proc. NII MPSS* 1: 1 (*in Russian*).
39. Vershkov, M.V., B.M. Levin and S.S. Fraiman. 1972. Antenna with meandering load. *Proc. CNII MF* 151: 73–80 (*in Russian*).
40. Levin, B.M. 1976. Meandering load with arbitrary number of wires. *Proc. CNII MF* 216: 130–139 (*in Russian*).

41. Levin, B.M. and A.F. Yakovlev. 1992. About one method of widening an antenna operation range. *Radiotechnics and Electronics Engineering* 1: 55–64 *(in Russian)*.

42. Hallen, E. 1962. *Electromagnetic Theory*. Wiley, New York. USA.

43. Rao, B.L.J., J.E. Ferris and W.E. Zimmerman. 1969. Broadband characteristics of cylindrical antennas with exponentially tapered capacitive loading. *IEEE Trans. Antennas Propagat.* 2: 145–151.

44. Wu, T.T. and R.W.P. King. 1965. The cylindrical antenna with non-reflecting resistive loading. *IEEE Trans. Antennas Propagat.* 3: 369–373.

45. Levin, B.M. and A.D. Yakovlev. 1985. Antenna with loads as impedance radiator with impedance changing along its length. *Radiotechnics and Electronics* 1: 25–33 *(in Russian)*.

46. Levin, B.M. 1990. Use of loads for creating a given current distribution along a dipole. *Radiotechnics and Electronics* 8: 1581–1589 *(in Russian)*.

47. Himmelblau, D. 1972. *Applied Nonlinear Programming*. McGraw-Hill, New York. USA.

48. Levin, B.M. and A.D. Yakovlev. 1988. Synthesis of antennas with loads by the method of the mathematical programming. *Radiotechnics and Electronics* 2: 254–262 *(in Russian)*.

49. Fradin, A.Z. 1977. *Antenna and Feeder Arrangements*. Sviyaz, Moscow. USSR *(in Russian)*.

50. Yaru, N. 1951. A note of super-gain antenna arrays. *Proceedings I.R.E.* 9: 1081–1085.

51. Bloch, A., R.G. Medhurst and S.D. Pool. 1953. A new approach to the design of super-directive aerial arrays. *Proceedings IEE—Part I* 67: 303–314.

52. Galichin, O.I. 1980. Optimum geometry of radiator. *Proc. Chelyabinsk Polytechnic Institute* 255: 110–114 *(in Russian)*.

53. Galichin, O.I. and V.P. Seregin. 1982. Synthesis of antenna shape for a given directivity pattern. *Proc. Chelyabinsk Polytechnic Institute* 273: 75–78 *(in Russian)*.

54. Levin, B.M. 2006. An antenna directivity calculation on the basis of main patterns. *Proc. 18th International Wroclaw Symp. on Electromagn. Compatibility*. Wroclaw (Poland): 64–67.

55. Vered, U. 2000. Estimation of intercardinal antenna pattern based on cardinal data. *Proc. 21st IEEE Convention of the Electrical and Electronic Engineers in Israel*. Tel Aviv (Israel): 37–41.

56. Ehrenspeck, H.W. and H. Poehler. 1959. A new method for obtaining maximum gain from Yagi antennas. *IRE Trans. Antennas Propagat.* 4: 379–386.

57. Chaplin, A.F. 1969. Synthesis of arrays of passive radiators. *Izvestiya Vuzov of USSR—Radioelectronics* 6: 559–562 *(in Russian)*.

58. Chen, C.A. and D.K. Cheng. 1975. Optimum element lengths for Yagi-Uda arrays. *IEEE Trans. Antennas Propagat.* 1: 8–15.

59. Chaplin, A.F., M.D. Buchazky and M.Yu. Mihailov. 1983. Optimization of director-type antennas. *Radiotechnics* 7: 79–82 *(in Russian)*.

60. Powell, M.J.D. 1964. An efficient method for finding the minimum of a function of several variables without calculating derivatives. *Computer Journal* 7: 155–162.

61. Popovic, B.D. 1970. Polynomial approximation of current along thin symmetrical cylindrical dipoles. *Proceedings IEE* 5: 873–878.

62. King, R.W.P., R.B. Mack and S.S. Sandler. 1968. *Arrays of Cylindrical Dipoles*. Cambridge University Press, New York. USA.

63. Mirolubov, N.N., M.V. Kostenko, M.L. Levinstein and N.N. Tichodeev. 1963. *Methods of Electrostatic Field Calculation*. Visshaya Shkola, Leningrad. USSR *(in Russian)*.

64. Levin, B.M. 1997. Calculation of electrostatic field in heterogeneous media. *Journal of Communications Technology and Electronics* 8: 852–857.
65. Bachvalov, Yu.A. and L.A. Panukov. 1982. Condition of static fields intensity invariance in the piecewise-homogenous media. *Izvestiya Vuzov of USSR—Electromechanica* 4: 408–410 (*in Russian*).
66. Bachvalov, Yu.A. and L.A. Panukov. 1983. Calculation of electrical capacitance in piecewise-homogenous media. *Izvestiya Vuzov of USSR—Electromechanica* 9: 26–31 (*in Russian*).
67. Levin, B.M. and V.G. Markov. 1998. Calculation of fields in coaxial chamber with two helical conductors. *Proc. 14th International Wroclaw Symp. on Electromagn. Compatibility.* Wroclaw (Poland): 215–219.
68. Kalantarov, P.L. and L.A. Zeitlin. 1986. *Calculation of Inductances.* Energoisdat, Leningrad. USSR (*in Russian*).
69. Levin, B.M. and V.G. Markov. 1998. Capacitance between crossing wires parallel to an interface. *Journal of Communications Technology and Electronics* 5: 478–483.
70. Carrel, R.L. 1958. The characteristic impedance of two infinite cones of arbitrary cross-section. *IEEE Trans. Antennas Propagat.* 2: 197–201.
71. Buchholz, H. 1957. *Elektrische und Magnetische Potentialfelder.* Berlin. FRG (*in German*).
72. Pistolkors, A.A. 1944. General theory of diffraction antennas. *Journal of Technical Physics* 12: 693–701 (*in Russian*).
73. Pistolkors, A.A. 1948. Theory of the Circular Diffraction Antenna. *Proceedings of IRE* 1: 56–60.
74. Mushiake, Y. 1992. Self-Complementary Antennas. *IEEE Antennas and Propagation Magazine* 6: 23–29.
75. Mushiake, Y. 1996. *Self-Complementary Antennas: Principle of Self-Complementarity for Constant Impedance.* Springer, London. UK.
76. Booker, H.G. 1946. Slot aerials and their relation to complementary wire aerials (Babinet's principle). *The Journal of the Institution of Electrical Engineers,* Part IIIA 4: 620–626.
77. Rumsey, V.H. 1966. *Frequency Independent Antennas.* Academic Press, New York. USA.
78. Levin, B.M. and V.G. Markov. 1997. *Method of Complex Potential and Antennas, Ship Electrical Engineering and Communication,* St. Petersburg. Russia (*in Russian*).
79. Cortes-Medellin, G. 2011. Non-planar quasi-self-complementary ultra-wideband feed antenna. *IEEE Trans. Antennas Propagat.* 6: 1935–1944.
80. Yang, J. and A. Kishk. 2012. A novel low-profile compact directional ultra-wideband antenna: the self-grounded bow-tie antenna, *IEEE Trans. Antennas Propagat.* 3: 1214–1220.
81. Korn, G. and T. Korn. 1961. *Mathematical Handbook for Scientists and Engineers.* McGraw-Hill, New York, Toronto, London. USA.
82. IEEE. Measurement Techniques. *IEEE Standards Coordinating Committee* 34 (1528TM), 2003. Recommended Practice for Determining the Peak Spatial-Average Specific Absorption Rate (SAR) in the Human Head from Wireless Communications Devices.
83. Feld, Ya.N. 1948. *Foundations of Slot Antennas Theory.* Sovetskoye Radio, Moscow. USSR (*in Russian*).
84. Perini, J. and D.J. Buchanan. 1982. Assessment of MOM techniques for shipboard applications. *IEEE Trans. Electromagnetic Compatibility* 1: 32–39.

85. Bank, M. and B. Levin. 2007. The development of the cellular phone antenna with a small radiation of human organism tissues. *IEEE Antennas Propagat. Magazine* 4: 65–73.

86. Levin, B., M. Bank, M. Haridim and V. Tsingauz. 2010. Dimensions of a dark spot produced by the compensation method. *Proc. 19th International Wroclaw Symp. on Electromagn. Compatibility*. Wroclaw (Poland) : 317–322.

87. Bank, M. and M. Haridim. 2009. A printed monopole antenna for cellular handset. *International Journal of Communications* 2: 54–61.

88. Johnson, R.C. 1993. *Antenna Engineering Handbook*. McGraw-Hill, New York. USA.

89. Menzel, W., D. Pilz and M. Al-Tikriti. 2002. Millimeter-wave folded reflector antennas with high gain, low loss and low profile. *IEEE Antennas Propagat. Magazine* 3: 24–29.

90. Tsai, F.-C.E. and M.E. Bialkowski. 2003. Designing a 161-element Ku-band micro strip reflect array of variable size patches using am equivalent unit cell waveguide approach. *IEEE Trans. Antennas Propagat.* 10: 2953–2962.

91. Vendik, O.G., D.S. Kozlov, M.D. Parnes, A.I. Zadorozhny, and S.A. Kalinin. 2013. Influence of mutual coupling and current distribution errors on advanced phased antenna array nulling synthesis. *Open Journal on Antennas and Propagat.* 3: 35–43.

92. Woodward, P.M. 1947. A method of calculating the field over a plane aperture required to produce a given polar diagram. *Journal I.E.E. (London)*, Part IIIA: 1554–1558.

93. Vendik, O.G. and D.S. Kozlov. 2012. Phased antenna array with a side lobe cancellation for suppression of jamming. *IEEE Antennas and Wireless Propagat. Letters*: 648–650.

94. Widrow, B., P.E. Mantey, L.J. Griffiths and B.B. Goode. 1967. Adaptive antenna systems. *Proceedings IEEE* 12: 2143–2159.

95. Widrow, B. and J.M. McCool. 1976. A comparison of adaptive algorithms based on the methods of steepest descent and random search. *IEEE Trans. Antennas Propagat.* 5: 615–637.

96. Compton, R.T., R.J. Huff, W.G. Swarner and A.A. Ksienski. 1976. Adaptive arrays for communication systems. *IEEE Trans. Antennas Propagat.* 5: 599–607.

97. Guan, N., H. Furuya, D. Delaune and K. Ito. 2008. Antennas made of transparent conductive films. *PIERS online* 1: 116–120.

98. Peter T., T. Abd Rahman, S.W. Cheung, R. Nilavalan, H.F. Abutarboush and A. Vilches. 2014. A novel transparent *UWB* antenna for photovoltaic solar panel integration and RF energy harvesting. *IEEE Trans. Antennas Propagat.* 4: 1844–1853.

99. Hautcoeur, J., F. Colombel, X. Castel, M. Himdi and E. Motta Cruz. 2009. Optically transparent monopole antenna with high radiation efficiency manufactured with silver grid layer (AgGL). *Electronics Letters* 20: 1014–1016.

100. Saberin, J.R. and C. Furse. 2012. Challenges with optically transparent patch antennas. *IEEE Antennas Propagat. Magazine* 3: 10–16.

101. Vershkov, M.V., T.A. Gurgenidze, A.G. Dadeko, A.B. Izraylit and B.M. Levin. 1970. Antenna-mast. USSR Patent #328824. *Bulletin of inventions*, 1973, no. 45.

102. Vershkov, M.V., E.A. Glushkovsky, T.A. Gurgenidze, A.B. Izraylit, B.M. Levin and E.Ya. Rabinovich. 1968. Ship antenna with inductive-capacitive load. *Antennas* 3: 9–16 (*in Russian*).

103. Smith, M.S. and M.J. Barrett. 1986. Experimental wire grid modeling for antennas on structures. *Electronic Letters* 18: 940–941.

104. Yakovlev, A.F. and B.M. Levin. 1996. Reducing of the cable downlead effect on the characteristics of vertical aligned radiators and the reciprocal coupling between them. *Proc.13th International Wroclaw Symp. on Electromagn. Compatibility.* Wroclaw (Poland): 123–126.

105. DuHamel, R.P. and D.E. Isbell. 1957. Broadband logarithmically-periodic antennas structures. *IRE National Convention Record*, Part 1: 119–128.

106. DuHamel, R.P. and F.R. Ore. 1958. Logarithmically-periodic antenna designs. *IRE National Convention Record*, Part 1: 139–151.

107. DuHamel, R.P. and D.C. Berry. 1958. Arrays of log-periodic antennas. *IRE Wescon. Convention Record*, Part 1: 161–174.

108. Isbell, D.E. 1960. Log-periodic dipole arrays. *IRE Trans. Antennas Propagat.* 3: 260–267.

109. Carrel, R.L. 1961. The design of log-periodic dipole antennas. *IRE Intern. Convention Record*, Part 1 : 61–75.

110. Yazkevich, V.A. and V.M. Lapizky. 1979. Exact and approximate methods of calculating log-periodic antennas. *Izvestiya vusov of USSR—Radioelektronika* 5: 69–72, (*in Russian*).

111. De Vito, G. and G.B. Strassa. 1973. Comments on the design of log-periodic dipole antennas. *IEEE Trans. Antennas Propagat.* 3: 303–309.

112. Yakovlev, A.F. and A.E. Pyatnenkov. 2007. *Wide-Band Directional Antennas Arrays From Dipoles.* S.-Petersburg. Russia (*in Russian*).

113. Doluchanov, M.P. 1972. *Wave Propagation.* Svijaz, Moscow. USSR (*in Russian*).

114. Wait, J.R. 1981. *Wave Propagation Theory.* Pergamon, New York. USA.

115. Doluchanov, M.P. 1970. Undeground wave propagation. *Radio* 1: 42–43 (*in Russian*).

116. Drabkin, A.L. 1987. Wireless transmission of information in underground conditions. *Radiotechnika* 5: 68–70 (*in Russian*).

117. Yazishin, V.I. 2011. *Underground Radio Communication in Mines (bibliographic index).* Kharkov. Ukraine (*in Russian*).

118. Hacker, P.S. and H.E. Schrank. 1982. Range distance requirements for measuring low and ultralow sidelobe antenna patterns. *IEEE Trans. Antennas Propagat.* 5: 956–966.

119. Polk, C. 1956. Optical Fresnel-zone gain of a rectangular aperture. *IEEE Trans. Antennas Propagat.* 1: 65–69.

120. Bleyer, D.J., R.C. Wittman, and A.D. Yaghjian. 1992. On Axis Fields from a Circular Uniform Surface Current. *Proc. Ultra-wideband Short-Pulse Int. Conf., Brooklyn*, NY: 285–292.

121. Kinber, B.E. and V.B. Ceytlin. 1964. On measurement error of a gain and pattern at short distances. *Radiotechnika and Electronika* 9: 1581–1593 (*in Russian*).

122. Fradin, A.Z. and E.V. Rizkov. 1972. *Measurement of Antennas-Feeders Devices Characteristics.* Svyaz, Moscow. USSR (*in Russian*).

123. Valenti, C. 2002. NEXT and FEXT models for twisted-pair North American loop plant. *IEEE Journal Select. Areas Commun.* 5: 893–900.

124. Cochrane, D. 2001. Passive cancellation of common-mode electromagnetic interference in switching power converters. M.S. Thesis, Virginia Polytechnic Inst. State Univ., Blacksburg.

125. Glushkovsky, E.A., B.M. Levin and E.Ya. Rabinovich. 1969. Thin magnetic impedance antennas. *Antennas* 5: 108-120 (*in Russian*).

Index

Milton Keynes UK
Ingram Content Group UK Ltd.
UKHW050454071024
449327UK00015B/381